W9-AQS-765

Lowrys' Handbook

OF

RIGHT-TO-KNOW

AND

EMERGENCY PLANNING

Handbook of Compliance for Worker and Community, OSHA, EPA, and the States

**George G. Lowry
and Robert C. Lowry**

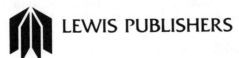 LEWIS PUBLISHERS

Library of Congress Cataloging-in-Publication Data

Lowry, George G.
  Lowrys' handbook of right-to-know and emergency planning:
handbook of compliance for worker and community, OSHA, EPA, and
the states
  George G. Lowry and Robert C. Lowry
    p. cm.
  Bibliography: p.
  Includes index.
  ISBN 0-87371-112-2
  1. Industrial hygiene—Law and legislation—United States.
2. Hazardous substances—Law and legislation—United States.
3. Industrial safety—Law and legislation—United States.
I. Lowry, Robert C. II. Title.
KF3570.L69   1988
344.73'0465—dc19
[347.304465]   88-8998

LEWIS PUBLISHERS, INC.
121 South Main Street, Chelsea, Michigan 48118

PRINTED IN THE UNITED STATES OF AMERICA

George G. Lowry received a PhD in Physical Chemistry from Michigan State University and an MS in Organic Chemistry from Stanford University. He also received a BS in Chemistry from California State University at Chico, and a Certificate in Industrial Relations from the University of California.

Dr. Lowry currently is Professor of Chemistry at Western Michigan University in Kalamazoo, where he has developed a graduate-level course in chemical laboratory safety. Before becoming a professor, he was an industrial research chemist for the Dow Chemical Company.

He is also a member of an American Chemical Society task force on laboratory hazardous waste disposal, and he is an independent consultant in chemical safety. In that capacity he has delivered safety instruction to industrial workers and supervisors and has served as an expert witness in liability cases involving alleged injury from chemical exposure. He has also spoken on chemical safety and Right-to-Know topics to industry, education, and professional groups in six states, and has served as an advisory consultant to numerous industrial companies.

Dr. Lowry is co-author of the book *Handbook of Hazard Communication and OSHA Requirements* (Lewis Publishers, Inc., Chelsea, MI, 1985).

Robert C. Lowry received a JD degree from Boalt Hall School of Law at the University of California at Berkeley, where he was an associate editor of *California Law Review*. He also received two BS degrees, one in Economics and the other in Civil Engineering with a concentration in Transportation Systems Analysis, from the Massachusetts Institute of Technology.

Mr. Lowry is an attorney who specializes in economic regulatory law. He is currently a member of the legal staff of the United States Postal Service in Washington, D.C., where he handles cases involving rates and classifications, contracts, labor negotiation, and consumer protection. He has been involved in litigation in federal courts in the D.C. Circuit, the 2nd Circuit, the 4th Circuit, and the 7th Circuit. He has worked for public interest groups as well as both public and private sector employers including the Union Pacific Railroad Law Department, the Massachusetts Public Interest Research Group, and the Police Review Commission of Berkeley, California.

Mr. Lowry is co-author of the book *Handbook of Hazard Communication and OSHA Requirements* (Lewis Publishers, Inc., Chelsea, MI, 1985).

Table of Contents

APPENDICES

List of Figures and Tables

FIGURES

TABLES

Preface

On November 25, 1983, the Occupational Safety and Health Administration of the U.S. Department of Labor promulgated a new Hazard Communication Standard (29 CFR 1910.1200). The purpose of this Standard (the HCS) is to

> ensure that the hazards of all chemicals produced or imported by chemical manufacturers or importers are evaluated and that information concerning their hazards is transmitted to affected employers and employees within the manufacturing sector.

Although the HCS originally applied only to manufacturing employers, it has recently been expanded to cover all private-sector employers, and many states have expanded its coverage to public-sector employers within their boundaries.

On October 17, 1986, President Reagan signed into law the Superfund Amendments and Reauthorization Act (SARA) of 1986, Title III of which is a stand-alone statute called the Emergency Planning and Community Right-to-Know Act (EPCRA), or SARA Title III. The purpose of SARA Title III is to

> encourage and support emergency planning efforts at the state and local levels and provide the public and local governments with information concerning potential chemical hazards present in their communities.

Many regulated firms are understandably uneasy about these laws. In addition to the usual negative reaction to new regulations and standards, there is concern for the complexity of the regulations and the ambiguity of certain provisions. SARA Title III also creates new responsibilities for local government bodies that may or may not be prepared to carry them out effectively.

Nevertheless, the concepts behind these "Right-To-Know" laws are supported by major labor organizations, major industrial groups, and most environmental and community groups.

This book is intended to help all affected parties work with the HCS and SARA Title III as smoothly as possible. Contained herein are explanations of what is required, as well as recommendations of how to comply, particularly in areas where details are not spelled out in the statutes and regulations. Numerous suggestions are offered that will help identify hazardous materials, plan for emergencies, obtain needed additional information, and comply smoothly with the many recordkeeping and reporting requirements. These are necessarily general statements and recommendations that deal primarily with federal laws. Whenever you have a question regarding your legal rights, responsibilities, and liabilities under Right-to-Know, you should consult your local attorney for interpretation of the law as it applies to your particular situation.

In Part A, the first three chapters of the book provide an overview of the basis and major elements of these laws. In Part B, the next five chapters provide specific information for those affected directly by the laws—chemical manufacturers and distributors, other firms using hazardous chemicals, and local emergency planning committees. In Part C, the final four chapters deal with legal relationships between the various directly affected parties, and between the Right-to-Know laws and other federal and state laws, and several related matters whose outcome is still to be determined. In the appendices are extensive tables of specific information about regulated chemicals, sources of information useful to those concerned with specific duties under the laws, and other useful forms and information.

In the end, whether Right-to-Know laws achieve the objective of enhanced worker and public safety with a minimum of burden and strife will often boil down to relations between the affected parties. Given a positive attitude, good labor relations, and good community relations, compliance should have real long-term benefits for all. We hope this book will make it somewhat easier to enjoy those benefits than if you were left to your own devices.

Recent Changes in the Hazard Communication Standard

OSHA has promulgated the statement that "The Office of Management and Budget has disapproved . . . the requirement that material safety data sheets be provided on multi-employer worksites. . . . " (53 FR 15035, April 27, 1988)

OSHA also promulgated the statement that "The Office of Management and Budget has disapproved . . . coverage [by the Hazard Communication Standard] of any consumer product excluded from the definition of 'hazardous chemical' under section 311(e)(3) of the Superfund Amendments and Reauthorization Act of 1986. . . ." (53 FR 15035, April 27, 1988) This serves to exempt " . . . any substance to the extent it is used for personal, family or household purposes, *or is present in the same form and concentration as a product packaged for distribution and use by the general public*." (Emphasis added.) This exempts any such material "whether or not it is used for the same purpose as the consumer product . . . [and thereby exempts] household or consumer products in commercial and industrial use. . . . " (52 FR 46078, December 4, 1987)

OSHA also promulgated the statement that "The Office of Management and Budget has disapproved . . . coverage of any drugs regulated by the U.S. Food and Drug Administration in the nonmanufacturing sector." (53 FR 15035, April 27, 1988) This is a significant expansion of the original drug exemption clause and is based on OMB's position that "Drugs for human consumption are heavily regulated by the Food and Drug Administration, which requires the transmittal of detailed information downstream from the manufacturer through professional package inserts and labels." (52 FR 46078, December 4, 1987)

On June 24, 1988, in response to a litigation action (*Associated Builders and Contractors, Inc. v. Secretary of Labor*, CA3, No. 88–3345), the U.S. Court of Appeals for the Third Circuit granted an emergency stay of the extension of the Hazard Communication Standard (HCS) pending further developments in the subject action. A temporary stay had already been granted on May 20. Various stages of written briefs were scheduled with due dates of July 8, July 22, and July 29, with oral arguments scheduled for August 22. On July 8, the Court lifted the stay for all employers except those in the construction industry.

Part A

Introduction and Overview of Legal Responsibilities

The Hazard Communication Standard (HCS) and the Emergency Planning and Community Right-to-Know Act (SARA Title III) complement each other by covering two sides of the potential hazards created by the use of dangerous chemicals: hazards to persons who work with and around the chemicals, and hazards to persons in the surrounding community. Part A of this book presents an overview of the basis and major elements of these laws.

Chapter 1 discusses the need to know information regarding chemical hazards, including the justification for federal, as opposed to state or local, requirements. Chapter 2 presents an overview of the HCS, and Chapter 3 does the same for SARA Title III. More specific information on the requirements and how to comply with them is provided in subsequent chapters.

The Need to Know

Hazardous chemicals, hazardous materials, or hazardous substances, as they are variously called, have long been used for many purposes in homes, business, and industry. Often people who use such materials are unaware of specific hazardous properties, and what should be done to protect themselves and others from illness or injury that those chemicals might cause. This often is both a cause and a result of a general apathy towards the potential hazards exhibited by such materials. After all, if a person has worked with a particular material for a long time and nothing alarming has ever resulted from that experience, it is easy to develop the feeling that there is nothing to worry about.

However, occasionally something alarming does happen as a result of working with some particular materials. When such an experience affects people other than the individuals working with the material, and possibly causes visible property damage as well, it becomes a matter of public concern. There are some journalists and many environmental activists who seem to be eager to attack all chemicals as a result of a few tragic accidents. In recent years, this phenomenon has resulted in a widespread epidemic of "chemophobia," or fear of chemicals.

1.1 EXTENT OF THE PROBLEM

In the late 1980s it has become commonplace to see references to "toxic chemicals," as if that were the principal matter of concern. Certainly toxicity of chemicals that enter the environment is a serious matter, but the simple fact is that in the quantities normally encountered, most toxic

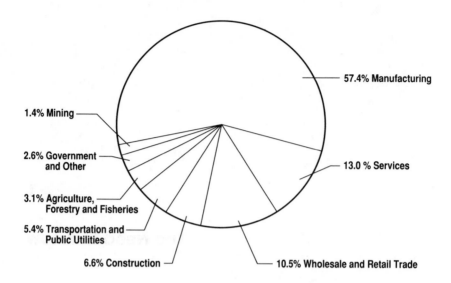

Figure 1. Distribution of chemical-source illnesses and injuries by industry groups. Source: 48 FR 53285 (Nov. 25, 1983).

chemicals are not lethal and do not cause serious harm (although some do). The most serious hazards connected with a large number of chemicals are connected with physical hazards such as fire and explosion rather than with health hazards such as toxicity and carcinogenicity.

Chemicals in the Workplace

The U.S. Department of Labor has gathered statistics showing the extent of illnesses and injuries resulting from chemical hazard exposure in the workplace. Some of those statistics were given in the *Federal Register* of November 25, 1983. For the years from 1976 through 1978 there were reported a total of 127,725 chemical-source illnesses and injuries in the United States, an average of 42,575 per year. And these figures represent only the reported cases. Certainly many others have gone unreported.

It should not be surprising that the incidence of such illnesses and injuries varies significantly from one industry group to another. Such a breakdown is shown in Figure 1, using Bureau of Labor Statistics figures as reported in the *Federal Register*. From this figure, it is obvious that the manufacturing sector bears the largest burden.

However, such percentage figures tell only part of the story, because different industry groups employ vastly different numbers of workers. For example, manufacturing firms employ only about one-third of American

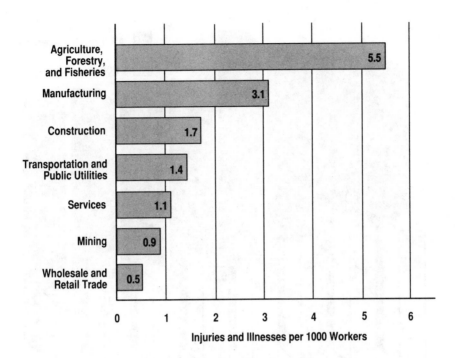

Figure 2. Frequencies of chemical-source illnesses and injuries by industry groups.
Source: 48 FR 53285 (Nov. 25, 1983).

workers, but this one-third experiences more than half of all the reported chemical-source injuries and illnesses. The frequencies of such cases are shown in Figure 2 for the same industry groups (with the exclusion of "Government and other") for the year 1977. Again, manufacturing stands out, but so does the agriculture, forestry, and fisheries group of industries.

In the absence of further statistics, one might guess that workers in the chemical industry would have among the highest incidences of chemical-source illnesses and injuries, simply because they work with such materials constantly. However, for many years, most large chemical manufacturing companies have had safety records that are far better than overall industry averages. This is illustrated in Figures 3 and 4, showing accident incidence and severity rates for the chemical industry compared to those for the average of all industries.

A total of 42 industrial categories were included in the studies on which these figures are based. In three of the seven years the chemical industry had the lowest incidence rate of the 42, and never did it rate poorer than fifth. In severity rates, the chemical industry ranked from second to fifth over the seven-year period, and no single industry ranked consistently better during this period. Before 1977, different classifications and report-

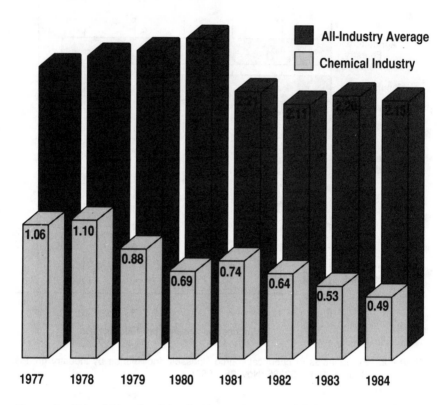

Figure 3. Industrial lost-time injury incidence rate per 100 full-time employees. Source: "Accident Facts," National Safety Council, 1978–85.

ing procedures were used, but qualitatively the same conclusions apply over a period of many years.

These records have been achieved in spite of the fact that chemical industry workers are more exposed to hazardous materials than are most workers in other industries. This probably is largely a result of better understanding of the hazardous properties by those whose main concerns are with the chemicals themselves rather than with some process or product that only incidentally involves chemicals.

It is almost axiomatic that if people who work with or near hazardous materials have a reasonable understanding of the properties of those materials, they will be better able, and often better motivated, to practice appropriate types of prevention and hence to work safely. Most employers provide equipment to protect their workers from material hazards, along with some training in its use. Information about the actual hazards involved has often been inadequate, though. Partly as a result of this, many workers simply will not wear the protective equipment provided them. In

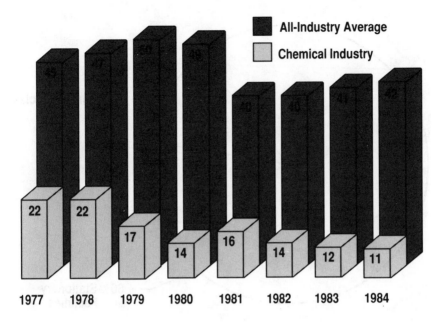

All-Industry Average

Chemical Industry

1977   1978   1979   1980   1981   1982   1983   1984

Figure 4. Industrial injury severity rate, days away from work per 100 full-time employees. Source: "Accident Facts," National Safety Council, 1978–85.

many cases, the employers themselves do not fully understand the nature of the hazards.

Chemicals in the Community

It is common to read in newspapers, and hear on radio and television news programs, about chemical accidents that cause injuries, deaths, evacuation of nearby residents and workers, or at least considerable concern. To gain some insight into the nature and severity of the effect chemical hazards may have on the greater community, it is appropriate to examine some additional statistics. Anything approaching a complete set of objective statistics on such matters, even for the United States alone, is difficult or impossible to obtain and interpret.

In his book *Bhopal: Anatomy of a Crisis,* author Paul Shrivastava presents a table of major industrial accidents (i.e., those causing more than 50 deaths each) worldwide between 1907 and 1984. He has excluded accidents in the USSR as well as those involving explosives, munitions, mining operations, gas distribution, or the commercial transportation of people.

The total number of deaths in all those major accidents over a 78-year

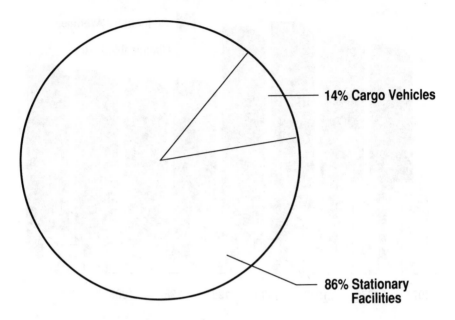

Figure 5. Distribution of deaths between accidents involving cargo vehicles and those
involving stationary facilities.

time span was only 6,936, a trivial number compared with the number of
deaths caused by automobile accidents in one year in the United States
alone. And yet, one would like to think that each of the accidents could
have been avoided, or at least the severity of results could have been
markedly decreased, if only the proper measures had been taken.

The numbers of deaths do not tell the whole story, however. Many
more people were injured or forced to flee their homes or workplaces as
a direct result of those accidents. Furthermore, the resulting property
damage represented a loss of capital to the owners and dislocated many
thousands of workers from their jobs at least temporarily, and sometimes
permanently.

The above figures, though, deal only with major disasters. They are
somewhat analogous to figures of casualties from major airline, bus, and
train casualties rather than from all transportation accidents together,
which eclipse the casualty figures from major disasters by a large factor.

It is instructive to break down Shrivastava's figures in various ways to
understand the situation and the problem a little better, and perhaps to find
better solutions. In Figure 5, it is seen that about one-seventh of all the
deaths in this sample involved cargo vehicles rather than stationary facilities such as warehouses and manufacturing plants. The concern about

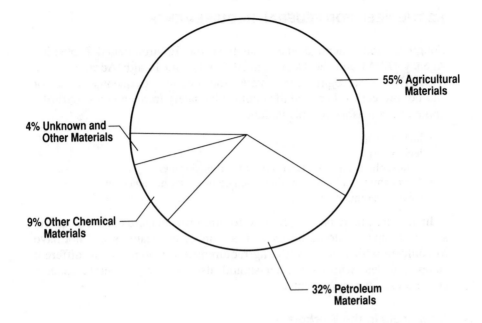

55% Agricultural
Materials

4% Unknown and
Other Materials

9% Other Chemical
Materials

32% Petroleum
Materials

Figure 6. Distribution of deaths from accidents by general categories of materials.

such incidents motivated the Hazardous Materials Transportation Act, and yet much needs to be done to bring this portion of the problem under adequate control.

In Figure 6 it is seen that over half of the deaths involved accidents in which agricultural materials (fertilizers, pesticides, etc.) played a key role. About one-third of the deaths involved accidents with petroleum materials such as crude oil, fuel oil, gasoline, or natural gas. Only about one-tenth of the deaths resulted from accidents involving all other types of chemicals.

Regardless of the type of facility or the type of material involved in industrial accidents, the tragedy usually could have been much less serious if proper steps had been taken. In order to do so, the people involved would have had to understand better what the hazards were and how to manage them to prevent an accident from becoming a major crisis—a true disaster.

When chemical accidents occur, it is important to know when to evacuate people, what steps are available short of evacuation, and how to choose among the options in order to minimize the casualties. Generally speaking, disaster personnel are not trained to evaluate such options as well as they should be. Apparently, neither are most of the personnel who manage the facilities in which such accidents could occur.

1.2 THE NEED FOR FEDERAL REQUIREMENTS

Congress, and federal agencies such as the Environmental Protection Agency (EPA) and the Occupational Safety and Health Administration (OSHA), have recognized that states and localities frequently cannot or will not protect workers and the public adequately from chemical hazards. Some of the reasons for this include:

- lack of funds
- lack of expertise
- philosophical attitudes favoring minimal government
- "company town" status or other such commercial influence on local government

In addition, there is a strong need for uniform standards in such matters so that companies doing business in a number of locations will not have to conform with many conflicting requirements. Companies in different states that deal with each other should also be able to operate under a common set of requirements.

Chemicals in the Workplace

In 1970, Congress enacted the Occupational Safety and Health Act, which in part required that any standard promulgated under the Act "shall prescribe the use of labels or other appropriate forms of warning as are necessary to insure that workers are apprised of all hazards to which they are exposed, relevant symptoms and appropriate emergency treatment, and proper conditions and precautions of safe use or exposure" [29 USC 655(b)(7)]. Since then, OSHA has developed standards for specific materials which comply with those requirements. Until recently the agency had not taken a systematic approach.

In 1983, OSHA announced its federal Hazard Communication Standard (29 CFR 1910.1200), known as the HCS, first published in 48 FR 53280 (Nov. 25, 1983). Part of the publication in the *Federal Register* is an extensive preamble, which gives a good background survey of the development of the HCS. The HCS is written as a standard that preempts any state or local laws dealing with the same considerations. This preemption was the subject of much displeasure and litigation, but it has been upheld in court.

Chemicals in the Community

To deal with the community problems that can arise from the presence of moderate to large quantities of chemicals, many legislative bodies had begun to consider and enact "Community Right-to-Know" laws by about

1980. By mid-1983, there were a dozen such state laws on the books. In early February of 1985, James Florio (D-NJ), who at the time was Chairman of the Subcommittee on Commerce, Transportation, and Tourism of the House of Representatives Committee on Energy and Commerce, introduced bills to require that certain types of steps be taken to prepare adequately for emergency response to chemical accidents.

Though many of the provisions Florio had in his original bills did not survive, the final outcome of the effort was Title III of the Superfund Amendments and Reauthorization Act of 1986 (SARA Title III), which is a stand-alone law entitled the Emergency Planning and Community Right-to-Know Act (EPCRA). Following experiences with the HCS, and to avoid some types of questions concerning preemption, SARA Title III includes a section stating specifically that the statute does not preempt any state or local laws dealing with chemical emergencies and disasters.

1.3 BASIC CONCEPTS OF HAZARD, RISK, AND SAFETY

Before considering the details of the HCS and of SARA Title III and how to comply with them, it is well to consider some critical ideas and definitions. These are of central importance to a good understanding of the whole subject of chemical hazard safety.

It is tempting to rely on the imposition by some authority of a set of safety rules to cover essentially all conceivable situations that could lead to unsafe acts or conditions. However, as is well known to experienced observers of such matters, there is a serious flaw in such an approach.

If workers are not convinced that they are likely to be hurt if they disobey a particular safety rule, many of them will conform grudgingly, and only when they have to. This is a key factor in any realistic concept of safety. An individual does not consider a situation unsafe, and therefore will resist forced compliance with safety rules, if he or she does not believe that the situation poses an unacceptable amount of risk.

A common definition of safety is "freedom from danger or harm." However, a little reflection leads to the recognition that in this sense nothing is really safe, because it isn't possible for anything to be completely free from danger or harm. This doesn't mean we must abandon the goal of safety, but it does mean we need to consider more complicated concepts if we are to deal effectively with the question.

In the next several paragraphs, these concepts are defined, and examples are given, to set the foundation for understanding the OSHA Standard and for using it most effectively. The same concepts apply to SARA Title III also.

Hazard

A hazard may be defined as any substance, situation, or condition that is capable of doing harm to human health, property, or system functioning. Note particularly that this definition does not say that the hazard will do harm, but merely that it has that capability. Also, it does not say how much harm might be done. Thus a situation or material that can only result in a slight irritation, and only under unlikely circumstances, is a hazard just as is a situation or material that can result in a fatality and that is very likely to do so. In other words, the term "hazard" does not discriminate very well about how serious a potential harm might be or how likely it is to occur.

Unfortunately, we commonly think of hazardous things as being likely to cause serious harm, but that simply is not so. Many things that the OSHA HCS defines as hazardous can only cause slight harm or irritation, and with proper precautions even that is very unlikely. This is a point that needs to be made strongly during the training required by the Standard—otherwise workers can become paranoid about perceived dangers.

Risk

Risk may be defined as a measure of the probability and severity of harm to human health, property, or system functioning. In other words, "risk" includes both a sense of how likely harm is to occur and an indication of how serious the harm may be if it does occur.

The severity of a collision between the earth and a planet from another solar system is about as great as anything imaginable. But if there is virtually no likelihood that such a collision will ever occur, then the risk, which is a composite of the two factors, is extremely low. The severity of a minor skin irritation that vanishes within a short time is very small. If it is almost sure to occur, though, the risk may be considered greater than that of the planetary collision.

Risks are very difficult to evaluate in any sense that we can reliably rate them numerically, though in some special cases ways have been devised to do just that. But we can all have our own feelings about how risky the presence of certain hazards might be. The fact that different people evaluate the risk of any particular hazard differently is natural and is the source of many problems. In some cases regulatory bodies, and even courts of law, have made an evaluation for us. But in many cases we do have some options ourselves.

Safety

Safety may be defined as a condition that is judged to be free of unacceptable risks. That is, once we have estimated how risky something is, we judge for ourselves whether we consider it safe and therefore whether we will voluntarily accept the risks.

Many people feel that the risks involved in riding a motorcycle or in skydiving are unacceptable, and those people will not consider such activities to be safe. For many fewer people, apparently, the risks involved in driving or riding in an automobile without seat belts are unacceptable and those people will not consider such activities to be safe. Similar statements can be made about cigarette smoking, heavy drinking of alcohol, eating heavily salted food, working in chemical plants, etc. In many cases, some people will judge a risk to be acceptable, and therefore the particular activity is, to them, safe; the same activity is unsafe to others who are unwilling to accept the same risk. Sometimes the reason is because of a difference of judgment as to how great the risk is, and sometimes it is primarily a matter of different attitudes of how much risk is acceptable. Statistical studies seem to indicate a range of a factor of about ten in the level of risk acceptable to the American public. That is, some people usually will accept risks that are about ten times as great as those that are barely acceptable to other people.

Sometimes people are unwilling to accept a situation because they think it is much riskier than it actually is. Other times, some individuals may be willing to accept a situation because they think it is much less risky than it actually is. A large part of the reason for the Right-to-Know movement is to provide workers and public officials with reliable information about material hazards so they can better understand the risks and consequently make better decisions.

A judgment of what is safe will not be the same for all individuals. However, most people resent being placed into a risky situation against their will or without their knowledge, regardless of what their decision might be if it were made freely. The idea that people should not be subjected involuntarily to risk is a philosophical basis for governmental regulation of safety in the workplace and the surrounding community, and for certain kinds of liability law as well, particularly the doctrine of "Strict Liability."

2

The Hazard Communication Standard

The purpose of the HCS is to remove as much as possible the mystery surrounding the risks that are due to chemical hazards in the workplace. It does so by creating a "right to know" for workers who are involved in the manufacture and handling of hazardous chemicals, or who are otherwise routinely exposed to them. Under other OSHA regulations, some of these risks are also regulated in terms of allowed exposure level. The HCS is thus designed to form one part of an integrated approach to safety in the workplace.

The HCS originally covered only manufacturing establishments in Standard Industrial Classification (SIC) codes 20–39, but applies to all private sector employers effective May 23, 1988. It requires that private sector employers who manufacture, import, distribute, or otherwise use hazardous chemicals institute risk management and safety programs. These programs are to inform workers of the hazards they work with and how they can minimize both the probability and the severity of potential harm.

Although the Standard is aimed at enhancing safety, OSHA did not attempt to define an unacceptable level of risk. Rather, the intent of the Standard is to provide workers with enough information to make their own safety judgments. The Standard covers labeling of containers, availability of material safety data sheets (MSDSs), development of a written hazard communication program, and training of workers. All of these requirements are aimed at the general goal of providing workers with reliable information about the various material hazards they may meet on the job.

The complete text of the current Standard is reprinted in Appendix C.1.

It is likely that many employers will experience a fair amount of difficulty in attempting to comply with the Standard, which is a performance-oriented standard written in rather vague language that tends to rely on both bureaucratic and technical jargon. This is particularly so for employers who do not have trained chemical safety experts on their staff and who may not even be sure which of the chemicals they use are considered hazardous. Similarly, workers and their representatives may be uncertain as to just what is and is not required under the Standard.

The remainder of this chapter provides an overview of the current scope and requirements of the OSHA HCS, as well as exemptions and enforcement procedures and sanctions. Subsequent chapters provide more detailed information on how to comply with the technical aspects of the Standard, and discuss further the legal and policy aspects of implementation.

2.1 EXPANDED COVERAGE

Nonmanufacturing Sectors

On August 24, 1987, OSHA published a final rule expanding coverage of the HCS to include all private employers in nonmanufacturing as well as manufacturing sectors of the economy (52 FR 38152). Employers in the nonmanufacturing sectors must be in compliance with all provisions of the standard as of May 23, 1988.

OSHA took this action following lengthy litigation concerning the scope of the original HCS. In May 1985, the U.S. Court of Appeals for the Third Circuit ruled in *United Steelworkers of America v. Auchter*, 763 F.2d 728, that OSHA had failed to demonstrate that inclusion of all workers within the scope of the Standard was not feasible. Although OSHA argued that over 50% of all reported cases of illness due to chemical exposure occur in the manufacturing sector, the Court found that

> some employees in specific nonmanufacturing categories, such as hospital workers, are exposed to a greater number of toxic substances than are typical workers in the manufacturing sector. Moreover, some workers in specific non-covered industries have higher reported rates of chemical source illness and injury than do workers in many covered industries.

The Court also rejected OSHA's argument that uncovered workers would benefit from "trickle down" protection as a result of hazard communication in the manufacturing sector. It therefore directed OSHA to either expand coverage of the HCS, or state reasons why such expansion would not be feasible. Two years later, OSHA still had not acted. (OSHA did issue an advance notice of rulemaking in November 1985, in which it requested comments on the feasibility of expanding the HCS to include all workers [52 FR 48794]. No further action was taken, however.)

In May 1987, the Third Circuit ordered OSHA to publish either an expanded HCS or a statement why expansion was not feasible, within 60 days (*United Steelworkers of America v. Pendergrass*, 819 F.2d 1263). The Court stated that its previous judgment in the *Auchter* case "did not contemplate going back to square one," and ordered OSHA to take action based on the record already developed. OSHA's request for rehearing was denied August 7, and, following refusal by the Department of Justice to appeal the case to the Supreme Court, OSHA issued its long-awaited rule.

In its final rulemaking notice, OSHA stated that it lacked evidence to demonstrate that the Standard would not be feasible for any part of the nonmanufacturing sector. It nevertheless invited persons to provide information on the practicality of the rule and any recommendations for further rulemaking within 60 days. On October 28, 1987, the Office of Management and Budget (OMB) directed OSHA to engage in additional rulemaking directed at reducing some of the paperwork requirements of the expanded Standard, and to develop a generic hazard communication program that could be used by all employers. OSHA has indicated that it plans to address the paperwork issue before the May 1988 effective date of the expanded Standard.

In addition, a total of four lawsuits have now been filed by various groups challenging the validity of the expanded standard. (See Appendix C.4 for a list of pending and decided cases concerning the HCS.) It is not clear when a decision will be reached in these cases.

Public Sector Employers

Although public sector employers are not technically within OSHA's jurisdiction, the Federal Advisory Council on Occupational Safety and Health has recommended that all federal agencies voluntarily comply with the key requirements of the HCS, including hazard identification and labeling, development of MSDSs and written communication plans, and worker training. As of August 1987, 38 federal agencies had agreed to participate in a uniform approach to hazard communication. A manual titled *Hazard Communication—A Program Guide for Federal Agencies* has been developed, and is available from the Government Printing Office. Plans have also been made to develop automated systems for MSDSs, standardized label forms, and training materials to be used throughout the federal government. Private sector employers are free to adopt portions of this uniform approach, but are not required to do so.

State and local public sector employers are not covered by the HCS, but may be covered by similar or identical provisions in state laws or regulations. See Section 10.2 and Table 11 for further details on state laws.

2.2 BASIC REQUIREMENTS

The provisions of the HCS which apply to each firm depend primarily on whether that firm is a chemical manufacturer or importer or a downstream employer. Figure 7 is a decision tree which traces the differences drawn in the standard.

Chemical Manufacturers and Importers

The basic requirements of the expanded HCS applicable to firms which manufacture or import and repackage covered materials remain the same as under the original Standard. These firms must first evaluate chemicals which they produce or import to identify those considered to be hazardous materials within the meaning of the Standard. Further details on how to accomplish this task are provided in Chapter 4. They must also ensure that each container of hazardous chemicals leaving the workplace is labeled, tagged, or marked with the identity of the hazardous chemical(s), appropriate hazard warnings, and the name and address of the chemical manufacturer, importer, or other responsible party. For further information on labeling, see Section 5.1.

Chemical manufacturers and importers must also obtain or develop an MSDS for each hazardous chemical they produce or import. The information required on MSDSs remains the same as under the original HCS, and is discussed further in Section 5.2. MSDSs must be provided with all shipments of hazardous chemicals to downstream suppliers or employers. They may be sent inside the package with the chemicals, with the shipping papers, or in a separate mailing, but they must be sent without waiting for the customer to request them. Whenever new hazard information relating to any of the covered chemicals is obtained, the MSDS must be updated within three months to include that information. The revised MSDS must be transmitted with the next shipment to each customer. If the chemical is not in production when the new information is received, the MSDS must be revised at the time of the first shipment when production resumes.

In addition to these requirements, chemical manufacturers and importers must also comply with the other provisions of the HCS applicable to downstream employers.

Downstream Employers

All private sector employers must be in compliance with the provisions of the expanded Standard as of May 23, 1988.

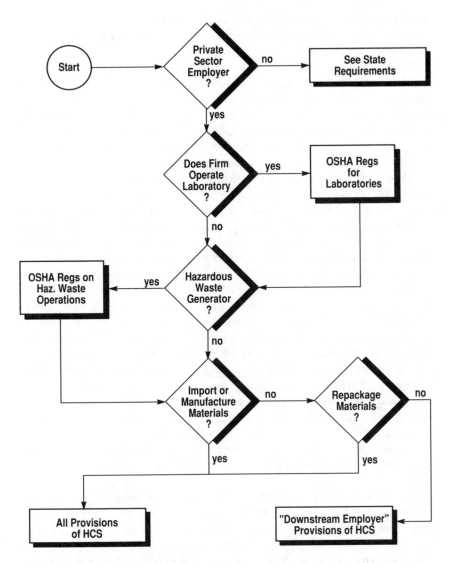

Figure 7.   Decision tree to determine applicable hazard communication requirements.

All firms are responsible for an initial assessment of the potential hazards present in the workplace. The same firm may simultaneously be considered an original manufacturer or importer for some operations and chemicals, and a downstream employer for others. Firms which supply chemical materials to other firms must assess the hazards of any materials they produce, import, or distribute, whether repackaged or not. Downstream employers may rely on the original manufacturer or importer for the hazard determination as well as the preparation of the labels and MSDSs, or they may do these jobs themselves. All three types of firms may also contract out these tasks.

Downstream employers who do not receive an MSDS with a shipment that has been labeled as a hazardous chemical should obtain one from the supplier as soon as possible. A prudent practice would be to include a routine request with each first-time order for a particular chemical or mixture. Similarly, downstream employers who receive an unlabeled shipment and are uncertain whether it contains hazardous chemicals should request this information as soon as possible. Those downstream employers who undertake an independent evaluation will assume responsibility for the adequacy and accuracy of the information they use.

All employers must also prepare a list of all the hazardous materials used in the workplace. For a large or complex operation, smaller lists should be made for each of the departments or areas in the facility. All firms where workers may be exposed to hazardous materials must also have a complete written hazard communication program in operation. Details of this program are covered in Section 7.1. The entire written program must be available for inspection upon request by workers or their designated representatives or OSHA. This requirement applies as well to distributors who simply repackage hazardous materials before reselling them. Even importers or distributors whose employees are not normally exposed because they do not repackage hazardous materials would be well advised to develop a program, because of the ever-present possibility of exposure resulting from accidental breakage or spills.

Copies of the MSDSs for all the hazardous materials used must be included in the written program. The written program must also include a statement as to how the hazards were assessed, which may be merely a statement that the assessments of the suppliers were adopted by the employer.

Hazardous materials lists and MSDSs must be available to all workers at all times. It is not enough to have them on file in an office unless that file is open and easily accessible to all. It is also doubtful whether having the lists and MSDSs on a computer will meet the requirements of the Standard, first, because there is always the possibility that the computer will be "down" at the time an accident occurs, and second, because not all workers may be computer literate.

Employers must also establish a training program to ensure that workers are aware of the provisions of the Standard, and that they know how to recognize the hazardous materials to which they may be exposed, what the hazards are, and how to use the materials safely. Important features of such a program are covered in Section 7.2. Training must also be provided for maintenance, security, and other such personnel who may be exposed occasionally. Office workers and other employees of the company who do not enter the regions of the facility where hazards exist do not need to be included in the training plan. However, hazardous materials are used in many offices, and where they are, the workers who use them on a regular basis must be included in the program. (For example, occasional use of a copying machine to make copies is not covered by the rule. However, if a particular worker is designated as a copying machine operator and is responsible for handling the chemicals associated with its use, he or she is entitled to information and training.) Customers or other visitors do not require training. A description of the training program and how it will be carried out must be included in the written program.

Construction Sites

One particular concern discussed by OSHA in its August 24, 1987 Notice of Final Rulemaking involved the difficulty of applying the provisions of the HCS to the construction industry, where several contractors and subcontractors may share a single work site, and where individual workers may shift from one site to another on an almost daily basis. OSHA had in fact received a recommendation from the Advisory Committee on Construction Safety and Health that employers in the construction industry be regulated under a separate standard, rather than being treated as downstream employers under the HCS.

OSHA decided, however, to include construction employers within the scope of the expanded Standard. It found the argument that construction sites are unique industrial workplaces to be unpersuasive where all that is required is the transmittal of information. OSHA noted that similar arguments regarding transient workers, mobile work sites, and other special problems can be made for other nonmanufacturing users of chemicals as well as the construction industry. Rather than preparing and cross-referencing completely separate standards for each such industry, it concluded that any special problems can be dealt with more effectively by future modifications to the expanded HCS.

Although it is likely that OSHA will eventually modify the HCS as it applies to the construction industry, it is uncertain when or how this modification will take place. On the one hand, at least two of the pending court challenges to the HCS mentioned above directly concern its application to construction sites and subcontractors. On the other hand, the Advisory Committee on Construction Safety and Health has revised its earlier recommendation, and advised OSHA *not* to exempt any industry sectors from the expanded HCS prior to its May 1988 implementation.

For now, employers in the construction industry must comply with all requirements applicable to other downstream employers.

Laboratories

Laboratories are subject to the provisions of the HCS to only a limited extent. Employers must ensure that labels on incoming containers are not defaced or removed, maintain any MSDSs that are received with incoming shipments of hazardous chemicals and make them readily accessible to laboratory employees, and provide a worker information and training program.

OSHA believes that this limited coverage for laboratories is appropriate for a number of reasons. First, laboratories commonly use small quantities of many different hazardous chemicals for short periods of time. Second, the conditions and purposes of the use of individual chemicals frequently change in ways that cannot be easily predicted. Third, many substances used by laboratories are of unknown toxicity. Finally, many laboratory workers are highly trained in the use of potentially hazardous chemicals.

It should be noted, however, that OSHA is currently considering a separate rule specifically addressing occupational exposure to toxic substances in laboratories, which may take effect in the near future (51 FR 26660 [July 24, 1986]). The proposed rule calls for laboratory employers to develop and implement written Chemical Hygiene Plans designed to protect laboratory workers from the health hazards associated with exposure to toxic substances. Each plan must incorporate standard operating procedures for all work involving toxic substances, must include criteria to invoke specific exposure control measures, and must identify procedures, activities, or operations requiring prior approval. Each plan must also include provisions for an exposure evaluation and medical consultation for workers who reasonably believe that they have been overexposed to a toxic substance.

2.3 EXEMPTIONS

Certain types of hazards are specifically excluded from coverage by the HCS, usually because they are regulated by other agencies of the federal government. Within the range of hazards covered, certain conditions of use are also wholly or partially excluded from coverage under one or more sections of the Standard.

Total Exemptions

Tobacco and tobacco products are excluded completely from the provisions of the HCS, as are food, drugs, or cosmetics which are intended for personal consumption. Food, drugs, cosmetics, and alcoholic beverages in retail establishments are also exempt if they are packaged for sale to consumers. Although some of these products may meet the definition of a "hazardous chemical," they do not pose any different or more significant hazards in the workplace than at home.

Wood and wood products are also excluded, even when the wood is treated with hazardous preservatives and other chemicals. Most such chemicals are covered by EPA regulations and are not included in the HCS. Wood dust, however, which is typically created as a byproduct of manufacturing operations involving wood and wood products, *is* considered a hazardous chemical, and is therefore covered by the Standard.

Also excluded are "articles," but this exclusion is a bit more complicated and has been subject to some interpretation. An article is defined as a manufactured item which is formed to a specific shape or design and which has functions dependent in whole or in part upon its shape or design during end use. Typical examples of articles are office products such as pencils, pens, typewriter ribbons, and the like. The purpose of this exemption is to avoid placing a regulatory burden on articles that are not actually hazardous during normal use. However, if the article releases, or would otherwise result in exposure to, a hazardous chemical under normal conditions of use, it is not exempt.

OSHA has sought to clarify this definition by stating that the exemption applies solely to the ultimate end use—intermediate uses which result in exposure are covered and trigger the requirements of the Standard. Thus, encapsulated asbestos insulation is covered by the HCS during installation, which involves hammering the material into openings, thus releasing the asbestos. Once the insulation is installed, however, it is considered an article, and is exempt. Another example concerns the use of paper in high-speed converting. Although paper is normally considered an article, paper dust is regulated, and if it is produced in an appreciable quantity, the paper is covered by the HCS.

Consumer products and hazardous substances covered by the Consumer Product Safety Act and Federal Hazardous Substances Act are exempt if they are used in the workplace in the same manner as normal consumer use, and if the exposure of employees is no greater than that experienced by consumers. An example given by OSHA involves abrasive cleaners which, if used intermittently to clean a sink, are exempt, but which are covered if used to clean out reactor vessels, or if used by the same worker to clean sinks all day long.

Finally, any drug regulated by the Food and Drug Administration (FDA) which is in solid form for direct administration to the patient (such as aspirin tablets) is exempt from the Standard.

Hazardous Wastes

Section (b)(6)(i) of the HCS states that any hazardous waste as defined by the Resource Conservation and Recovery Act (RCRA) and subject to the regulations issued under that act by the EPA is exempt from all aspects of the Standard. (Further information on RCRA can be found in Section 10.1.) In the Superfund Amendments and Reauthorization Act (SARA) of 1986, however, Congress directed OSHA to adopt additional standards for protection of the health and safety of workers engaged in hazardous waste operations. These regulations must specifically address the following areas:

- site analysis
- training by contractors
- medical examination, monitoring, and surveillance
- personal protective equipment, clothing, and respirators
- engineering controls
- maximum exposure limits
- informational programs regarding likely exposure
- handling, transporting, labeling, and disposal
- introduction of new equipment or technologies
- decontamination procedures
- emergency response and protection of workers

OSHA has published an interim final rule, and a notice of proposed rulemaking and public hearings in order to comply with this direction (29 CFR § 1910.120, 52 FR 29620 [Aug. 10, 1987]). Final regulations are in effect as of August 10, 1988.

Mixtures

If a mixture is tested as a whole and found to be hazardous in any of the defined ways, it is considered to be hazardous. Any mixture that contains at least 0.1% of a carcinogen or at least 1% of any other health hazard is considered to be a hazard, and the identity of the specific hazard must be given on the label and in the MSDS. Even mixtures containing smaller amounts are covered if it is reasonable to expect that vapors might be released in excess of applicable threshold limit values® (TLVs) or permissible exposure limits (PELs). If a mixture is not tested as a whole to determine physical hazards, evaluation of the physical hazard potential of the mixture can use any scientifically valid data that may be available.

If any mixture is found to be hazardous, the identity of the compound or compounds imparting that property to the mixture must be indicated on the label and the MSDS unless exempt under the trade secret provisions of the Standard. (See Chapter 9 on trade secrets).

Modified Labeling Requirements

Many items are exempt from the labeling requirements of the HCS, although still subject to its other provisions, because they are covered by labeling requirements of other agencies which are even more stringent than those of OSHA's Standard. These items include:

- food, food additives, drugs, or cosmetics, including flavors and fragrances, subject to labeling requirements and regulations issued by the Food and Drug Administration under the Federal Food, Drug, and Cosmetic Act
- alcoholic beverages intended for nonindustrial use which are subject to labeling requirements and regulations issued by the Treasury Department's Bureau of Alcohol, Tobacco, and Firearms
- consumer products or hazardous substances defined in the Consumer Product Safety Act and the Federal Hazardous Substances Act, provided they are subject to requirements or regulations issued by the Consumer Product Safety Commission
- pesticides defined in the Federal Insecticide, Fungicide, and Rodenticide Act (FIFRA) which are subject to regulations issued by the EPA under FIFRA

Further information on these related federal statutes is provided in Section 10.1.

Process tanks and reactors whose contents may vary from batch to batch do not have to be labeled in the same way as original packages or storage tanks. The information normally required on the label must be available, but it can be in the form of placards, batch process tickets at the process control desk, or some other form selected by the employer. Any such alternative method of providing warning information must clearly indicate the containers to which it applies, and must be available to all workers affected throughout their shift.

The HCS does not require that piping be labeled within the plant. Note, however, that some state Right-to-Know laws do require this, so each employer should check to see what requirements apply. The HCS does require that the information that would normally have to be included on a label must be available to the workers at all times, and some way of identifying what is in the particular pipes must be provided. This may be a piping layout diagram, or the pipes may be color coded or marked with some identifiable symbol, or any other effective method may be used.

Portable containers into which hazardous chemicals are transferred from labeled containers are exempt from labeling requirements under certain conditions. At all times, such "transfer vessels" must be under the control of the person who originally filled them, and the label exclusion only applies for the immediate use of the material in the vessel, i.e., within the work shift in which it is transferred. OSHA has stated that the actor in interpreting the scope of this exemption is whether the contents of the transfer vessel are known to the workers exposed to them. Thus, a worker who takes a meal break, for example, should transfer any hazardous chemicals back into labeled containers before taking his or her break, so that other workers who are not aware of the contents of transfer vessels will not be exposed.

Containers of chemicals in laboratories do not have to meet the label requirements of the Standard, provided that proper labels originally on the containers cannot be defaced or removed while any chemical remains in the container. In effect, this exemption applies primarily to chemicals synthesized or in process within the laboratory itself. Transfer vessels used in laboratories do not need to be labeled.

2.4 ENFORCEMENT AND SANCTIONS

Civil and Criminal Penalties

The HCS was issued under the authority of the Occupational Safety and Health Act (the OSH Act). The OSH Act provides both civil and criminal penalties for violations of standards, rules, or orders issued under its authority (29 USC § 666).

An inspector who discovers such a violation is required to issue a citation. First-time violations generally result in civil fines of up to $1,000 for each violation. Failure to correct a cited violation may result in a civil fine of up to $1,000 for each day during which the violation continues. Willful or repeated violations are punishable by fines of up to $10,000. A willful violation which results in the death of a worker is a criminal violation punishable by a fine of $10,000 and six months in jail. A second conviction raises the penalties to $20,000 and one year in jail.

Rules Governing Inspections and Enforcement Proceedings

OSHA has adopted rules governing the proper conduct of all inspections and enforcement proceedings carried out under its authority. These rules, which are described in general terms below, are published in 29 CFR Part 1903.

OSHA inspectors may enter any factory, plant, establishment, construction site, or other area where work is performed, during regular hours and at other reasonable times, for the purpose of conducting inspections. Advance notice of these inspections is specifically *prohibited,* except where notice may enable an employer to abate an apparent, imminent danger as quickly as possible. Employers have the right to refuse to allow an inspector to enter the site or any part of the site or to question certain persons, after which the inspector may seek a court-issued warrant. Once a warrant is obtained, the inspection must be allowed to take place.

Inspectors are authorized to take environmental samples, obtain photographs, and use other reasonable investigative techniques, including asking questions in private. Representatives of both the employer and the workers shall be allowed to accompany the inspector during the physical inspection. At the beginning of the inspection, the employer may identify areas containing trade secrets, and may restrict access to those areas to certain worker representatives.

At the conclusion of the inspection, the inspector must confer with the employer or his representative and review any apparent safety or health violations. If the inspector believes there is an imminent danger of death or serious physical harm, a citation may be issued immediately, followed by a court order to correct the dangerous condition. Otherwise, citations may only be issued following review of the inspector's report. Any citation issued must describe the nature of the violation, fix a reasonable time for correction of the violation, and inform the employer of his right to appeal the citation to the OSH Review Commission. The employer must also be notified of the proposed penalty, which shall become final unless an appeal is made.

All citations, whether appealed or not, must be posted in a prominent place where they will be readily observable at or near the place of the alleged violation, and must remain posted for three working days or until the violation has been corrected, whichever comes later. If the citation is being appealed, the employer may also post a notice to that effect.

Although OSHA has authority to assess civil penalties under the procedures outlined in its rules, no criminal penalty may be imposed without an opportunity for a jury trial in a federal district court. Employers who are cited for a violation by OSHA may want to consult an attorney concerning the specific issues of their case, and should definitely do so if the violation may lead to a criminal penalty.

Civil Tort Liability

Aside from the possibility of OSHA citations, compliance or noncompliance with the HCS can affect a firm's civil tort liability in the event of injury to a worker caused by hazardous chemical materials. Civil tort liability generally is the result of a finding that a person or firm is in some way responsible for an injury to the health or property of one or more other persons. The type of responsibility involved may vary depending on the relationship between the parties and the nature of their activities. In general, if a firm or employer fails to exercise reasonable care under the circumstances to protect its workers from harm, then it may be negligent. If negligence is proven in court, then damages may be awarded, normally by a jury. These awards are often substantial, and may involve punitive as well as compensatory damages.

How well a firm or employer complies with the HCS can have an influence on negligence liability in two ways: (1) it can affect the standard to be applied in determining whether negligence has occurred, and (2) it can influence the decision as to whether that standard has been satisfied. Because the exact nature of this relationship will depend on all the facts of a particular situation, as well as the law of each jurisdiction, specific questions should be addressed to a local attorney.

Many workers are covered by workers' compensation plans which provide an exclusive remedy for job-related injuries. Unintended lapses in compliance with the HCS are not likely to affect the employer's liability under such a plan. Where the employer is also a manufacturer or importer, however, there may be a question as to whether the injury is job-related, depending on the cause of the injury and the work performed. Intentional noncompliance may also result in a suit for punitive damages, regardless of whether the injury itself was intended. Thus, a workers' compensation plan does not automatically insulate the employer from civil liability to injured workers.

For injured employees not covered by workers' compensation, any failure to comply with the HCS may create an implied proof of negligence on the part of the employer. This is particularly so since one purpose of the HCS is to ensure that employees are adequately informed of the hazards of the workplace. Even if the court does not consider the failure of compliance itself to be proof of negligence, expert witnesses will probably testify that the prevailing standard of the industry includes compliance. Failure to meet the prevailing standard of the industry in any particular situation is likely to result in a finding of negligence.

Chemical manufacturers, importers, and other suppliers could also be subject to claims for injuries in downstream firms if they fail to comply with requirements for labels and MSDSs. Liability could result either from a failure to supply labels and MSDSs (although here the downstream employer also has a duty to request documentation), or from the provision of misleading information which results in an injury.

Compliance with the HCS does not ensure that a firm will not be found to be negligent, though, if it can be shown that the prevailing standard of the industry is to provide more information than is actually required by OSHA, or if the firm involved is subject to other federal, state, or local laws or regulations requiring additional information. (See Chapter 10 on related laws.) Finally, claims may be brought against an employer for injuries caused by the failure of a contractor to comply with the Standard under a theory of joint responsibility. No generalized statements can be made regarding the possible success of such actions, which will depend on the facts of each case and the law in the local jurisdiction.

Each firm must therefore be aware of its own legal requirements and prevailing industry standards, and seek to disclose hazard information in accordance with them. Moreover, any questions regarding potential liability resulting from a specific incident should be directed to an attorney familiar with the laws applicable to that jurisdiction.

3

The Emergency Planning and Community Right-to-Know Act

The Emergency Planning and Community Right-to-Know Act of 1986 was signed into law by President Reagan on October 17, 1986. It is a self-contained Title III of the Superfund Amendments and Reauthorization Act of 1986 (SARA Title III). It is codified as Chapter 116, Title 42, United States Code (42 USC § 11101 *et seq*.).

SARA Title III was initially proposed in Congress following the December 1984 chemical disaster in Bhopal, India, where more than 2000 people died following a leak of methyl isocyanate from a plant owned by Union Carbide (India), Ltd., a partially-owned subsidiary of Union Carbide Corp. Its purpose is to "encourage and support emergency planning efforts at the state and local levels and provide the public and local governments with information concerning potential chemical hazards present in their communities." Although the Act is codified as Title III of SARA, it is not part of the Superfund law itself.

The Act is organized into three subdivisions: subtitle A deals with community emergency preparedness and notification in the event of a hazardous chemical accident; subtitle B contains reporting requirements regarding hazardous chemical inventories and toxic releases, and establishes a national toxic release data base; and subtitle C includes various provisions regarding administration, enforcement, and trade secret protection. The provisions of SARA Title III are administered by two offices within the EPA: the Office of Solid Waste and Emergency Response is responsible for administering the emergency preparedness portions of the law, and the Office of Pesticides and Toxic Substances is responsible for the reporting portions. Actual implementation, however, is left up to state and local bodies, which are free to adopt stricter standards applicable within their jurisdiction.

The complete text of SARA Title III is reprinted in Appendix C.2. Regulations proposed and adopted by the EPA as of February 1988 are listed in Appendix C.3.

3.1 EMERGENCY PLANNING AND NOTIFICATION

Subtitle A of SARA Title III requires the establishment of a national network of local emergency response plans to deal with the release of hazardous chemicals into the environment from establishments where those chemicals are used. It forms a logical counterpart to the National Contingency Plan established by Congress for dealing with oil spills and hazardous substance releases on a national level. (See Section 8.4.)

Subtitle A is similar to the HCS in that it is a performance-oriented statute which contains few specifics. Instead, it describes a general administrative structure and lists various factors to be considered. Chapter 8 provides a more detailed discussion of how individual planning committees might go about fulfilling their duties under SARA Title III.

State Commissions, Planning Districts, and Local Committees

Administration of the emergency planning and notification provisions of SARA Title III is left up to a system of state emergency response commissions and local emergency planning committees. State commissions were required to establish emergency planning districts no later than July 17, 1987. Districts are defined geographically, and may be identical to existing political subdivisions. There is no limit to the number of districts within each state. Where appropriate, multistate districts may be established by agreement of the state commissions involved. Local committees were to be established by August 17, 1987. Each committee is responsible for a single emergency planning district. Local committees, like state commissions, must include representatives of groups or organizations listed in the statute.

Although the provisions of subtitle A are federal requirements, Congress made it clear that they are to be implemented solely at the state and local level. The EPA also has stated that implementation of Title III is primarily a state and local responsibility: EPA does not intend to oversee the operation of individual commissions and committees. Each local committee is required to establish its own rules and procedures, and is responsible for obtaining the resources necessary to carry out its duties under SARA Title III.

Further discussion of state commissions, planning districts, and local committees appears in Section 8.1.

Comprehensive Emergency Response Plans

Each local committee must have completed preparation of an emergency response plan no later than October 17, 1988. Plans must be reviewed annually, or as circumstances require. Each plan must include, at a minimum:

- identification of facilities and transportation routes likely to be involved
- methods and procedures for responding to a release
- designation of community and facility coordinators
- procedures for public notification
- methods for determining the occurrence of a release, and the area or population likely to be affected
- a description of emergency equipment and facilities
- evacuation plans
- training programs
- methods and schedules for exercising the plan

These elements are discussed further in Section 8.2.

The emergency response planning process must be open to public participation. Once developed, the plans are reviewed by the state commissions, and may also be reviewed by regional response teams established under the hazardous waste provisions of Superfund. See Section 8.3.

Substances and Facilities Covered

All facility owners or operators initially covered by the Act must have notified their state commission that they are covered by May 17, 1987, and must have appointed an emergency preparedness coordinator to participate in the local planning process by September 17, 1987. Covered facilities must also comply with requests from local planning committees for basic information necessary to develop emergency response plans.

A facility is covered if it contains a quantity of an extremely hazardous substance which exceeds the threshold quantities established by the EPA, or if it has been specially designated by the state commission. Appendix D.2 reproduces the list of extremely hazardous substances covered by subtitle A, along with the threshold planning quantities (TPQs) established by the EPA. Local planning committees are encouraged to add additional hazardous substances to their own lists of substances requiring notification, so facility owners and operators should contact their local committee to make sure they have a complete list.

Facilities which are not covered initially, but which either contain sufficient quantities of a chemical added to the list at a later time, or which increase the quantity maintained of an extremely hazardous chemical above the TPQ, must notify the state commission and local committee within 60 days.

The state commission must notify the EPA of each facility within its jurisdiction subject to the requirements of subtitle A.

Notification of Releases

Once a local emergency response committee has been established, facilities within that committee's jurisdiction must notify it of most releases of a reportable quantity (RQ) of a chemical which is on both the SARA Title III extremely hazardous chemicals list and the hazardous substances list established under Superfund. If the release exceeds the RQ for a substance on only one list, it must generally still be reported if it occurs "in the same manner" as a reportable release under Superfund. Appendix D.2 lists RQs for all chemicals covered under either statute.

The EPA announced plans to simplify these requirements by producing a single list of hazardous substances, each with its own RQ, applicable to both SARA Title III and Superfund. This list was to have been published by April 30, 1988.

Notice, when required, must be provided immediately by whatever means is available (usually telephone) to the local community emergency coordinators, and to any state commission likely to be affected. This notification must be updated by written notice, as soon as practicable. The written notice shall include information on actions taken, known or anticipated health risks, and medical advice where appropriate.

A more detailed discussion of when notification is required and what information must be provided is contained in Section 6.4.

3.2 REGULAR FACILITY REPORTING REQUIREMENTS

In addition to the emergency planning and notification provisions of subtitle A, SARA Title III also imposes a number of regular reporting requirements on facility owners and operators. The purpose of these requirements is to increase the availability of information to state and local officials and the general public so as to improve the ability to prevent chemical emergencies, to react quickly and effectively when they do happen, and to understand and address their long-term effects. Sections 6.2, 6.3, and 6.5 present a detailed discussion of methods for collecting and providing the information required under subtitle B.

Transportation, including storage incident to transportation, is exempt from the reporting requirements of subtitle B. Emergency notification must still be provided under subtitle A, but this may generally be fulfilled by dialing 911 or the operator, since many transportation operators on the road will not know the telephone numbers of the relevant state and local officials.

Hazardous Substance Lists or MSDSs

The owner or operator of each facility required to maintain MSDSs under OSHA's HCS must either submit copies of the MSDSs or a list of chemicals for which MSDSs are kept to the local planning committee, the state commission, and the local fire department. The expansion of the HCS on May 23, 1988 to include employers in nonmanufacturing sectors means that this requirement applies to virtually all private sector facilities where hazardous chemicals are present in amounts exceeding the threshold limits established by the EPA. Hazardous substance lists or MSDSs must be submitted no later than three months after the facility is first required by OSHA to maintain MSDSs.

The EPA is encouraging facility owners and operators to use the option of providing a list of chemicals in lieu of MSDSs. This will reduce the information management burden on recipients of the information, as well as providing a list for indexing of the hazardous chemical inventory forms discussed below. Lists submitted in lieu of the actual MSDSs must include the chemical name or common name of each chemical, grouped according to health and physical hazards, and the hazardous component of each. Lists must be updated within three months following a revision of what is covered by the HCS. Should the owner or operator elect to submit a list of hazardous substances in lieu of MSDSs, the local planning committee may obtain a copy of the MSDS for any listed chemical upon request. The local planning committee is authorized to make MSDSs available to the public within 30 days of a request.

For purposes of this requirement, "hazardous chemical" is defined as in the OSHA regulations, except that it does not include substances regulated by the FDA; solids used in manufactured items; substances used for personal or household purposes; substances used in research laboratories, hospitals, or medical facilities under direct supervision; and fertilizers and substances used in routine agricultural operations.

Hazardous Chemical Inventory Forms

Facility owners or operators who must submit MSDSs or hazardous chemical lists must also prepare an inventory form covering those chemicals for which MSDSs are maintained and which are present in excess of specified threshold reporting quantities. Standard inventory forms have been prepared by EPA and must have been completed with respect to

"Tier I" information and submitted to the local planning committee, the state commission, and the local fire department no later than March 1, 1988, and must be updated annually thereafter. Tier I information consists of the estimated daily and maximum quantities of hazardous chemicals present at the facility during the preceding year, and their general location. The information may be aggregated according to health and physical hazards as defined in OSHA regulations.

"Tier II" information, consisting of specifics pertaining to particular chemicals, shall be provided upon request. State commissions and local committees may make requests on behalf of state or local officials acting in their official capacity, and may also request information on behalf of the general public. Requests by the general public to the state commission must generally be granted, but may be denied if they concern a chemical which a facility has stored in an amount less than 10,000 pounds during the preceding year.

The EPA has adopted threshold quantities below which this requirement does not apply. These RQs will be phased in over a period of three years. Minimum RQs are 10,000 pounds effective for two years as of March 1, 1988, and any amount of a chemical for which an MSDS is required in subsequent years. For extremely hazardous substances, the threshold in the first two years is 500 pounds or the TPQ, whichever is lower. Appendix D.2 identifies all chemicals specifically regulated under SARA Title III and lists TPQs and RQs. As with hazardous substance lists, these thresholds may be changed, and states or local communities may require additional information under their own laws in addition to that required under SARA Title III.

Copies of Tier I and Tier II inventory forms are reproduced in Appendix E.2. The information required and procedures for filling out these forms are discussed in greater detail in Section 6.3.

Toxic Chemical Release Inventory Forms

Owners or operators of facilities that have ten or more employees and that are in SIC codes 20 through 39, as well as other facilities specifically identified by the EPA, must also have submitted a toxic chemical release form to the EPA and a designated state official on or before July 1, 1988 and annually thereafter. These forms, which are discussed in greater detail in Section 6.5, must include, at a minimum:

- the name, location, and principal business activities of the facility
- an appropriate certification
- whether each toxic chemical is manufactured, processed, or otherwise used
- the general category of use of each chemical

- the estimate of the maximum amounts of each toxic chemical present at the facility at any time during the preceding year
- waste treatment or disposal methods employed for each waste stream and an estimate of the treatment efficiency typically achieved
- the annual quantity of each toxic chemical entering each environmental medium

Congress specifically stated that information required under this section may be collected from readily available data, and that no monitoring is to be required beyond that necessary to comply with provisions of other laws or regulations.

The EPA issued its final rule implementing this requirement on February 16, 1988 (53 FR 4500). A separate list of toxic chemicals has again been established, subject to threshold reporting quantities, which are listed in Appendix D.2. For purposes of submitting toxic release inventory forms, the threshold quantity is stated in terms of the minimum quantity manufactured, imported, processed, or otherwise used at a facility during a calendar year. The threshold quantity for toxic chemicals which are manufactured, imported, or produced is 50,000 pounds in 1988 and 25,000 pounds in 1989 and thereafter. For toxic chemicals which are otherwise used, the threshold is 10,000 pounds for the applicable calendar year. Covered facilities must submit an annual form for each toxic chemical which satisfies the established threshold.

Although the EPA may modify the frequency with which reports must be submitted, it may not do so earlier than July 1, 1993.

3.3 PUBLIC DATA BASES AND TRADE SECRET PROTECTION

Emergency response plans, MSDSs or hazardous substance lists, and Tier I inventory forms must be made available by local planning committees for public inspection at a designated location during normal business hours. Information contained on Tier II inventory forms may generally be obtained by filing a request with the state commission. In addition, the EPA is required to make the information submitted on toxic release inventory forms accessible through telecommunications from a national computer data base on a cost reimbursement basis. More information about these data bases and how to obtain access is contained in Section 11.2.

Facility owners or operators required to submit MSDSs or hazardous substance lists, inventory forms, and toxic release inventory forms (as well as other information required to assist in emergency planning) may withhold a specific chemical identity if it is a protected trade secret. Specific chemical identities must be provided to health professionals, nurses, and doctors when needed to provide medical diagnosis or treatment or to carry out the preventive purposes of Sara Title III. Except in medical emergencies, requests for this information must be in writing, and the information obtained must be held in confidence. Trade secrets and the release of protected information under both the HCS and SARA Title III are discussed further in Chapter 9.

3.4 ENFORCEMENT AND SANCTIONS

Administrative and Criminal Enforcement Procedures and Penalties

The enforcement procedures and penalties for SARA Title III vary widely, depending on which provision of the Act is at issue and the specific violation alleged. These procedures were set forth by Congress in section 325 (42 USC § 11045).

The EPA is authorized to order any facility owner or operator to comply with the emergency planning provisions of subtitle A, and to seek enforcement in federal district court. Failure to respond to a court order can result in a civil fine of up to $25,000 per day. No fine can be assessed, however, until a court has issued an order with which the owner or operator has failed to comply in a reasonable time.

The EPA is authorized to directly assess a civil fine of up to $25,000 per violation for an inadvertent failure to provide notification of an emergency release to local committees and fire departments. No fine may be assessed without notice and an opportunity for an informal hearing, however, and owners or operators may appeal a fine to federal district court. A civil fine of up to $25,000 may be assessed for each additional day during which notification is not provided, and for up to $75,000 in the case of a second or subsequent violation. Penalties based on continuing violations may only be assessed following a formal administrative hearing conducted according to procedures published in 40 CFR, Part 22. They include, among other things, the right to submit evidence, to present and cross-examine witnesses, and to be represented by an attorney.

A civil fine of up to $25,000 may be assessed by the EPA for failure to submit inventory forms and toxic chemical release inventory forms as required under subtitle B. Failure to provide MSDSs or hazardous substance lists, and failure to provide specific chemical identities to health professionals when required, may result in a fine of up to $10,000. Finally, the EPA may assess a civil fine of up to $25,000 per claim for a frivolous trade secret claim. These penalties may be assessed either directly or by bringing an action in district court.

Persons who knowingly or willfully fail to provide emergency notice may be subject to criminal penalties of up to $25,000 and two years in jail for a first offense, and up to $50,000 and five years in jail for subsequent offenses. Persons who knowingly or willfully disclose protected trade secret information may be subject to criminal penalties of up to $20,000 and one year in jail. Criminal penalties may not be assessed until a conviction has been attained in a criminal trial brought by a federal prosecutor.

Civil Enforcement of Statutory Requirements

In addition to EPA enforcement actions, Congress has also authorized a number of ways for private citizens and state and local government officials to enforce the provisions of SARA Title III.

Any health professional who is denied access to specific chemical identities following a proper request may bring an action to obtain that information in federal district court. This procedure will obviously be of little use during medical emergencies, but may be used to obtain information required for preventive measures or long-term monitoring or treatment of persons injured by toxic releases.

Congress has also authorized civil suits by private citizens against facility owners or operators, state and local governments, or the EPA to enforce the requirements and deadlines of SARA Title III. This authorization is in addition to any other rights possessed by private citizens. Each such suit must be preceded by 60 days' notice, and no suit may be brought against a facility owner or operator if the EPA has commenced and is diligently pursuing its own action. The court may award litigation costs to the substantially prevailing party.

Because no suits have yet been filed under this provision, it is somewhat difficult to determine what standards will be applied or how effective they are likely to be. Congress has included similar provisions in a variety of other environmental laws, however, and a brief summary of issues in light of those laws is provided in Section 11.2.

State and local governments and planning bodies are similarly authorized to file civil actions against facility owners or operators or the EPA, except that they are not subject to the limitations applicable to citizen actions regarding notice and diligent action by the EPA. Finally, a state may file suit against the EPA for failure to provide information submitted in support of trade secret claims which has been requested by the state.

Civil Tort Liability

Just as compliance or noncompliance with the HCS may affect civil tort liability to injured workers, compliance or noncompliance with SARA Title III may affect liability for injuries to persons or property in the surrounding communities following a release of hazardous chemicals. The fact that a substance is on one or more of the lists compiled by the EPA or the local planning committee may be offered as evidence that a duty exists to take reasonable measures to protect against harm. Failure to comply with provisions of SARA Title III may likewise be offered as evidence of negligence in carrying out that duty and may be sufficient in and of itself to establish civil liability, depending on the jurisdiction.

Depending on the size of the release, the prevailing weather conditions, and the location of the facility, the damage following a hazardous chemical release could be enormous. Once negligence has been established, a primary issue in any case will be whether all of the alleged injuries and property damage have been "proximately caused" by the release, i.e., whether there is an uninterrupted chain of causation stretching from the negligent action to the injury or damage, which is not superseded by a subsequent, intervening cause. Issues of causation are likely to be highly complex and will depend on the particular facts of each case.

Again similar to the HCS, compliance with SARA Title III does not guarantee immunity from negligence liability—liability for failure to provide warning or notification may still exist if there are local laws or prevailing industry standards which require more than the federal law. Moreover, no amount of notification can erase liability for negligent actions which cause a hazardous chemical release to occur in the first place.

It is also possible that claims could be brought based on failure to develop emergency response plans or negligent performance of response actions following a release. Again, the standards to be applied and the likely outcome will vary with the jurisdiction and the facts of each case.

Finally, the mere fact that more information is available to the public regarding the nature of chemicals in the community may lead to an increase in civil suits based on personal injury claims from persons who otherwise would not have identified a hazardous chemical as a possible cause. Similarly, it is possible that suits to enjoin certain activities based on public nuisance claims or other environmental statutes will increase for the same reason. In either case, SARA Title III does not affect existing duties to exercise reasonable care and comply with other statutes and regulations—it only results in a "higher profile" which may lead indirectly to more claims.

Part B

Carrying Out the Responsibilities

Compliance with the provisions of the HCS and SARA Title III is a complex task which will probably create significant difficulties for many chemical manufacturers and distributors, other firms using hazardous chemicals, and local emergency planning committees. Part B of this book is addressed to those persons.

Chapter 4 deals with identification and characterization of hazardous chemicals as required under the HCS. It covers both physical and health hazards. Chapter 5 deals with documentation of hazards and warnings under both the HCS and SARA Title III. It covers label content and design and preparation of MSDSs. Chapter 6 addresses facility recordkeeping and reporting requirements under SARA Title III. It covers the development of hazardous substance lists, completing emergency and hazardous chemical inventory forms, emergency notification of hazardous material releases, and completing toxic chemical release inventory forms. Chapter 7 deals with preventive measures in the workplace under the HCS. These include the written hazard communication program as well as worker information and training. Finally, Chapter 8 deals with the planning requirements faced by local communities under SARA Title III. Each of the required elements of the plan is addressed briefly, and information on where to obtain more detailed guidance is provided.

4

Identification and Characterization of Hazardous Chemicals

Both health hazards and physical hazards are covered by the HCS. Some of these hazards are potentially severe, and some are relatively mild, but all are covered equally.

The first step in identifying the hazardous materials is to make a complete inventory of all the chemicals and mixtures of chemicals on the premises. The second step is to prepare a list of suspects for more detailed investigation.

Following this, individual materials can be characterized as to the hazards they present, and organized by hazard and by location in the facility. This inventory, characterization, and organization provides the basis for the entire written program and information and training system.

4.1 HAZARD IDENTIFICATION

The Chemical Inventory

The initial inventory should include all chemicals and other materials on the premises. Thus, everything that is a containerized gas or liquid, or is a powder or granular solid, should be included if it is in a can, bottle, box, drum, cylinder, or tank. In addition, raw materials that will be processed in a manufacturing operation, such as metals, wood, paper, and plastics, should be included on the original inventory.

Chemical Manufacturers and Distributors

The inventory list for detailed examination of potential hazards should include all individual chemicals, all mixtures prepared from individual chemicals or other mixtures, and the raw materials used to produce finished products. Because manufacturers or importers have the primary legal responsibility to characterize all the material hazards connected with any of their products, they must conduct an exhaustive search for hazardous properties.

If a particular product is not covered under the Standard, a form letter of disclaimer should be prepared for distribution to those customers who request it. Figure 8 illustrates such a letter for nonhazardous materials such as plastics molding products. Figure 9 is a similar letter for such articles as plastic sheet or film or insulating board, paper, composition building panels, etc. However, the use to which such materials will be put by those who purchase them must be taken into account. For example, if a manufacturer purchases paper stock which will be run through cutting machinery that generates airborne paper dust, that is by definition a hazardous material and the manufacturer of the paper stock is obligated to furnish MSDSs and other information. Similar comments apply to wood, cloth, and other materials.

Because of variations in the provisions of the HCS adopted in some state-plan states (see Section 10.3) and approved by OSHA, a local attorney should be asked to check these letters of disclaimer for both the state of origin and the state of destination. After this has been done, a sentence might be added to the effect that the manufacturer "believe(s) that statutes and standards in {state of destination} are similar with respect to such materials."

Employers in All Other Industries

For an employer in any sector other than one producing hazardous materials, the first step is to survey the facility to identify all possible materials that might be classified as hazardous. Later, the results of this survey will be used to determine which materials are in fact hazardous, and the specific hazards involved. But the first step is to make an inventory from which a list of prime suspects will be prepared.

Purchase orders for the most recent 12-month period should be examined. (In some facilities, a longer period may be more appropriate.) Both materials that are ordered regularly and those that are only ordered occasionally should be included. Where each material is used in the facility should be determined and noted on the inventory. Those that prove to be hazardous will need to be listed by location anyway, so this is a chance to get that job partly done.

GEORGE LOWRY ASSOCIATES, INC.   5878 SCENIC WAY DR., KALAMAZOO, MI 49009   (616) 375-4368

April 1, 1988

{ Inside }
{ Address }
{ Here   }

Dear { Name }:

The following information is supplied in response to your request
for a Material Safety Data Sheet on our { Name of Product }.

This material is not a listed carcinogen, does not have a TLV value
assigned, and is not previously regulated by OSHA. Furthermore, a
diligent search of the available literature has failed to establish
that this material is hazardous under the Hazard Communication
Standard as defined in 29 CFR 1910.1200 Section (c). Hence a
Material Safety Data Sheet is not available.

Thank you for your interest in our products.

Sincerely yours,

George G. Lowry
President

Figure 8.   Sample disclaimer for nonhazardous materials.

GEORGE LOWRY ASSOCIATES, INC.   *(616) 375-4368*
5878 SCENIC WAY DR., KALAMAZOO, MI 49009

April 1, 1988

{ Inside }
{ Address }
{ Here   }

Dear { **Name** }:

The following information is supplied in response to your request
for a Material Safety Data Sheet on our { **Name of Product** }.

This item is an "article" as defined in 29 CFR 1910.1200 Section
(c). Further, this article does not release or otherwise result in
exposure to a hazardous chemical under the conditions of your
intended use. As stated in Section (b), Subsection (5), the Hazard
Communication Standard does not apply to such articles. Hence a
Material Safety Data Sheet is not available.

Thank you for your interest in our products.

Sincerely yours,

George G. Lowry
President

Figure 9. Sample disclaimer for "articles."

Either before or after the purchase order survey, a walk-around inspection of the premises should be conducted to find anything that might not have appeared on the recent purchase orders. This would include those materials that might have been around longer than a year, free samples provided by sales representatives, and materials purchased with petty cash. All these should be added to the inventory, with a notation of their locations.

Also, during the survey of materials present in the facility, particular attention should be paid to Department of Transportation (DOT) hazardous material labels on any drums or other containers. Some of the hazard definitions used by DOT are different from those used by OSHA, but anything classed as hazardous by DOT will usually be classed as hazardous by OSHA too. The DOT class will also provide a starting point for identifying the OSHA hazard classes. Also, some chemical suppliers provide hazard information on package labels in addition to what DOT requires, so look for these as well. If hazardous waste disposal pickups are made regularly, the manifest files should be examined. These will serve as clues of where the hazardous waste originated and will lead to the source materials, which in most cases will be hazardous also.

Once the initial survey has been completed, the vendors of all the materials should be asked to send either MSDSs or disclaimers stating that specific materials are not classified as hazardous or are not covered by the Standard. While awaiting a response from your vendors, a preliminary check should also be made using various reference sources.

Materials Covered by Reference

All substances already regulated by OSHA are included by reference in the HCS. These substances are listed in Tables Z-1, Z-2, and Z-3 located at 29 CFR 1910.1000–1048 in the OSHA regulations.

Also covered by reference are all substances listed in the latest edition of *TLVs: Threshold Limit Values for Chemical Substances . . .* published by the American Conference of Governmental Industrial Hygienists (ACGIH).

All the substances in the TLV list and in the Z-tables (between 600 and 700 of them) are listed together in Appendix D.1, as well as an indication of whether they are known carcinogens and whether they present a skin absorption hazard. In this table, substances are listed by Chemical Abstracts Service Registry Number (CAS No.), a widely used identification system. The table in Appendix D.4 permits finding the CAS number for any given substance, as well as commonly used synonyms and trade names.

The materials inventory should be checked against the list of substances in Appendix D.1. Any suspect chemicals that actually appear on this list are definitely regulated. Those facilities that don't employ a chemist may want to enlist the help of one for this purpose.

Only the compounds listed in the OSHA Tables Z-1, Z-2, and Z-3 have an OSHA-specified limit of workplace air concentration (PEL), and only those listed in the ACGIH booklet have an assigned TLV value. The term TLV is a registered trademark of ACGIH, and for anyone to assign their own value and call it a TLV, however well-intentioned that act might be, may be a violation of laws protecting trademarks. If a TLV is not given, it usually means that ACGIH has not evaluated its hazard thoroughly enough to assign a TLV.

The employer, manufacturer, distributor, or importer still must evaluate the specific hazards associated with each substance on these source lists, however. That is, either ACGIH or OSHA or both may say that something is hazardous and that its concentration in the workplace atmosphere should be kept below a certain level, but their tables usually do not say just what the hazards are. Some substances may have several different hazardous properties, and these must all be identified.

Probably the best sources for identification of physical hazards are the latest edition of the *Fire Protection Guide on Hazardous Materials*, published by the National Fire Protection Association (NFPA), and the current edition of the DOT *Hazardous Materials* book, an emergency response guidebook. In some respects the DOT book is the better of the two because it gives more complete information, but it does not include as many different substances. The NFPA book also has information on trade name materials that are not listed in the DOT book. For information on materials that might be dangerously reactive, Bretherick's *Handbook of Reactive Chemicals Hazards* is a key reference source.

In order to identify health hazards associated with specific chemicals, a key reference is the latest edition of the *Registry of Toxic Effects of Chemical Substances* published by the U.S. Department of Health and Human Services. This book also includes references to the carcinogens listed by the National Toxicology Program (NTP) and the International Agency for Research on Cancer (IARC).

In addition, a number of other references, including computer data bank services, are included in Appendices B.1 and B.2. However, computer data banks should not be used as the only source of information. Many such banks are incomplete—their data are obtained from research literature that only goes back a few years. For some materials, important information was obtained many years ago and hasn't reappeared in the more recent technical literature. Such information is missing from some computer data banks.

Materials Covered by Definition

All chemicals are covered by the Standard if there is scientifically valid evidence that they have one or more of the characteristic properties listed in Table 1. This probably includes about 70,000 to 80,000 commercial materials, as well as mixtures containing those materials.

In the following pages, the hazards covered by the HCS are described in terms of their definitions and one or more examples of materials exhibiting each hazard. In several cases, a more detailed discussion of the hazardous properties is included as well as information about how to deal with them safely.

As in the text of the HCS itself, these hazard properties are grouped into physical hazards and health hazards. In line with SARA Title III reporting requirements, they are further divided into three subgroups of physical hazards and two subgroups of health hazards.

Any chemical, or mixture of chemicals, can be assumed to be possibly hazardous in one or more ways. But certain groups of materials are more often hazardous than others. A list of the more likely types of these materials is given in Table 2. Any material that is in one or more of these categories should be considered as probably hazardous until an exhaustive search proves otherwise.

The presence of certain elements in the chemical names of materials will suggest the probable existence of hazards. A list of the key elements for this purpose is given in Table 3. Not all compounds containing these elements are hazardous, but many are. Also, some of the elements in this table are hazardous themselves, in addition to imparting hazardous properties to compounds that contain them.

Finally, the presence of certain words and word fragments in the names of chemicals can indicate possible hazards. Table 4 is a useful list of these. Not every chemical compound whose name contains these words or word fragments is hazardous, but most are.

Thus Tables 2, 3, and 4 point to materials on an inventory list that should be considered hazard suspects. Not all hazardous materials will be located with the aid of these tables, but most will.

4.2 PHYSICAL HAZARDS

The harmful effects from physical hazards generally do not affect health through chemical action on the body. Rather, they interfere with the functioning of a system or the integrity of property. They may also damage health through some physical means, such as by inflicting cuts or bruises, breaking bones, or causing internal injuries.

Physical hazards are conveniently classified as fire hazards, explosion

Table 1.   Material Hazards Covered by the Hazard Communication Standard

Physical Hazards	Health Hazards
Fire Hazards	*Acute Hazards*
Combustible liquids	Corrosive materials
Flammable aerosols	Irritants
Flammable gases	Sensitizers
Flammable liquids	Toxic agents
Flammable solids	Highly toxic agents
Oxidizers	Agents affecting target organs:
Pyrophoric materials	Blood
	Skin
Explosion Hazards	Eyes
Compressed gases	Central nervous system
Explosives	
	Chronic Hazards
Reactive Hazards	Carcinogens
Organic peroxides	Agents affecting target organs:
Unstable reactive materials	Hematopoietic system
Water-reactive materials	Lungs
	Liver
	Kidneys
	Central nervous system
	Reproductive system

Table 2.   Types of Materials that Usually are Hazardous

Adhesives	Monomers
Aerosols	Office copier chemicals
Anodizing chemicals	Paints
Battery fluids	Pesticides
Catalysts	Photographic chemicals
Cleaning agents (all types)	Photoresists
Degreasing agents	Pickling agents
Detergents	Printing inks
Duplicating machine fluids	Process chemicals
Electrolytes	Resin ingredients
Electroplating chemicals	Rubber chemicals
Etching baths	Shellacs
Foaming resins	Soaps
Foundry mold materials	Solvents
Fuels (all types)	Surfactants
Industrial oils	Varnishes
Janitorial supplies	Wastewater treatment
Lacquers	Water treatment

hazards, and reactive hazards, though some hazards fall into more than one of these classes. Below is a detailed discussion of each of these terms.

Fire Hazards

Fire is an ever-present possibility when chemicals, particularly liquid organic chemicals, are being used. Materials that can burn are conveniently

Table 3.   Elements Whose Presence Signals Potential Hazards

Aluminum	Chromium	Manganese	Silver
Antimony	Cobalt	Mercury	Tellurium
Arsenic	Copper	Molybdenum	Thallium
Barium	Fluorine	Nickel	Tin
Beryllium	Hafnium	Platinum	Tungsten
Bromine	Indium	Rhodium	Uranium
Cadmium	Iodine	Selenium	Yttrium
Chlorine	Lead	Silicon	Zirconium

Table 4.   Words and Word Fragments that Signal Potential Hazards

acid	brom	hydroxide	nitro
acryl	caustic	isocyanate	nitroso
alcohol	chlor	ketone	perox
aldehyde	chrom	mercaptan	phenol
allyl	cyan	nitrate	sulfide
amine	epoxy	nitrile	thio
amino	ether	nitrite	vinyl
anhydride	glycol		

divided into those that are *ready to burn* and those that are *almost ready to burn.*

Those substances that are *ready to burn,* often called flammable, merely need a source of ignition to produce a flame. Such flammable materials include liquids whose flash points are at or below normal room temperature, as well as gases and aerosols. Note that the older term *inflammable* is now considered to be obsolete. It is ambiguous, and hence does not serve a useful purpose.

The flash point of any material is the lowest temperature at which a flame can be produced when an ignition source is introduced near the material's surface. Note that it DOES NOT MEAN that a material WILL BURST INTO FLAME when above its flash point, but merely that it CANNOT BURN at lower temperatures.

Examples of ignition sources include very hot surfaces (even pipes carrying superheated steam), exposed glowing electrical filaments, flames, glowing tobacco, and electric sparks, as in a static discharge or an arcing connection within an electric circuit.

It is also important to note that if the concentration of fuel vapors in air is below the lower flammability limit (LFL), the mixture CANNOT be ignited. If the concentration of fuel vapors in air is above the upper flammability limit (UFL), the mixture CANNOT be ignited, either. Both the LFL and the UFL are characteristic properties that differ from one substance to another. Note that the LFL is sometimes called the *lower explosion limit* (LEL), and the UFL is sometimes called the *upper explosion limit* (UEL). Both sets of terms have identical definitions, but LFL and UFL are the preferred terms.

Combustible Liquids

A combustible substance is something that is *almost ready to burn*. That is, it has a flash point somewhat above room temperature and needs to be heated before it is ignitable. It is a liquid that has a flash point between 100°F (38°C) and 200°F (93°C). Thus, a combustible liquid presents a danger of fire at slightly elevated temperatures, but not at or below a normal room temperature. Examples include 10% ethyl alcohol, jet fuel, methyl cellosolve, mineral spirits, No. 1 fuel oil, phenol, and pine oil.

If 99% (by volume) of the components in a mixture have flash points above 200°F (93°C) the mixture is not considered combustible (under the HCS). Many mixtures containing more than 1% of a combustible liquid have flash points high enough that they are not classified as combustible. If in doubt, they should be tested to determine whether they have to be rated combustible.

Flammable Aerosols

A flammable aerosol is defined in two different ways:

- A flammable aerosol yields a flame projection of more than 18 inches at full valve opening.
- Alternatively, a flammable aerosol yields a flame extending back to the valve at any valve opening.

All aerosols are mixtures. Whether a particular aerosol is flammable often depends on the particular propellant formulation, so general examples cannot be given for this category of hazard.

It should be noted that an aerosol consists of a dispersion of microscopic liquid or solid particles in gas or air. One property of such a dispersion is that it may be flammable even if the flash points of its contents would be too high to be classified as flammable liquids.

Flammable Gases

A flammable gas is defined in two different ways:

- a gas whose LFL is less than 13% by volume in air (for example, butane)
- a gas whose UFL is more than 12% higher than its LFL, regardless of the value of the latter (for example, producer gas)

Note that some gases will burn but are not classified as flammable by OSHA, presumably because they do not ignite easily enough to present a significant hazard. However, if they are present when a fire breaks out they may increase the intensity of the fire. Ammonia is such a gas—its LFL is 16% and its UFL is 25%, a difference of only 9%.

Flammable Liquids

A flammable liquid is one whose flash point is below 100°F (38°C). It is a fire hazard if present in open containers near a source of ignition at or below normal room temperatures. Examples include acetone, ethyl acetate, 95% ethyl alcohol, gasoline, and turpentine.

If at least 99% (by volume) of the components of a liquid mixture have flash points above the specified range, it is not considered flammable under the HCS. However, many mixtures containing more than 1% of a flammable liquid have flash points high enough that they are not classified as flammable, or sometimes even as combustible. If in doubt, they should be tested to determine whether they have to be rated flammable or combustible.

Flammable Solids

A flammable solid ignites and burns with a self-sustained flame at a rate of at least 0.1 inch per second along its major axis. This category of hazard does not include blasting agents or explosives. Examples include magnesium metal and nitrocellulose film.

Oxidizers

An oxidizer is a chemical, other than a blasting agent or an explosive, that can initiate or promote combustion in other materials. Thus it may cause fire either of itself or through the release of oxygen or other gases. Examples include chlorine, fluorine, hydrogen peroxide, nitric acid, and oxygen. Organic peroxides would also come under the definition of oxidizers, but they have a special classification all their own.

Potential fuels should not be stored close to strong oxidizers, including organic peroxides. Potential fuels include such things as alcohols, aldehydes, ketones, organic acids, and petroleum products.

If a small amount of fuel is mixed with a relatively large amount of oxidizer, an explosion, or at least a fire, can easily result. This condition can exist if, for example, sawdust is used as an absorbent while cleaning up a spill of a strong oxidizer.

Many strong oxidizers, such as chlorates, chlorine dioxide, chromates, perchlorates, permanganates, and peroxides, are fairly stable when pure. But in the presence of very small amounts of impurities (including water, sometimes), they tend to decompose violently.

Pyrophoric Materials

A pyrophoric material will ignite spontaneously in air at temperatures below 130°F (54°C). Examples include white (or yellow) phosphorus and some catalysts used in the chemical and petroleum processing industries.

Pyrophoric materials must be handled in such a way as to avoid exposure to air. They must be stored out of contact with air to prevent ignition, usually under oil or water, or in a hermetically sealed container. The potential for a serious burn should always be kept in mind when working with such materials.

Explosion (Sudden Release of Pressure) Hazards

Technically, an explosion is defined as a sudden release of pressure. Thus, the rupture of a toy balloon is an explosion, though it is usually not considered to be hazardous. The mechanical failure of any compressed gas container results in a potentially dangerous release of pressure.

If the failure results from the formation of a small hole, such as when a valve, gage, or pipe fitting breaks off, and if the container is not adequately fastened in place, the result can be the same as the launching of a rocket, i.e., a missile is born. If the failure is more generally catastrophic, flying shrapnel can result. In either case, the velocity of the escaping gas can also harm nearby persons or equipment. Two types of explosion hazards are compressed gases and explosives.

Compressed Gases

A compressed gas is defined in three different ways:

- It may be a confined gas or mixture of gases having an absolute pressure of at least 40 psi at 70°F (21°C);
- It may be a confined gas or mixture of gases having an absolute pressure of at least 104 psi at 130°F (54°C); or
- It may be a liquid having a vapor pressure of at least 40 psi at 100°F (38°C).

None of the definitions depend on any other properties of the gas, such as toxicity or flammability. The fact that the gas is stored under pressure results in a hazard, as it could cause harm if the storage were disrupted improperly. Examples include acetylene, argon, carbon dioxide, nitrogen, oxygen, and propane.

Explosives

An explosive is any chemical that causes a sudden, almost instantaneous release of pressure, gas, and heat when subjected to shock, pressure, or high temperatures. Examples include diacetyl peroxide, gunpowder, and nitroglycerine.

Reactive Hazards

These hazards can produce excessive heat or poisonous or obnoxious products. Many unstable compounds can also decompose explosively under some circumstances.

Some highly reactive substances are shipped and stored with "inhibitors" or "stabilizers" present. While these inhibitors prevent the dangerous reactions from occurring uncontrollably, at least for a time, the inhibitors normally are gradually consumed during storage. Thus, there is usually a limited "shelf life" for any material that has a stabilizer or an inhibitor present.

Organic Peroxides

An organic peroxide is a derivative of hydrogen peroxide in which one or both hydrogen atoms have been replaced by an organic radical or radicals. This definition also covers the class of compounds known to chemists as organic hydroperoxides. Examples include benzoyl peroxide, cumene hydroperoxide, and methyl ethyl ketone peroxide.

Organic peroxides are usually stored refrigerated, but they should never be stored together with fuels. Combustion can result if the two materials should come into contact. All of the precautionary statements given about oxidizers apply to organic peroxides as well.

Unstable Reactive Materials

An unstable material is a chemical which in the pure state, or as produced or transported, will vigorously polymerize, decompose, condense, or become self-reactive under conditions of shock, pressure, or temperature. Examples include acrylonitrile, benzoyl peroxide, and butadiene.

Acrylic compounds, vinyl compounds, and some other types of compounds readily undergo self-reaction to form high polymers. Such reactions commonly generate troublesome amounts of heat. Monomers should be stored in a cool place to minimize the likelihood of polymerization. It should never be assumed that the polymerization inhibitor initially supplied by the manufacturer will protect against unwanted polymerization for more than six months to a year.

Water-Reactive Materials

A water-reactive material interacts with water to produce a flammable or toxic gas. Examples include calcium carbide and sodium metal.

Some substances react with water to produce much heat, sometimes reacting violently and spattering corrosive solutions. Examples include acid anhydrides and acid chlorides, both of organic acids and of "mineral acids."

4.3 HEALTH HAZARDS

Potential health hazards exist for virtually everything found in any facility using chemicals. The specific nature of each hazard depends not only on the identity of the particular material, but also on the amount that is present and the form in which it exists. Furthermore, the potential for harm depends on the type and amount of exposure.

The key reference for identification of health hazards associated with specific chemicals is the latest edition of the *Registry of Toxic Effects of Chemical Substances* published by the U.S. Department of Health and Human Services. This book also includes references to the carcinogens listed by NTP and IARC. A number of other selected references are given in Appendices B.1 and B.2.

Effects of Exposure

Amount of Exposure

Whether harm will occur to a person exposed to a hazardous material, and sometimes even the type of harm that occurs, depends upon the amount of exposure. On the one hand, literally everything is harmful to health if too much of it is introduced into the body in some way. Even pure water is lethal if it is injected into the bloodstream in large doses. On the other hand, even materials considered to be extremely toxic will not cause any harm if present in small enough quantities. In fact, some mineral elements that are lethal in high concentrations are actually essential to life in small concentrations.

As a method of dealing effectively with the matter of exposure level, OSHA established its PELs for atmospheric concentrations of hazardous materials to which workers may be exposed. These PELs are established as levels below which normally healthy workers are not expected to have health problems as a result of exposure, and have been developed mostly from clinical data collected in connection with actual exposures. ACGIH established TLVs for a number of materials, with essentially the same implication as the PELs.

The differences between TLVs and PELs are several, and should be kept in mind. PELs have regulatory status—if a substance is present in the atmosphere at a higher level than its stated PEL, the atmosphere is not in compliance with OSHA requirements. TLVs are advisory in nature, without regulatory status. The first PELs established were simply the TLVs in effect at the time, but since then many differences have evolved.

TLVs are constantly under review. In any given year a significant fraction of them are in the process of being changed in light of new experience. These reviews are handled by committees of experts who carefully evaluate all pertinent data.

The PEL values, which normally are changed only through a formal rule-making process, don't necessarily keep up with new scientific information as do the TLVs, but PELs may be changed by court orders or possibly as a result of the influence of pressure groups. OSHA is said to be considering the possibility of eliminating the PELs and simply adopting the TLVs by reference in their regulations. However, there is known to be strong opposition to such a move, and at the time of this writing the fate of such proposals is uncertain.

In general, the prudent action is to maintain workplace air clean enough that nothing is present at higher concentration than the PEL or the TLV (whichever is lower) for the particular substance.

Extent of Effect

Contact (localized). Once a material has made contact with the body, it may produce a harmful effect at the site of entry. This is known as a contact effect or a local effect. Common local effects include corrosive attacks or irritation of the skin, eyes, respiratory tract, or gastrointestinal tract. Other types of localized effects occur when an agent accumulates selectively in a particular organ and affects that organ.

Systemic. Alternatively to a local effect, the material may be transmitted through the blood to other parts of the body, with a more generalized harmful effect. The harmful effect may occur more or less immediately, or the material may be accumulated over time with a gradual increase of effect.

Time Effects

Health hazards are conveniently divided between acute health hazards and chronic health hazards. Some hazards have characteristics of both, but generally they are one or the other.

Acute. The effects of acute hazards are manifested soon after a single, brief exposure. If death does not occur, many acute effects disappear soon and do not linger permanently. Some may also show permanent effects, and therefore can be considered both acute and chronic.

The word acute can also mean sharp or severe, as with an acute pain. However, when used to describe a hazard, the word does not imply any sense of severity—merely that it has a short-term effect.

Chronic. Chronic hazards have a long-term effect, sometimes over the lifetime of the affected individual. The effects may be slow to develop, often as a result of repeated or continuous low-level exposure over a long period of time. Such effects may result from a gradual deterioration of function or a degradation of organ systems; they usually are irreversible.

Other chronic effects may result from a gradual buildup of the concentration of a toxin in some organ or portion of the body. A person can be exposed to something for months or years before experiencing any alarming symptoms. On continued exposure, the symptoms get progressively worse, sometimes culminating in death. If the exposure is removed, the symptoms sometimes may gradually become less serious as the concentration of the substance in the organ decreases through the body's normal elimination functions.

Routes of Entry

An important classification of material hazards is the route of entry to the body, which must be included on the MSDS and on the package label as well. If the route of entry is known with certainty, it can be guarded against more effectively. The usual routes of entry are through the skin, the eyes, the respiratory tract, or the digestive tract.

Entry Through the Skin

In addition to contact materials that cause localized damage to the skin, some materials, usually liquids, vapors, or gases, are absorbed through the skin into the bloodstream. In some cases, protection against such materials is relatively simple; in other cases it can be very difficult.

Absorption of systemic toxins through the skin can be even more dangerous than by ingestion. Substances that are absorbed into the bloodstream through the walls of the digestive tract pass through the liver, which detoxifies many poisons, before being circulated to other organs. A toxin that is absorbed through the skin into the bloodstream, though, is distributed throughout the body without first being detoxified by the liver. Thus, most organ systems may be exposed to whatever effect it has.

If a chemical is absorbed through the skin, the amount entering the bloodstream is proportional to the area of skin covered. If an attempt is made to wash off the material with alcohol, for instance, it may be dissolved, spread over a larger area than initially, and sometimes even be absorbed faster than in the absence of the alcohol. Cases have been reported in which such a procedure resulted unnecessarily in serious poisoning.

If a solid or viscous liquid material is spilled on the skin, the recommended first step is to wipe off as much as possible. The second step is to wash the contaminated area thoroughly with cold running water, generally for 15–30 minutes. Soap and water should then be used, followed by a final rinse with cold running water.

When the skin has been scratched, cut, abraded, or lacerated, its barrier function is defeated. Until a wound begins to heal, the break in the skin provides a more direct path for harmful materials to enter the bloodstream and be transmitted throughout the body. Thus, workers with scratches, cuts, and abrasions to their skin must be particularly careful to use protective equipment when working with hazardous materials. If a person should be wounded while handling chemicals, the recommended first response is to cleanse the wound by washing thoroughly and to allow the blood to flow briefly to flush contaminating chemicals from the interior of the wound.

Entry Through the Eyes

The covering of the eyeballs is normally the least important route of entry for systemic toxins, mainly because of the small surface area exposed, though it is very important in the case of contact hazards. However, the blood vessels of the eyes are more closely connected with the blood supply to the brain than are those in most parts of the skin. Therefore, the eyes may sometimes be a serious, important route of entry for neurotoxins and other systemic toxins, especially gases and liquids.

Entry Through Breathing

A very common route of entry is by inhalation. If a material can enter the bloodstream through the lungs, or if it damages the lungs or bronchia locally, then the inhalation hazard may be significant. Any harmful gases or vapors that are in the air we breathe are quickly and efficiently transmitted into the bloodstream. Except by direct injection into a blood vessel, this is often the most efficient way for toxic materials to enter the bloodstream for distribution throughout the body.

Inhalation hazards constitute a very common problem with chemicals in industry. When materials are handled in the open, they often become dispersed in the air so that anyone nearby will inhale them. Properly selected and fitted respirators can prevent inhalation of many harmful materials. But whenever practicable, it is best to keep materials that present a respiratory hazard in closed systems as much as possible.

Entry Through Swallowing

Some hazardous materials cause localized damage to the lining of a portion of the gastrointestinal tract. In addition, ingestion is a dangerous form of exposure to many systemic hazards. Such materials, even if highly toxic when ingested, do not present a significant risk to workers if they simply use care to avoid getting the material in their mouths. If the material is finely dispersed in the workplace air either as a dust or as a liquid mist or vapor, this may merely involve the use of an adequate dust mask.

Acute (Immediate Effect) Hazards

Corrosive Materials

A corrosive material causes visible localized destruction or irreversible alteration of living tissue by chemical action at the site of contact. Such materials commonly affect skin, eyes, gastrointestinal tract, or respiratory tract. Examples include boron trifluoride, caustic soda (sodium hydroxide), hydrofluoric acid, liquid peroxides and other oxidizing agents, phenol, phosphorus halides and oxyhalides, and sulfuric acid.

A substance is considered to be corrosive if, when tested on the intact skin of albino rabbits by a prescribed test procedure, it destroys tissue or irreversibly changes its structure at the site of contact following an exposure period of four hours.

Strong mineral acids and some halogenated organic acids change the structure of the skin by precipitating protein and forming a barrier layer, which becomes part of a resultant scar tissue. However, the barrier layer also prevents further penetration into the flesh by the acid.

Concentrated solutions of strong bases are also corrosive; in a sense they are even more dangerous than acids because they do not precipitate a protein barrier. They continue to destroy tissue until they have become so diluted that they can do no more. Reportedly they do not cause pain as they do their damage. This combination of properties makes them serious insidious hazards; i.e., their harmful actions may not be recognized until it is too late to prevent serious harm.

Any material that is corrosive at room temperature is even more dangerous at higher temperatures. At steam temperature, a corrosive material often will act so fast that it is literally impossible to prevent damage. In most cases the same material can be washed away without much harm being done if contact occurs at room temperature.

Irritants

An irritant is defined as a material that causes localized reversible inflammation at the site of contact by chemical action. Irritants commonly affect skin, eyes, gastrointestinal tract, or respiratory tract. Examples include acrolein, ammonia vapors, calcium hypochlorite, chlorine, ethyl alcohol, formaldehyde, hydrogen fluoride, nitric oxide, nitrogen dioxide, organic selenides, ozone, particulate sulfates, peroxyacetylnitrites (PANs), sodium hypochlorite, stannic chloride, sulfur dioxide, and sulfuric acid mists.

The inflammatory effect caused by a skin irritant is commonly called dermatitis and is characterized by redness, heat, swelling, pain, or some combination of those symptoms. As with corrosive agents, an irritant will usually cause a more severe effect when hot than when cold.

There are officially prescribed tests for irritants in which the substance is applied to the intact skin of an albino rabbit for four hours. The appearance of the exposed area of the skin is then evaluated by a numerical scale which grades redness, swelling, and extent of affected area. If the score exceeds a prescribed value on this scale, the substance is considered to be an irritant. Even though the grading scale is empirical and its application may be somewhat subjective, dermatologists consider it to be reliable.

Sensitizers

A sensitizer is a material that causes a substantial portion of exposed people or animals to develop an allergic reaction in normal tissue after repeated exposure. Examples include 2,4-D, 2,4,5-T, acridine, acrolein, ammonia, arsenic, benzoyl peroxide, beryllium salts, carbamate herbicides, chromic acid and chromates, cobalt, creosote, epoxy compounds, formaldehyde, furfural, hydroquinone (present in many photographic developers, and as a polymerization inhibitor in some vinyl and acrylic monomers as well), isocyanates (widely used in the production of urethane plastics and foams), nickel, nitrobenzene, organophosphate insecticides, ozone, phthalic anhydride, platinum compounds, pyridine, and tetracyclines.

Some skin irritants also act as sensitizers. After repeated exposure to such a material, allergic people react much more strongly than they would to a simple irritant. As with any allergy, these reactions can occasionally be life-threatening. Not everyone will become sensitized to a particular substance, but the possibility exists when a substance is a known sensitizer, so its effects must be monitored.

Note that a substance is generally termed a sensitizer only if it affects a substantial portion of the exposed population. However, some individuals may become allergic to substances that are not generally classified as sensitizers. For them the situation is just as threatening as if the substance were classified as a sensitizer.

Toxic Agents

Toxic agents are those that affect the proper functioning of an organism, such as a human, resulting in a change in physiology through a chemical rather than a physical process. In the extreme, the effect of an acute toxic agent is death, and this is the basis of definition of toxicity within the HCS. Most often a small, brief exposure results in some type of physiological effect short of death. In many cases the effect of such a small exposure is considered to be essentially trivial, even by the person affected, and recovery may be complete with no lingering damage.

A toxic agent is defined in three different ways, depending on route of entry to the body. In the definitions given below, the terms LD_{50} and LC_{50}, also known as median lethal dose and median lethal concentration, refer to the doses at which one-half of the test animals died. At higher doses more than half of the animals died, and at lower doses, fewer.

Frequently the values of LD_{50} or LC_{50} given in the literature for a particular compound have been measured for a different animal species or under different exposure rates than are specified in the following definitions. In such cases, approximate translations can be made, but they always contain significant experimental uncertainty. See Appendix F for a discussion of lethal dose equivalencies.

Entry by ingestion (swallowing). A toxic agent has an LD_{50} for oral doses in rats of between 50 milligrams per kilogram of body weight (mg/kg) and 500 mg/kg. Examples include 2,4-D, acrylonitrile, aniline, and epichlorohydrin.

Entry by absorption through the skin. A toxic agent has an LD_{50} for rabbit skin in a 24-hour exposure of between 200 mg/kg and 1000 mg/kg. Examples include acrylonitrile and epichlorohydrin.

Entry by inhalation (breathing). A toxic agent has an LC_{50} for inhalation doses administered for one-hour durations in rats of between 200 parts per million (ppm) in air and 2000 ppm. Examples include ammonia, boron trifluoride, ethylene oxide, and nitrogen dioxide.

Highly Toxic Agents

A highly toxic agent is also defined in three different ways, depending on the route of entry.

Entry by ingestion (swallowing). A highly toxic agent has an LD_{50} for oral doses in rats of less than 50 mg/kg. Examples include Aldrin, ethyleneimine, and hydrogen cyanide.

Entry by absorption through the skin. A highly toxic agent has an LD_{50} for rabbit skin of less than 200 mg/kg. An example is mustard gas.

Entry by inhalation (breathing). A highly toxic agent has an LC_{50} for inhalation doses administered for one-hour durations in rats of less than 200 ppm. An example is dimethylnitrosamine.

Agents with Target Organ Effects

Some materials affect only one or a few specific body organs rather than producing a general effect. The resulting effects are called target organ effects; some are acute effects and some are chronic. Only the acute hazards are listed here; chronic hazards with target organ effects are listed later. Some target-organ agents can be either chronic or acute, but except for neurotoxins and circulatory system toxins, they are listed in only one of the two categories.

Agents which act on the blood. This type of agent is a substance that decreases the hemoglobin function and deprives the body tissues of oxygen. The resultant acute affects include cyanosis, dizziness, and loss of consciousness as typical symptoms. Examples include aniline, arsine, carbon monoxide, chloroprene, cyanides, dinitrobenzene, hydroquinone, mercaptans, metal carbonyls, nitrobenzene, 2-nitropropane, sodium nitrite, and TNT.

Other types of blood toxins affect the blood's clotting ability or its ability to fight infection. Examples include arsine, aspirin, boron hydrides, magnesium oxide, naphthalene, tetrachloroethane, and Warfarin.

Cutaneous hazards. A cutaneous hazard is a material that will affect the dermal layer of the body, such as by defatting of the skin, drying it out and reducing its elasticity, at least temporarily. This can lead to irritation or to greater susceptibility to corrosive materials than healthy skin. Examples include acetone, benzyl chloride, carbon disulfide, chlorinated organic hydrocarbons, chromic acid and other soluble chromium compounds, ethylene oxide, hydrogen chloride, iodine, MEK, phenol, phosgene, picric acid, styrene, sulfur dioxide, toluene, and xylene.

Other skin hazards include corrosive agents, irritants, or sensitizers. A skin irritation may seem to be a minor problem, and often it is. Nevertheless, statistics show that in industry there is more time lost from work as a result of skin irritation than from any other single type of injury or illness. Economically at least, these are extremely important types of injuries to deal with.

Eye hazards. An eye hazard is a material that affects the eye or visual capacity, for example by causing corneal damage or conjunctivitis (an inflammation of the mucous membrane lining the inner surface of the eyelids and covering the front part of the eyeball). Examples include acetone, ammonia, calcium oxide (lime) and calcium hydroxide (slaked lime), ethyl alcohol, hydrogen chloride, methyl alcohol, naphthalene, phenothiazine (an insecticide), picric acid, potassium hydroxide, sodium hydroxide, sulfuric acid, tannic acid, thallium, and toluene.

Materials that commonly cause such effects usually are also irritants, corrosives, or sensitizers for the skin. Often the effect is more serious with the eyes than with the skin, because of the important visual function of the eyes. Many materials cause the same types of problems with both skin and eyes, though.

Lachrymators constitute an additional type of hazard whose effects are specific to the eyes. A lachrymator is an agent which causes an excessive formation of tears, usually accompanied by a stinging or burning sensation in the eyes. Examples include benzoyl chloride, bromoacetone, chloroacetophenone (tear gas), isocyanates, isothiocyanates, xylyl bromide, and wood smoke.

As with skin contact, eye contact with hazardous materials is best treated by washing extensively with cold running water. First aid eyewash solutions should never be used for this purpose. Washing thoroughly with running water for 15–30 minutes is the only treatment to be applied within the facility or in the field. This should be done with the assistance of another person if at all possible. When a foreign substance contacts the eyes, the natural reflex tendency is to clamp the eyelids shut to prevent further contamination. For this reason, the eyelids should be held back by two fingers throughout the washing procedure, and this is best done by someone other than the person who suffered the contamination. The victim may be in some pain, or at least discomfort, and will find it difficult to function under the stress condition as rationally as he or she normally would.

If any residual irritation or other effect remains, professional medical attention should be obtained immediately. Health professionals may then use medicated solutions as part of the treatment, but lay persons should never do so.

Anything that is a contact hazard to the eyes at room temperature is a much more serious hazard at higher temperatures. In fact, many substances can be flushed away without lasting damage if they contact the eyes at room temperature. Often the same substances will cause permanent damage if they contact the eyes at temperatures above about 50°C (122°F)—it is not possible to wash them out fast enough to prevent damage.

Neurotoxins. A neurotoxin is a substance that causes primary toxic effects on the central nervous system, such as narcosis, behavioral changes, decrease in motor functions, etc. Narcosis ranges from a mild light-headed feeling to a strong anesthetic effect. Often this is an acute toxic effect from which recovery is complete, rather than chronic. However, even a mild narcosis can be dangerous because of a diminution of attentiveness, muscle reflexes, and motor coordination. Examples include acetone, acetylene, controlled narcotic drugs, ether, ethyl alcohol, nitrous oxide, some hydrocarbons, and many volatile chlorohydrocarbons.

Chronic (Delayed Effect) Hazards

Carcinogens

A carcinogen is defined by OSHA in the HCS as a substance that is listed as a carcinogen in one of the following three sources:

- NTP *Annual Report on Carcinogens*, latest edition
- IARC *Monographs*, latest edition (but only those categorized as Groups I and II by IARC, not Group III)
- OSHA's 29 CFR 1910 subpart Z

Examples include asbestos, benzene, benzidine, beryllium, bis-chloromethyl ether, cadmium oxide, carbon tetrachloride, condensed polynuclear aromatic hydrocarbons (PNAHs, or PAHs), formaldehyde, nickel carbonyl, radium, radon and daughters, strontium-90, trichloroethylene, trivalent arsenic compounds, vinyl chloride, and zinc chromate.

The effects of carcinogens may appear long after exposure (often 10–30 years later), even if there is no exposure in the meantime. It is now known that usually exposure to two different substances, the second being called a cocarcinogen, is required before a malignant tumor develops. Sometimes the identity of such a cocarcinogen is known, and sometimes not. Sometimes it may be any of a family of chemical compounds rather than a single one that triggers the tumor formation.

In addition to carcinogens and cocarcinogens, hereditary factors are undoubtedly involved. Some people seem to be particularly susceptible to carcinogenesis, and others seem to be resistant. At present, the state of knowledge is not advanced enough to be able to identify such individuals in either category.

It is prudent to minimize the use of known carcinogens in the workplace whenever possible. When it is necessary to use them, rigorous precautions should be exercised to minimize exposure, even when it is expensive to do so.

In addition to known carcinogens, there are many suspected carcinogens. If there is scientific evidence to support such a suspicion, the HCS requires that the MSDS indicate that fact, even though the evidence is still so weak that they are not identified as known carcinogens. The prudent action in such cases is to minimize contact with such materials as much as is practical, but not to become so concerned about the hazard as to become paranoid.

Cancer is a much-dreaded disease that used to be nearly always fatal, though many types of cancer now have excellent remission rates if detected early enough. Being informed that a substance is even suspected of being a carcinogen can trigger a strong reaction in many people. When training workers about carcinogens, their possible effects should not be glossed over, but should be explained carefully, frankly, and with no attempt to cover up the risks.

Agents with Target Organ Effects

The agents in the following categories cause mainly chronic effects. Short exposures, or even rather extended exposures to low concentrations, may not produce identifiable symptoms. Over very long exposure times, even at low concentrations, many of these agents produce cumulative effects that are fatal or at least debilitating.

Often the buildup of such agents in the body can be detected by clinical tests on blood, urine, or exhaled gases. When workers are exposed to such target-organ agents, their presence or absence in the bodies of the workers should be monitored periodically so that exposure can be stopped before any serious symptoms appear.

Agents which damage the blood and hematopoietic system. These agents affect the bone marrow and lymphatic system, thereby producing a long-term effect on the circulatory system. Examples include arsenic, benzene, chloroform, and ethylene dibromide.

Agents which damage the lungs. These agents irritate the pulmonary tissue, resulting in cough, tightness in the chest, and shortness of breath. Examples include asbestos, coal dust, cotton fibers, organic isocyanates, silica, and components of tobacco smoke. In addition to irritation, asbestos, coal dust, tobacco smoke, and perhaps others can cause lung cancer as well.

Irritation of the respiratory tract is similar in nature to the irritation effect of eyes and skin, except that the lungs cannot be irrigated with running water to flush away the offending agent. Some irritants will dissolve in the mucous in the upper respiratory tract and then be swept upward and out by the action of specialized (ciliated) cells. Generally, though, any irritant that is not so removed may cause a lingering irritation effect. Often such effects are reversible, but not always.

Emphysema is a lung disease that is caused generally by exposure to certain kinds of chemicals or particulate matter in the air. Emphysema is essentially similar to a corrosive effect in which the membranes lining the air sacs actually become scarred and less elastic than in a healthy lung. In extreme cases, many of the air sacs become bridged across by scars, permanently removing significant portions of the lungs' gas transport membrane from useful functioning. People with emphysema may be doomed to a lifetime of shortness of breath, constant pulmonary therapy, or worse. Examples of agents that cause emphysema include asbestos, beryllium oxide fumes, cellulose, coal dust, cotton, flax, graphite dust, hemp, iron oxide fumes, kaolin (a form of clay) dust, silica, talc, tin oxide fumes, and tobacco smoke.

Some agents cause an edema (a swelling and accumulation of fluids within the tissues) of the membranes lining the lungs. This results in a loss of efficiency of transfer of oxygen into the blood and of carbon dioxide out from the blood. This, too, may be a temporary effect, but of somewhat longer duration than in the case of simple respiratory irritation. Chlorine is an example of such an agent.

Some substances cause pneumonia-like symptoms, often after a delay of several hours or a few days following exposure. The pneumonia-like symptoms involve an accumulation of fluid in the lungs, but potentially even greater than with chronic edema, possibly leading to fatality in extreme cases. The nitrogen oxides display such effects.

Hepatotoxins. A hepatotoxin is a substance that can cause liver damage such as enlargement, cirrhosis, or jaundice. Examples include aflatoxin, arsenic, carbon tetrachloride, chlorobenzene, chloroform, ethyl alcohol, nitrosamines, vinyl chloride, and trichloroethylene.

Some types of liver damage are cumulative. That is, as a portion of the liver is damaged, it may not be regenerated. It is thus no longer able to detoxify foreign substances in the body and excrete them. As the toxic process continues, less and less of the liver function is available, until finally the individual dies. This is the principal cause of death by alcohol overindulgence.

Nitrosamines, vinyl chloride, and some other substances are implicated in liver cancer.

Nephrotoxins. A nephrotoxin is a substance that can cause kidney damage such as edema or proteinuria. Examples include 2,4,5-T, arsenic compounds, cadmium compounds, carbon disulfide, carbon tetrachloride, chloroform, chromium compounds, ethyl alcohol, ethylene dibromide, ethylene glycol, gold compounds, lead compounds, mercury compounds, silver compounds, trichloroethylene, uranium compounds, and vinyl chloride.

Neurotoxins. A chronic neurotoxin is a substance that, received even in small doses over a long period of time, results in loss of motor coordination, palsy, behavioral changes, and a degeneration of mental capacity, leading to insanity in the extreme. Examples include lead compounds, mercury, manganese, and thallium.

Reproductive toxins. There are three types of reproductive toxins that are of general concern. One type causes sterility, infertility, or a significant reduction in fertility. Examples include 2,4-D, 2,4,5-T, anesthetic gases, benzene, boron, cadmium, carbaryl, chlordane, DDT, dibromo-chloropropane, diethylstilbestrol (DES), dioxin, ethyl alcohol, ethylene dibromide, hexafluoroacetone, kepone, lead, mercury, methyl mercury, paraquat, parathion, PCBs, toluene, vinyl chloride, and xylene. In addition, natural hormones and structurally similar synthetic compounds can cause sterility in both males and females.

A second type of reproductive toxin is a teratogen, i.e., a substance that can cause birth defects. Examples include 2,4,5-T, anesthetic gases, cadmium sulfate, dioxin, organic mercury compounds, phenylmercuric acetate, phthalic acid esters, sodium arsenate, and thalidomide.

A third type of reproductive toxin is a mutagen, i.e., one that can cause permanent genetic changes. Examples include 2,4-D, 2,4,5-T, benzene, cadmium sulfate, DDT, dioxin, ethylene oxide, ethyleneimine, hydrazine, hydrogen peroxide, lead salts, nitrites, ozone, and sodium arsenate.

5

Written Forms of Hazard Identification

5.1 LABEL CONTENT AND DESIGN

Any substance that is covered either by definition or by reference must be properly labeled with a warning of known hazards. The label warning must contain information about the significant hazards known to be associated with the substance. For some materials this will involve several hazards.

Content

Initially, OSHA set out to develop a labeling standard for hazardous materials, largely because of a realization that there was a high incidence of work-related accidents and illness that could be attributed to mishandling of hazardous materials. But that effort developed into the current HCS, partly because there is a very significant problem with putting too much information on a label. The average person will not read a label, even the most important part of it, if it contains too much information.

A proper balance between too much information and too little is difficult to achieve. Too little information carries the risk of injury because of a lack of knowledge about the material. Too much information carries the risk that it will not be heeded, and hence there is still a lack of knowledge about the material. OSHA only requires a minimum amount of essential information on the labels, but requires that much more complete information be included on the MSDS, which is the key information source that is to be available to all workers at all times.

OSHA Requirements

In general, the only items of information required on the labels by this Standard are:

- the identity of the material
- the hazard warnings
- the name and address of a responsible party from whom additional information can be obtained if needed

However, OSHA has published complete standards for certain materials, including requirements of specific wording on labels. A list of those particular materials is given in Table 5. The indicated regulation should be consulted for the actual text of the required wording and for other requirements.

ANSI Standards

Partly for the purpose of reducing exposure to liability cases brought with the allegation of inadequate warning of hazards, the American National Standards Institute has published a voluntary labeling standard (ANSI Z129.1–1982). It suggests including the following nine items on labels of containers of hazardous materials:

- the identity of the product or its hazardous components. This, of course is also included in the requirements of the HCS.
- a signal word. CAUTION, WARNING, and DANGER are perhaps the most commonly used signal words, in order of increasing severity of the hazard represented. This is not required by the HCS, but probably should be included even if no other voluntary information is provided.
- a statement of the actual hazards present. This too is required by the OSHA Standard.
- precautionary measures to prevent physical harm when using the product, such as wearing rubber gloves, goggles, respirators, etc.
- instructions in case of contact or exposure, such as to induce or not induce vomiting in case of swallowing, to wash with cold water, etc.
- antidotes to be used in case of poisoning
- notes to physicians as to emergency treatment recommended
- instructions in case of fire and spill or leak, i.e., type of fire extinguisher to be used, how to clean up spill without undue risk, etc.
- instructions for container handling and storage, such as to keep it in a cool place, away from fires, away from strong acids, etc.

The HCS requires information similar to the last six of these items on MSDSs, but not on labels. Note, however, that following the ANSI standard alone would not satisfy the OSHA HCS, as the name and address of a responsible party are not included among the nine items.

Table 5.   Substances for Which OSHA Has Issued Standards Specifying Wording on Labels, Among Other Requirements

Substance	Location of Standard
2-Acetylaminofluorene	29 CFR 1910.1014
Acrylonitrile	29 CFR 1910.1045
4-Aminodiphenyl	29 CFR 1910.1011
Asbestos	29 CFR 1910.1001
	& 29 CFR 1910.1101
Benzene	29 CFR 1910.1028
Benzidine	29 CFR 1910.1010
bis-Chloromethyl ether	29 CFR 1910.1008
Chloromethyl methyl ether	29 CFR 1910.1006
Coke oven emissions	29 CFR 1910.1029
1,2-Dibromo-3-chloropropane	29 CFR 1910.1044
3,3'-Dichlorobenzidine	29 CFR 1910.1007
4-Dimethylaminoazobenzene	29 CFR 1910.1015
Ethyleneimine	29 CFR 1910.1012
Ethylene oxide	29 CFR 1910.1047
Formaldehyde	29 CFR 1910.1048
Inorganic arsenic	29 CFR 1910.1018
Lead	29 CFR 1910.1025
alpha-Naphthylamine	29 CFR 1910.1004
beta-Naphthylamine	29 CFR 1910.1009
4-Nitrobiphenyl	29 CFR 1910.1003
N-Nitrosodimethylamine	29 CFR 1910.1016
beta-Propiolactone	29 CFR 1910.1013
Vinyl chloride	29 CFR 1910.1017

A new revised version of the ANSI Standard, Z129.1–1988, has been in development for some time and is being reviewed for final approval at the time of this writing. Those who wish to obtain copies of the new Standard, available as of June 1988, should contact ANSI at 1430 Broadway, New York, NY, 10018, (212) 354–3300.

General Considerations

In addition to the required items, either the employer or the manufacturer or supplier may find it desirable to add information considered helpful to anyone using the material. If it were used only by workers at facilities that are in full compliance with the HCS, then presumably nothing more would be needed, as the MSDS and the training program should suffice.

Because most commercially available hazardous materials are sometimes used by others, though, the supplier might decide to give more information on the label. It is largely a matter of judgment as to how much information to include on the label beyond what the HCS requires, if any. Perhaps an important consideration in such a judgment will be to decide just how the material in question is used and distributed, and how likely it is to come into the hands of persons who are not instructed under the HCS requirements.

Language

Reading Level

The statements on the label should be worded as simply as can be done without sacrificing accuracy. To be effective, a hazard warning label must be interpreted easily and quickly. Most people will read a label only once, if that, prior to an emergency need. They usually will have forgotten the contents of the label by the time there is an emergency, and will then have to read it, probably quickly, under stressful conditions. Thus, the wording should be simple and practical, and it should be aimed at avoiding significant hazards that might reasonably be expected to be a problem if handling, storage, and use are improper or are accidentally disrupted.

Because hazard warning labels must be easily understood, they should use as few polysyllabic, pedantic words (such as "polysyllabic" and "pedantic") as possible. Such instructions as "wash the affected area with copious amounts of water" should be avoided. Many people don't know the meaning of copious; even those who do cannot really say just how much is meant by copious in the particular case. It is better to say "wash (or rinse) with running water for at least 15 minutes." This language provides both plainer and more definite instructions.

Also, if instructions are given beyond the requirements of the HCS, be sure they are unambiguous and useful. Useless instructions will normally be disregarded, as they should be, and sometimes ambiguous instructions are actually misleading. An example of poor, inadequate information: one often reads labels indicating that one is to "avoid contact with skin and eyes." Such an instruction doesn't indicate the type of problem that would be incurred if such contact were not avoided. It also does not indicate HOW to avoid such contact, i.e., whether simple safety glasses or chemical splash goggles are needed, and what type of gloves to wear to protect the hands.

Other Than English

The HCS requires that labels must be in English. If a significant number of workers do not read or speak English well, it may be desirable to add a second label with the necessary information translated into the appropriate language. This may actually be required in some jurisdictions. A local attorney should be consulted if this is a concern. If labels in languages other than English are used, they should be written by a skilled interpreter rather than by someone who simply can speak and write a little of the language. These are legally required means of communicating critically important information to workers. If they are written in such a way that they fail to communicate precisely the correct information, they fail their purpose and therefore could be more trouble than they are worth. The fact that the initial language is in English, and properly written, will not serve the purpose if a defective translation is relied on by some workers.

Design

Color

It is tempting to color code labels on hazardous materials in some way, but this involves a serious risk of misinterpretation. Unless the same colors are used as on DOT labels, for instance, there can be confusion between the two. However, in some cases the DOT specifies a white background, which might be interpreted as an absence of hazard. If for that reason some other color were selected by the label designer, then it could lead to further confusion.

In addition to these problems, such color coding would be of limited use in communicating hazards to color-blind workers. About 8% of all men, and less than 1% of all women, suffer some defects in color vision. For such workers, relying on color coding can be worse than useless. Different inks that look the same to persons with perfect color vision can look very different to a person who suffers partial color blindness. To such a person, color coding can be very confusing, and hence dangerous, where dealing with hazard information. Still other persons with imperfect color vision may see two different colors as though they were the same. This, too, leads to confusion.

Finally, even a person with perfect color vision loses the ability to distinguish between different colors under very poor light conditions, which will often prevail in an emergency situation. In summary, reliance on color coding can conceivably lead to tragic consequences. If such coding is to be used at all, it certainly must be accompanied by other visual forms of communication that do not rely at all on color.

Risk Rating

It would seem to be very valuable to include on the label information regarding the amount of risk involved with the hazards rather than simply to indicate what the hazards are. The signal words "caution," "warning," and "danger" do this to a degree, but they are too general unless repeated for each type of hazard indicated. If there is a respiratory health hazard indicated, the person who reads the label generally may not be able to tell whether it might be a fatal material like carbon monoxide or simply a minor irritant. Two private associations have developed such risk ratings for use on labels, and they seem to be enjoying increasingly widespread adoption in industry and elsewhere.

The NFPA has developed its 704M System which rates risks from 0 (minimum) to 4 (maximum) in each of three areas. The National Paint and Coatings Association (NPCA) has developed its Hazardous Materials Information System (HMIS) which is very similar.

Unfortunately, OSHA inspectors have found that many workers in facilities that use either the NFPA or HMIS systems don't know whether a rating of "0" or "4" represents the greatest hazard. This emphasizes the need for careful, thorough training of workers, regardless of what system is used to communicate information.

NFPA 704M system. The NFPA 704M system is explained in detail in the NFPA book *Fire Protection Guide on Hazardous Materials*. It has the distinct advantage of giving an indication of the severity of risk in three different hazard categories: flammability, dangerous reactivity or explosivity, and general health hazard. A sample 704M symbol is illustrated in Figure 10.

This type of label does not give a breakdown of subclasses of hazard to the extent required by the HCS, but it does provide a good quick visual warning of how seriously hazardous a material is. The numerals used in each of the three hazard rating areas range from 0 to 4, with the numeral "0" representing no significant hazard of the particular type, and the numeral "4" representing the highest level of hazard. In addition, when appropriate, a symbol representing special hazard problems is placed in the bottom section of the label. In addition to the special symbols for water reactivity, radioactivity, and etiologic hazard, the following additional symbols are used: "P" for dangerous polymerizability, "Ox" for oxidizer, "acid" or "alk." for alkali or strong base, "corr." for corrosive, etc.

The meanings of the numerical ratings for health hazard identification, in terms of type of possible injury, are as follows:

0: materials which on exposure under fire conditions would offer no hazard beyond that of ordinary combustible material
1: materials which on exposure would cause irritation but only minor residual injury even if no treatment is given
2: materials which on intense or continued exposure could cause temporary incapacitation or possible residual injury unless prompt medical treatment is given
3: materials which on short exposure could cause serious temporary or residual injury even where prompt medical treatment is given
4: materials which on very short exposure could cause death or major residual injury even where prompt medical treatment is given

The meanings of the numerical ratings for flammability identification, in terms of susceptibility of materials to burning, are as follows:

0: materials that will not burn
1: materials that must be preheated before ignition can occur

**Flammability Hazard
(Red Background)**

**Health Hazard
(Blue Background)**

**Stability Hazard
(Yellow Background)**

**Special Hazard
(White Background)**

(Water Reactivity Symbol Shown)

Figure 10.   Design of the NFPA 704M hazard identification symbol.

2: materials that must be moderately heated or exposed to relatively high ambient temperatures before ignition can occur

3: liquids and solids that can be ignited under almost all ambient temperature conditions

4: materials which will rapidly or completely vaporize at atmospheric pressure and normal ambient temperature, or which are readily dispersed in air and which will burn readily

The meanings of the numerical ratings for reactivity (stability) identification, in terms of susceptibility of release of energy, are as follows:

0: materials which in themselves are normally stable, even under fire exposure conditions, and which are not reactive with water

1: materials which in themselves are normally stable, but which can become unstable at elevated temperatures and pressures or which may react with water with some release of energy but not violently

2: materials which in themselves are normally unstable and readily undergo violent chemical change but do not detonate; also, materials which may

react violently with water or which may form potentially explosive mixtures with water

3: materials which in themselves are capable of detonation or explosive
reaction but require a strong initiating source or which must be heated
under confinement before initiation or which react explosively with water

4: materials which in themselves are readily capable of detonation or of
explosive decomposition or reaction at normal temperatures and pressures

While the above definitions are very descriptive, they are long enough
and complex enough that it should not be surprising to learn that workers
are often confused about their meanings. For this reason, some safety
supply companies furnish blank, self-adhesive NFPA labels that contain
brief indications beside the square-on-point, such as those in Table 6. A
facility operator needs merely to write the appropriate numeral in each of
the three colored areas with indelible ink and apply the label to the container. A worker who reads such a label also can read the reminders of the
meanings of the numeric ratings if necessary, so he or she doesn't have
to rely on memory so much.

If NFPA labels are applied at the workplace where the materials are
used, they are excellent once the workers are familiar with their meaning.
It is strongly recommended that they be given consideration for use.
However, because of their square-on-point (diamond) format, the NFPA
labels are easily confused with DOT shipping labels and are forbidden to
be used in such a way that they are visible during shipment.

HMIS. A similar system, without the possibility of confusion with DOT
labels, is the HMIS, developed by the NPCA. In this system, the ratings
used involve the same numerical values, with virtually identical meanings, as the NFPA 704M system. The HMIS system includes a descriptive
means of evaluating the hazards applicable to any given material from
basic property data such as LD_{50}, flash point, etc. Using this information,
a knowledgeable person can apply hazard ratings to materials that have
not already been rated by NFPA. However, not all the information needed
to determine the numerical values in such cases will be found on MSDSs,
so it requires some further literature searching, usually by a health or
safety professional.

Another feature of the HMIS system is the use of a single-letter symbol
to indicate the appropriate combination of personal protective equipment.
Each of several combinations of different eye protection, different gloves,
boots, aprons, and other protective clothing, is represented by a single
letter (A, B, C, etc.). However, there are several such combinations, and
the letters bear no simple relationships to the equipment combinations they
represent. Thus it is doubtful whether such a system will communicate the
pertinent information quickly and accurately, except in facilities where
only about two or three different combinations are needed, and hence

Table 6.   Brief Descriptions of NFPA and HMIS Hazard Ratings

	Interpretation for:		
Rating	Health Hazard	Fire Hazard	Reactivity
4	Deadly	Flash point below 73°F	May detonate
3	Extreme danger	Flash point below 100°F	Shock and heat may detonate
2	Hazardous	Flash point between 100°F and 200°F	Violent chemical change
1	Slightly hazardous	Flash point above 200°F	Unstable if heated
0	Normal material	Will not burn	Stable

workers only have to memorize the meanings of two or three letter symbols. To deal with this problem, suppliers of HMIS materials provide explanatory posters as well as wallet cards to be distributed to workers. Entire packets of training materials are also available for the HMIS.

Pictorial Symbols

For quick reference, it is appropriate to use pictorial representations of various types of information, both for the protection of those workers who cannot read well and for the convenience of everyone. A graphic symbol that can be understood at a glance is much more effective than the simplest sentence that gives the same information, even to a highly educated, literate person.

Hazard types. A suitable form of graphic symbol is the type that is specified as part of the DOT labels and placards used for shipment of hazardous materials. Such symbols correspond to shipping labels that are already familiar to industrial personnel, and for the most part they require no training to interpret their meaning.

Several useful symbols are shown in Figure 11; most are commonly used already, but some are not yet so widely recognized.

For example, the symbols for "Carcinogen" and for "Etiologic agent" are becoming more widely used in some locations, but are not yet generally recognized. Thus, it is important that any pictorial symbols used be thoroughly discussed during training to be certain all workers understand them. It has been said that although the use of pictorial graphics to replace written warnings creates greater efficiency in communication, it also creates a new class of functional illiterates.

The symbol for "Radioactive" is well known, but radioactive and etio-

Flammable   **Explosive**   **Oxidizer**

Corrosive   **Water-Reactive**   **Radioactive**

Toxic   **Carcinogenic**   **Etiologic**

Figure 11.   Graphic symbols used to indicate material hazards.

logic hazards are not covered by the OSHA Standard. It is recommended, though, that whenever appropriate they be applied and referred to as a part of the overall HCS program, even though OSHA does not regulate them.

Several of these hazard symbols apply only to physical hazards, and not even to all of those. Other symbols could possibly be developed to warn of some of the other HCS-specified hazards, but it seems doubtful how useful they would be in many cases. It is hard to imagine specific graphic symbols for such hazards as hepatotoxins, nephrotoxins, mutagens, hematopoietic agents, etc. Thus, the use of pictorial symbols to warn of specific hazards has limitations, but it can be a great advantage where appropriate.

Routes of entry. Figure 12 depicts pictorial symbols designed to show important routes of entry. Route-of-entry information must be included on a proper label anyway, and such graphics can greatly improve the efficiency of communication of that information.

Personal protective equipment. Figure 13 shows several pictorial graphics for recommended personal protective equipment. While the presence of such information on the label is not required by the HCS, it is desirable. There is a slight problem here, though. Many manufacturers recommend extremes of protective equipment that may be needed only under unusual emergency conditions, apparently in order to protect themselves from liability. If workers sense that all the recommended equipment is not needed under normal working conditions, they will soon come to ignore such recommendations regardless of how appropriate they might be. This does not suggest that personal protective equipment should not be recommended on labels, merely that careful thought must be given to ascertain just what information will and will not be included.

Placement. The actual placement of symbols on a label is a matter of design preference that probably does not make much difference in terms of effectiveness. The most obvious placement is along a horizontal bar at the top or bottom edge of the label, or a vertical bar along the left or right edge of the label. Perhaps the hazard symbols should be in one location and personal protective equipment symbols in another.

5.2 PREPARING MSDSs

A chemical manufacturer, importer, or distributor must establish a system to deliver the MSDS to each purchaser of his or her products. MSDSs must be prepared for use even with hazardous materials that are exempted from labeling requirements, such as those regulated by the FDA, the Consumer Product Safety Commission (CPSC), FIFRA, and the Bureau of Alcohol, Tobacco, and Firearms (ATF). The delivery of MSDSs should be triggered automatically and could probably be handled most appropriately by the order-processing department.

Format

Careful consideration should be given to the form to be used for the MSDS. The HCS does not specify any particular format, but there are several factors involved. Some suppliers have still been using the OSHA

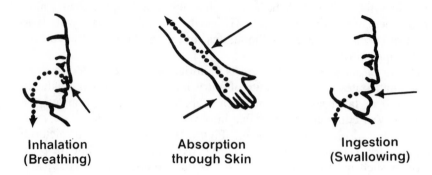

Inhalation	Absorption	Ingestion
(Breathing)	through Skin	(Swallowing)

Figure 12. Graphic symbols used to indicate routes of entry.

Form 20 for their MSDSs. It seems to carry the official sanction of OSHA, but it is obsolete, and therefore it should not be used.

OSHA has published a new Form 174 which, when properly completed, satisfies the requirements of the HCS. A copy of this form is given in Appendix E.1. Suppliers who have been using Form 20 might well consider changing to Form 174.

Some chemical companies furnish MSDSs in the form of a computer printout of the required information. While that practice may meet the formal requirements of the Standard, it has some problems of its own.

The first (and very practical) problem is that if an employer has this type of printout from several different suppliers, there will be no standardized place on all the MSDSs to find a particular type of information. This can lead to confusion, frustration, and an unwillingness to make use of the available data.

Second, NO ITEMS ON THE MSDS CAN BE LEFT BLANK. If information in any specific category is either not available or not applicable for a given material, the entry must be "n/a" rather than a blank space. With a computer printout, it may be difficult to tell whether the particular data sheet meets this requirement.

Third, the computer printout sometimes runs to eight, ten, or even more pages. This requires that someone other than the supplier sort through the data to find the information needed to protect workers. Unprocessed data and refined information are greatly different things, a fact that is all too often overlooked by professional people.

Employers might wish to transfer the data from their suppliers' MSDSs onto a single form, such as OSHA Form 174, and use it as a basis for their HCS programs. The original forms from the suppliers should then be retained on file with the master copy of the written program, as documentation. The standardized form then will be most useful to workers, as it will always have the same type of information in the same location.

Safety Glasses **Splash Goggles** **Face Shield**

Protective Gloves **Protective Apron** **Protective Boots**

Half-Mask Respirator **Full-Face Respirator** **Self-Contained Breathing Apparatus**

Figure 13. Graphic symbols used to indicate forms of personal protective equipment.

However, this suggestion must be accompanied by a caveat. If data from all manufacturers are copied to a standard form, the employer making the copies then becomes responsible for the content of the new MSDS. While that may be acceptable to some employers, many will lack the professional staff needed to deal with it adequately.

As an alternative procedure, some very large corporations send blank copies of MSDS forms of their own design to suppliers with a request that the indicated data be supplied and the form be signed and returned to the sender. But the Standard does not require that suppliers cooperate with such a request, and whether small firms can exert enough economic pressure to obtain the necessary cooperation is uncertain.

Both labels and MSDSs must be in English, though an additional copy in another language may be used also if appropriate. For example, if a large portion of a work force consists of Spanish-speaking people whose understanding of English is poor, their employer may want to have an additional MSDS file just for those people. In such a case, though, a skilled technical interpreter must be engaged to provide the necessary translation of information into the second language.

Identity of Material

The identity of the material must be included on all MSDSs and the form of identity must make it easy to locate the sheet for any particular material. The name that is commonly used by the workers must be on the sheet as well as a chemically specific name. A specific chemical name is one that will permit any qualified chemist to identify the structure of the compound(s) present to enable further searching of the literature for desired information. This identity will usually be a common or systematic scientific name, often accompanied by its unique CAS number.

Mixtures

If the hazardous material is a mixture that has been tested for hazards, the chemical name(s) of the substance(s) which contribute(s) those known hazards to the mixture must be given, as well as the common name of the mixture itself if it has one. If the hazardous material is a mixture that has not been tested as a whole for hazards, the ingredients that contribute to the actual or presumed hazards of the mixture must be listed, but the actual composition need not be.

Trade Secrets

If the composition of the hazardous mixture is a trade secret, the chemical identities of the ingredients need not be disclosed on the MSDS, but the hazards they present must be specified as well as appropriate protective or precautionary measures to be used with the material. More specific data about the composition of trade secret mixtures can be obtained by qualified health professionals or by OSHA under appropriate conditions and for certain types of reasons, such as a need for monitoring workplace air or for medical surveillance of exposed workers. Such composition information must also be made available to workers or their designated representatives under certain conditions. See Chapter 9 for a more complete discussion of trade secrets and the legal complications they involve.

Control of Hazards

Hazard Specifications

TLVs or OSHA PELs or both must be listed if they have been determined. However, remember that these values have not been assigned for the vast majority of hazardous materials—only those listed in Appendix D.1. If a TLV has not already been determined for a given material, it could be a violation of ACGIH's trademark for anyone else to assign a value and call it a TLV. A PEL can be assigned only by OSHA using a formal rulemaking procedure, or occasionally in response to a court order.

Safe Usage Data

Generally applicable control and monitoring methods as well as general precautions for safe use are to be given on the MSDS also. This information should include appropriate measures to be taken during repair and maintenance of contaminated equipment.

Emergency Response Data

Emergency and first aid procedures that are appropriate for workers exposed to the material must be included.

Applicable control and monitoring methods to be used during clean-up of spills and leaks must also be included in the information on the MSDS.

If the material has an RQ or RCRA Hazardous Waste Identification Number assigned by EPA, these might well be included on the MSDS also, though OSHA does not require their inclusion.

Characteristics of Material

Physical Hazard Data

The physical hazards of the material must be indicated, including potential for fire, explosion, and reactivity.

Health Hazard Data

The health hazards of the material must also be indicated, including signs and symptoms of exposure as well as any medical conditions known to be aggravated by exposure to the particular material. Along with the health hazards, the principal routes of entry must be given (skin, eyes, GI tract, respiratory system). *Indication must also be made as to whether the compound is a listed carcinogen.* This indication is one that could easily be missing from a computer-generated MSDS without its absence being noticed by many readers.

Physical and Chemical Properties

The MSDS must include physical and chemical characteristics of the hazardous chemical such as density relative to water, density of gas or vapor relative to air, boiling point, melting point, flash point, flammability range, vapor pressure at one or more temperatures, etc.

Supplier or Manufacturer Data

An essential part of the MSDS is the date of its preparation and the name, address, and telephone number of a responsible party who can provide additional information if necessary.

5.3 ADDITIONAL FORMS OF WRITTEN INFORMATION

Hazardous Substance Fact Sheets

Even MSDSs that are legally correct and adequate often are in such sketchy form that they are not as useful to health and safety professionals as might be desired. In addition, the fact that many different formats may be present in a given facility's file may make it difficult to locate critical information quickly in the event of an emergency.

A reasonable solution to these problems is to use a supplementary document that is uniform in format for all materials at a facility. Such a document would be filled out in enough detail to guide health and safety professionals in all aspects of any foreseeable emergency. Such a document would not replace MSDSs. It would be too involved and too complex to meet the needs of most of the workers.

The state of New Jersey makes available such a Hazardous Substance Fact Sheet for use by facility owners and operators in that state. Some other states or localities may use comparable forms. EPA has announced that it intends to furnish fact sheets modeled on the New Jersey documents for some, but probably not all, of the hazardous chemicals specifically covered by SARA Title III.

As a minimum, such a fact sheet should include the following types of information in sufficient detail that safety, health, and emergency response personnel can normally perform their essential duties with respect to the material without having to consult other references:

- identity
  - common chemical name
  - CAS No. and DOT No.
  - physical characteristics
  - brief summary of associated hazards
- exposure information
  - PELs or TLVs or other recommended limits
  - means of monitoring and reducing exposure
- health hazard information
  - health effects
  - medical testing procedures
- workplace information
  - engineering controls
  - storage and handling precautions
  - exposure preventive measures
  - recommended respiratory protection and other personal protective equipment
- emergency information
  - fire hazard and control information
  - spill or leak control and response
  - first aid

Hazardous Material Safety Guide

As has been discussed above, the wide variety of MSDS formats reduces the usefulness of such documents for their primary purpose—communication of health and safety information to workers. In addition, much of the information required for MSDSs, or included without being required, is "background noise" that gets in the way when a worker wants to find some information.

A worker wants to know whether a liquid is flammable, combustible, or neither—not what its flash point is, or what method was used to determine it, or what its evaporation rate is relative to ethyl ether or any other particular reference material. He or she wants to know whether a material is toxic, what symptoms of overexposure are, and whether protective equipment is needed—not whether it is a "CNS depressant," or a nephrotoxin, or what its PEL or TLV or LD_{50} are. He or she certainly isn't interested in the name and telephone number of a contact person in the manufacturer's or distributor's company. All of these things have their place, and they really need to be available. But even with a good training program, they will not be useful to the vast majority of workers.

It is recommended that a facility use a Hazardous Material Safety Guide (HMSG) as a supplement to MSDSs, labels, and training. The form shown in Figure 14 might be used, or another might be developed to meet the particular needs of the facility. The HMSG would have essentially all

HAZARDOUS MATERIAL SAFETY GUIDE

Material ID No.:

Name:

	HMIS/NFPA Hazard Rating
Health Hazard:	
Fire Hazard:	
Stability Hazard:	
Special Hazard:	
	Health Hazards

Routes of Entry:

Acute Hazards:

Symptoms:

Immediate Treatment:

Chronic Hazards:

Personal Protective Equipment:

Normal Conditions:

Emergency Conditions:

Figure 14.   Sample hazardous material safety guide.

Physical Hazards

Fire Hazards:

Explosion Hazards:

Stability Hazards:

Engineering Controls

TLV: PEL:

Ventilation:

Special Controls:

Spill Cleanup and Disposal

Fire Response

Figure 14, con't.

the hazard information that would normally be sought by most workers. About the only information on the sheet that might not be consulted by workers is the ID number used to key the HMSG to the facility's list of hazardous material. But it contains essentially nothing that doesn't directly tell the worker about hazards or precautions he or she needs to know. In addition, a uniform format for the HMSGs makes them easier for the workers to use, and thus they are more likely to be used.

HMSGs do not replace MSDSs. They should be maintained, probably in a separate loose-leaf binder, at each location where MSDSs are kept. Then workers can easily access and consult the basic, original MSDSs also if they need to or want to.

6

Keeping Records and Filing Reports

6.1 REQUIREMENTS

The HCS has extensive requirements for record keeping, but no requirements for reporting. The Standard has the purpose of protecting workers, and the only required communication to official agencies is through site inspections and responses to official requests for specific information or documentation.

SARA Title III, though, has the purpose of providing information and assistance to official agencies to aid in the management and control of emergencies arising from accidental release or spill of materials that could cause harm to the surrounding community. Thus, it has extensive reporting requirements but very few requirements for record keeping. It is not concerned with what goes on inside a facility except when it can harmfully affect the surrounding community.

Reporting

There are five types of reports required under SARA Title III. Each of these reports must be made by owners or operators of certain facilities. The requirements as to which facilities must report depend on the particular reporting function. They are specified in the following paragraphs.

If a facility is required to file a particular type of report, it is the responsibility of either the owner or operator of the facility to do so. The owner may in many cases be the same person as the operator. In many other cases different persons may be involved, and often the owner is a corporation but the operator is an individual employed by the corporation.

Either the owner or the operator may file the report, so long as one of them does. As explained in Section 3.3, there are penalties authorized by the statute: fines, jail sentences, or both. If a required report is not filed in timely fashion, a penalty may be levied on either the owner or the operator, but normally on both. The responsibility falls on both, and it is up to them, rather than to the EPA or other officials, to decide who will perform the function.

Emergency Planning Status

Chronologically, the first reporting requirement involves notification of the authorities that the facility falls under the emergency planning provisions of SARA Title III. (See Section 3.1.) This notification must be made if the facility has on hand, at any time, one of the extremely hazardous materials in excess of the TPQ. These materials are identified, and their TPQs at the time of this writing are given, in Appendix D.2.

Lists of Hazardous Materials

All facilities required to maintain MSDSs under the HCS are required to submit to their local planning committees either copies of the MSDSs or a list of hazardous materials kept or used in the facility. It is generally recommended that lists be submitted rather than the actual MSDSs, but some local committees may request the actual MSDSs. Any facility owner or operator may optionally submit copies of the MSDSs themselves.

If lists are submitted, they are to be organized in groups of materials having each of the general categories of hazards: fire hazard, sudden release of pressure hazard, reactivity hazard, immediate (acute) health hazard, or delayed (chronic) health hazard. Note that an individual chemical or mixture may exhibit more than one of these types of hazardous properties and will therefore appear in more than one of these categories on the list submitted. Further details about this reporting requirement are covered in Section 6.2.

The hazardous materials list required by the HCS contains much of the same information required on the lists to be submitted to local planning committees. Depending on how the list is organized, a single list may serve both purposes, or both may be generated from a single computer data base as described in Section 6.2. Thus, there is much overlapping information that doesn't need to be repeated as a separate project.

Hazardous Material Inventories

Facilities having or using more than specified quantities of hazardous materials are required to submit actual inventories of these materials. The inventories are organized in two categories: Tier I and Tier II inventories. Only the Tier I inventory is actually required to be submitted automatically by facilities with the specified quantities of materials, but Tier II inventories may be submitted instead. The Tier II inventories are actually required to be submitted only when requested.

Tier I inventory reports are to be available to any member of the public who asks to see them, and for this reason, many owners or operators may elect not to submit Tier II reports except on request. The Tier II reports contain far more detailed information, which owners and operators may prefer not be made available to the public for various reasons. Tier II reports, when requested, may be made available to the public under some conditions, but not generally. However, the most convenient way to prepare the reports is to prepare the Tier II report first and use it as a worksheet for preparing the Tier I report. Therefore, some owners and operators will elect to simply submit the Tier II report rather than do the slight amount of extra work to prepare the Tier I report.

Information required on the Tier II report, in addition to identification of individual chemicals and their physical and health hazards, includes the maximum and average daily amounts of the material present, the number of days it was present on site during the reporting period, and where on the premises, in what type of container, and under what temperature and pressure conditions it was stored. The Tier I report, though, merely requires reporting of the total amounts of materials in each of the five hazard categories stored, the average and maximum daily amounts, and numbers of days on site for each of the five categories, without identifying locations or a breakdown of individual materials within each of the categories. Further details regarding developing and submitting the Tier I and II inventories are given in Section 6.3.

Individual Spills or Emergency Releases

If quantities of certain materials in excess of specified amounts are accidentally spilled or released, that information must be reported to the proper authorities. In order to determine which materials are subject to such requirements and the specified amounts, refer to the RQ listed in the columns of the table in Appendix D.2. Some of these reports are required under CERCLA, or Superfund, and some under SARA Title III. The RQs under these two statutes are different for some materials, and both are indicated in the cited table. EPA has announced its intention of publishing a single consolidated list of materials with RQs. This should help to relieve some of the confusion regarding what to report and to what office or agency to report it. Note that if a spill or release covered by this section is not reported properly, it may be a violation of both CERCLA and of

SARA Title III, and penalties may be assessed under both statutes simultaneously.

Details of what to report when spills and emergency releases occur, and how to report it, are covered in Section 6.4.

Annual Toxic Chemical Release Inventory Report

Finally, Congress required EPA to develop an inventory of releases of certain toxic chemicals from manufacturing facilities having ten or more full-time employees. The chemicals subject to this annual reporting requirement are also given in the table in Appendix D.2. Further details regarding the annual spill reporting are given in Section 6.5.

Required Record Keeping

Required Under the HCS

The HCS requires that MSDSs be kept on file and readily accessible at all times to employees working in locations where they might be exposed to the particular materials. In addition, a copy of the written hazard communication program, including a list of chemicals, must be maintained and available to workers and their representatives as well as to OSHA or state compliance personnel when requested. It is recommended that this list be in a matrix form that includes an indication of the location(s) in the facility where each material is used or stored and of the type(s) of hazard associated with each material.

Required Under SARA Title III

The only records that SARA Title III requires facilities to keep and maintain are copies of the annual toxic chemical release inventory forms, together with copies of all supporting materials used to develop the information included in the submission. These materials must be kept on file for a period of five years following submission of the form.

Recommended Optional Record Keeping

There are a number of types of records that are advisable to be maintained by facilities even though they are not required by OSHA, EPA, or other federal agencies. One of these types of records involves documentation of compliance efforts under the HCS and under SARA Title III. For example, when MSDSs are requested from suppliers but not forthcoming, or when requests for more complete and proper MSDSs are made, copies of the letters making such requests should be kept on file as evidence of good faith effort on the part of the facility owner or operator.

There are no requirements to maintain records of training under the HCS, but in view of the fact that compliance officers are finding that many workers seem not to have learned the material they should have learned during training, it is almost essential that complete records be kept of training given. These records should report who received what training, when and for how long, and what evaluation method, if any, was used to determine the effectiveness of the training at the time.

There is no requirement to keep MSDSs for materials that are no longer in use in a particular facility. However, it is advisable to keep for an indefinite period of time (essentially permanently) a file of copies of obsolete MSDSs, together with an indication of the dates during which the material was used or stored. Also, if monitoring devices are used in the facility to determine the level of potential exposure of workers to particular materials, records of the results of such monitoring should be kept on file essentially permanently.

One reason for keeping such records is to protect against the possibility of future lawsuits alleging exposure to certain materials and damage resulting from the exposure. If the MSDSs and monitoring results pertinent at the time of the alleged exposure are not available, it may be difficult to defend such a case. Another reason for keeping such records is to aid in diagnosis and treatment of problems that appear years after exposure in current or former workers.

Though SARA Title III does not require keeping records other than those associated with the annual release reports, it is advisable to keep copies of all submitted reports and notifications on hand for an indefinite period of time. Again, such information may be useful in case of lawsuits filed years later.

6.2 DEVELOPING HAZARDOUS SUBSTANCE LISTS

As the lists of hazardous materials requiring MSDSs are developed under the requirements of the HCS, it is desirable to include considerably more information than actually required. At a bare minimum, each material entry should indicate where in the facility it is stored or used as well as the material hazards associated with it.

In addition, though, it is desirable to include the types of information that will be useful in meeting the various notification and reporting requirements under SARA Title III. For example, a given material's TPQ and RQ should be indicated, as well as whether it falls under the annual release reporting requirements. Note that the particular quantities associated with some of these requirements certainly will change from time to time. It is important to remain alert to changes in the reporting requirements, and when they occur to enter the changes in the lists. In order to meet Tier II inventory requirements, it would be well to indicate the kind of container and the temperature and pressure conditions of storage.

In addition, it is advisable to keep a record of the quantity of material on hand. Thus, the amount on hand at the beginning of the year would be recorded. When the material is depleted (if it is), the date should be recorded, and then the average daily amount would be simply one-half of the amount initially present, the maximum daily amount would be the initial amount, and the number of days present would be recorded. If the material is not depleted but more is ordered, then the amount remaining at the time the next shipment arrives would be recorded. One-half the sum of this amount and the amount at the beginning of the period would be the average daily amount. The amount at the beginning of the next period would be the amount remaining plus the amount received, and the process would be continued until the end of the year, until more is received, or until all is used up.

By keeping these records current along with the inventory of hazardous materials, the preparation of the inventory reports will be greatly simplified. At present it is not known whether there will be a minimum quantity to be reported on Tier I inventories in future years, so it is prudent to assume that within three years *everything* must be reported in this manner.

If the total number of hazardous materials at a facility is no more than about 20 to 50, these records can be kept on file cards or in loose-leaf notebooks fairly readily. However, as the number of different materials reaches about 100 or more, cross-indexing and other aspects of data filing and retrieval become extremely cumbersome for paper and ink methods. It then becomes appropriate to consider seriously the institution of a computerized inventory data base to manage all the Right-to-Know information.

If the facility has a mainframe computer with a staff of computer scientists, it may be most appropriate and convenient to develop facility-specific software to do the job. If a mainframe computer is not in use, or perhaps even if it is, then desktop computers, networked together if appropriate, may be the more reasonable way to go.

Several firms have designed and made available for sale computer software packages to manage Right-to-Know data. Some of these include MSDS data bases, which may or may not be useful. Some of them include data management packages to handle the types of information and data indicated in the preceding paragraphs. A number of firms reported to have such software packages available are listed in Appendix B.5. The authors of this book have not examined the specifications for any of these software packages and therefore cannot make any recommendations. If a facility owner or operator is interested in obtaining such a package, it would be advisable to write or phone several of the listed vendors to obtain the specifications and examine them carefully. It is possible that one or more would be exactly what is needed. However, it may be that none are exactly appropriate—they may include unwanted peripheral capabilities, or they may not include some capabilities that are important to the particular facility.

It is possible that for many facilities, a generalized relational data base software program will handle the needed functions and information as well as (or perhaps even better than) a packaged program. It is up to the person(s) who will be using the software to evaluate its suitability for the individual facility. Certainly if a packaged software program will do what is needed, it will be the most economical way to proceed. However, a program that doesn't meet the needs is never economical.

6.3 HAZARDOUS CHEMICAL INVENTORY FORMS

Copies of both Tier I and Tier II inventory forms are reprinted in Appendix E. Note, however, that EPA has indicated that these forms merely include the required information, but that the actual form can be changed by individual states. Reportedly, some states have already done so. For the most part, the changes apparently involve merely the addition of a facility identification number for use within the state and the community, but there could be other changes as well.

Any facility owner or operator who is required to make such inventory reports should check with the individual state agency to determine exactly what form is to be used. Also, it is possible that some states or communities will also have different levels at which submission of these inventories is required, so that, too, should be checked before submitting or deciding not to submit the reports.

Tier II Forms

This form is to be submitted within 30 days of receiving a written request for it from an authorized official. If it has not been requested, it is not required to be submitted at all, but it provides a convenient worksheet for preparing Tier I forms. Thus, it is suggested that it be completed on a regular basis for use in preparing the Tier I form. Then, if the Tier II form is requested, it will be simply a matter of submitting a report that is already on hand rather than having to interrupt other operations to prepare such a form on request.

Facility Identification

This portion of the form (see Figure 15) asks for the full name and street address of the facility. If there is no street address, then the location should be given in any way that will permit an unambiguous identification of where it is.

The primary SIC code of the facility must be given. If this is not known, it can usually be determined through the reference section of a local library. Also, the Dun & Bradstreet number of the facility must be given. If it is not known, it can be obtained from the state or regional office of Dun & Bradstreet. If the facility does not already have such a number, one will be assigned.

The identity, mailing address, and phone number of the owner or operator must also be entered. Finally, the name, title, and work phone number of at least one, but preferably two, person(s) or office(s) who can assist emergency responders to a chemical accident at the facility must be given. This must also include a telephone number where the individual(s) will be available for contact 24 hours a day, every day. This is why it is preferable to give the information for two persons rather than just one. If one person is ill or out of town or otherwise unavailable at a given time, then presumably the other will be available in case of emergency. Even a few minutes lost in trying to contact such a person can be critical in times of emergency.

Chemical Description

For each material, the CAS number must be given if one is available. If the material is a mixture, then the CAS number of the mixture may be used, if one is available. If one is not available, then this item can be left blank, but it is better to enter the CAS numbers of all the known components of the mixtures for which such numbers are known.

If the identity of the chemical is being withheld (legally) the box marked "Trade Secret" must be marked (see Figure 16), and the generic chemical class entered. For example, it might be "aliphatic alcohol," "organic isocyanate," "organic peroxide," etc. Also, the chemical name and *all* applicable state descriptors are to be marked, i.e., whether it is a pure compound or a mixture, and whether it is a solid, a liquid, or a gas. Thus, there should generally be exactly two of these boxes checked.

Under "Physical and Health Hazards," all appropriate boxes should be checked. Often two or more will apply. This information should already be available as a part of your list of hazardous chemicals, whether on paper or on a computer data base.

Under "Inventory," for each chemical enter the maximum and average daily amount on hand, which should already be available on your data

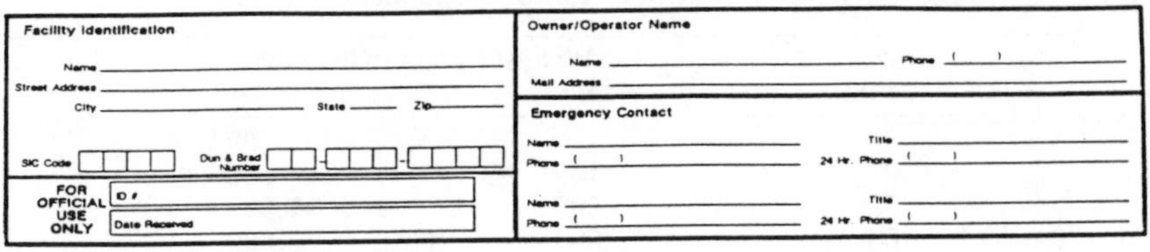
Facility Identification
Name
Street Address
City State Zip
SIC Code Dun & Brad Number
FOR OFFICIAL USE ONLY ID # Date Received
Owner/Operator Name
Name Phone ( )
Mail Address
Emergency Contact
Name Title
Phone ( ) 24 Hr. Phone ( )
Name Title
Phone ( ) 24 Hr. Phone ( )

Figure 15. Facility identification portion of Tier II inventory form.

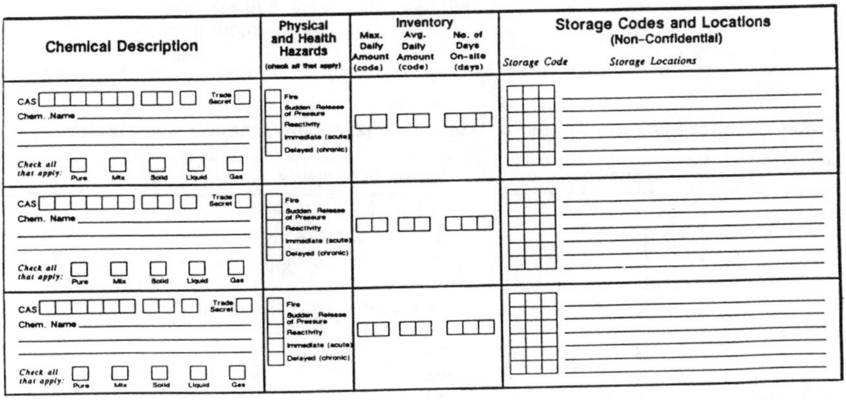
Chemical Description
CAS Trade Secret
Chem. Name
Check all that apply: Pure Mix Solid Liquid Gas
Physical and Health Hazards (check all that apply)
Fire
Sudden Release of Pressure
Reactivity
Immediate (acute)
Delayed (chronic)
Inventory
Max. Daily Amount (code)
Avg. Daily Amount (code)
No. of Days On-site (days)
Storage Codes and Locations (Non-Confidential)
Storage Code
Storage Locations

Figure 16. Chemical description portion of Tier II inventory form.

system. These entries are coded in two-digit numbers, using the reporting range values shown in Table 7 for both entries. The actual number of days the material was on site (rather than a coded value) is entered here also.

Under "Storage Codes and Locations," the type and location of storage of the material is entered. Under "Storage Codes" there are three boxes for each material. In the first box, the code for type of storage container is entered, as obtained from Table 8. In the second box, the pressure condition of storage is entered, using a code number from Table 9, and in the third box, the temperature condition of storage is entered, using a code number from Table 10.

For "Storage Location," there are two possible ways of presenting the required information. The first is to give the building name or number, or the tank number in a tank farm, or other identification such as "drum rack east of paint shop," etc. If possible, the room within the building should be given. The alternate way of giving the location information would be to give coordinates on a site plan. In either case, for the benefit of emergency planners, it is desirable, though optional, to attach a site

Table 7. Range Values for Use with Hazardous Chemical Inventories

Range Value	Weight Range in Pounds	
	From	To
00	0	99
01	100	999
02	1,000	9,999
03	10,000	99,999
04	100,000	999,999
05	1,000,000	9,999,999
06	10,000,000	49,999,999
07	50,000,000	99,999,999
08	100,000,000	499,999,999
09	500,000,000	999,999,999
10	1 billion	more than 1 billion

Table 8. Storage Type Codes for Use with Hazardous Chemical Inventories

Codes	Types of Storage
A	Above-ground tank
B	Below-ground tank
C	Tank inside building
D	Steel drum
E	Plastic or nonmetallic drum
F	Can
G	Carboy
H	Silo
I	Fiber drum
J	Bag
K	Box
L	Cylinder
M	Glass bottles or jugs
N	Plastic bottles or jugs
O	Tote bin
P	Tank wagon
Q	Rail car
R	Other

plan to the Tier II inventory form. The site plan may contain coordinates by which materials may be located, or the buildings and other structures on the site plan can be indicated by name.

Facility owners or operators may elect to withhold information on the locations of specific materials from public disclosure. There are many cases in which it may seem desirable to do so, and it may not always be known whether such a case will develop. Once such information has been released, it cannot be withheld later, except by moving things around and withholding the new locations.

The procedure for withholding such information is as follows. Under "Storage Codes and Locations (Non-Confidential)" the word "confidential" is entered. On a separate Tier II Confidential Location sheet, the

Table 9. Pressure Conditions for Hazardous Chemical Inventories

Codes	Storage Conditions
1	Ambient pressure
2	Greater than ambient pressure
3	Less than ambient pressure

Table 10. Temperature Conditions for Hazardous Chemical Inventories

Codes	Storage Conditions
4	Ambient temperature
5	Above ambient temperature
6	Below ambient temperature but not cryogenic
7	Cryogenic conditions

name and CAS number of each such chemical are entered together with the locations and storage information, in the same manner as described above. This sheet is then attached to the first sheet, and the location and storage information will be available to authorized emergency planners and responders but not to the public at large.

Finally, the correctness and accuracy of the inventory must be certified by a signature of the owner or operator, or the authorized representative thereof.

Tier I Forms

Tier I inventory forms must be submitted annually (as of March 1, 1988) by facility owners and operators to each of the following organizations:

- the state emergency planning commission
- the local emergency planning committee
- the local fire department

Only those materials present at any time during the reporting year in amounts greater than specified below need be included on the inventories:

- January to December 1987 (or first year of reporting), 10,000 lbs
- January to December 1988 (or second year of reporting), 10,000 lbs
- January to December 1989 (or third year of reporting), 0 lbs, or whatever final threshold EPA publishes for the third and later years, following additional analysis
- for extremely hazardous substances, 500 lbs or the TPQ, whichever is less, for all years of reporting

Reporting is required, within the above guidelines, for all materials for which a facility is required to maintain MSDSs under the HCS. The following materials are also specifically excluded from inventory reporting requirements under SARA Title III:

- foods, food additives, color additives, drugs, or cosmetics regulated by the FDA
- solid substances used in manufactured items if exposure to the substance does not occur under normal use
- substances used for personal, family, or household purposes, or substances present in the same form and concentration as a product packaged for distribution and use by the general public
- substances used in research laboratories, hospitals, or other medical facilities under the direct supervision of technically qualified individuals
- substances used in routine agricultural operations or fertilizers held for sale by retailers to the ultimate customers

In the "Physical and Health Hazards" portion of the form, aggregate information on chemicals by hazard categories is required. For each hazard type, the total amounts and general locations of all chemicals of that type present in the facility during the reporting year are to be included. The total actual weights, in pounds, for the maximum and average daily amounts are to be computed for all the materials of each hazard type. The codes to be entered are those shown in Table 7, just as for Tier II forms.

Note that a material may have more than one hazard type, and will contribute to the totals in each of the hazard types reported. Thus, the grand total of amounts for all five of the hazard types may well be far greater than the actual total amounts of material present. Thus, ethyl alcohol is a flammable liquid and an acute health hazard and a chronic health hazard. Its quantity would contribute to the total amounts in all three types. Completing this information will be a simple matter of adding up the figures from the various materials in the Tier II form for each of the five hazard types, provided the Tier II form has been completed.

The only other information required for the Tier I reports, in addition to the usual certification section, is the location information. For each hazard type, all the general locations are to be entered at which such materials are stored or used at the facility. Again, a site plan may be attached to make the location information easier to file and to use, using either names of buildings and structures or coordinates.

6.4 EMERGENCY NOTIFICATION OF RELEASES

If there is a release of a covered substance in excess of the listed RQ, then the owner or operator of the facility must immediately notify the community emergency coordinator for the affected area and the State Emergency Response Commission. The following information must be provided to the extent known at the time of notice, and so long as no delay in notice or emergency response results:

- chemical name or identity of substance released
- indication of whether substance is extremely hazardous
- estimate of quantity released
- time and duration of release
- medium or media into which substance is released
- known or expected health risks associated with emergency, and advice concerning medical attention needed for exposed individuals where appropriate and available
- proper precautions to be taken as a result of release, including evacuation if appropriate
- name(s) and telephone number(s) of person(s) for further contact

This notification can be made by telephone, by radio, or in person. Although transportation operations are exempted from many provisions of SARA Title III, spills and releases from vehicles transporting covered substances must still be reported. In transportation incidents, notification can be made by calling 911 or the telephone operator.

As soon as practicable after a release, follow-up written notice must be submitted by the owner or operator. This notice must include the following information:

- actions taken in response to the release
- known or expected health risks associated with the release
- advice regarding medical attention for exposed individuals where appropriate and available

There are several exemptions from notification provided under the statute. The following types of releases do not require notification under SARA Title III, but some of them may still have to be reported under CERCLA Section 103. The exemptions are as follows:

- federally-permitted releases as determined under CERCLA Section 101
- releases which result in exposure only within the boundaries of the facilities (such releases are not exempted under CERCLA reporting requirements)
- releases from a facility that produces, uses, or stores no hazardous chemicals
- continuous releases, stable in quantity and rate, as defined under CERCLA Section 103
- releases of pesticides registered under FIFRA

6.5 TOXIC CHEMICAL RELEASE INVENTORY FORMS

Facilities that are covered under the reporting requirements of SARA Title III, Section 313 include those that:

- perform any manufacturing functions under SIC codes 20–39
- employ ten or more persons full-time
- manufacture, process, or otherwise use a listed chemical in amounts that exceed threshold quantities during the reporting calendar year

The regulations covering this reporting requirement are located in 40 CFR Part 372, initially published as a notice in 53 FR 4500 (Feb. 16, 1988).

If the facility is not primarily a manufacturer in SIC codes 20–39 but includes one or more manufacturing operations, the text of the regulation should be consulted for interpretation of how the requirements apply. This is also true if the number of full-time workers is less than ten, but part-time workers are also employed. Also, certain laboratory "establishments" within manufacturing facilities are exempted from the reporting require-ments. The preamble in the *Federal Register* notice includes a discussion of these subjects that may be found useful in determining whether a facility is required to report.

In cases involving mixtures of unknown composition, the facility owner or operator is required to make a reasonable attempt to identify reportable chemicals by contacting the suppliers, if the information was not already known. If the supplier indicates the presence of a covered material but not the composition, it must be dealt with for Section 313 purposes as if it were 100% pure; otherwise, the percentage present in the mixture may be applied to the total amount of the mixture to determine whether the amount present places it in the reportable category.

Beginning with 1989, however, suppliers of mixtures are required to report to their customers the presence of reportable materials in the mix-tures, as well as their percentage compositions. If a mixture is trade-secret protected, the presence of the reportable materials must be stated in terms of chemical family names. If the percentage composition of such a mix-ture is trade-secret protected, its composition must be reported in terms of maximum amounts, such as "less than 10%," "less than 35%," etc. This information must be incorporated in or attached to MSDSs covering such mixtures, and must be transmitted with the first shipment of the material in 1989 and each succeeding year. For details of these reporting requirements, including the possibility that a material may be exempted, the actual regulation should be consulted.

If the facility manufactures or processes a listed chemical (see Appen-dix D.2 for their identity), it must be reported if manufactured in excess of the following thresholds:

- 75,000 lbs for calendar year 1987
- 50,000 lbs for calendar year 1988
- 25,000 lbs thereafter

For facilities "otherwise using" the listed toxic chemical, the threshold amount is 10,000 lbs for all reporting years. If a listed chemical is present only in a mixture to the extent of less than 1% of the composition (0.1% in the case of carcinogens), then it does not need to be reported. Releases of less than 1000 lbs of a specific chemical may be reported as ranges, rather than specific amounts, for the years 1987–89. Beginning with the 1990 reporting year, actual amounts released must be reported, though that provision of the regulation is scheduled to be reviewed by EPA and may be changed.

Toxic Chemical Release reporting forms must be submitted to the EPA and to the state. The form should be submitted either to the State Emergency Planning Commission or to the governor's office unless the state has specified another recipient. All reporting deadlines are July 1 of the year following the reporting year.

The EPA published a preliminary version of Form R (EPA Form 7740–20 [6/87]) which was to be usable for reporting 1987 results. However, in the February 16, 1988 *Federal Register* notice, a revised Form R (EPA Form 9350–1 [1/88]), reproduced in Appendix E, was published together with detailed instructions for completing and submitting the form. This new form must be used for reporting results for 1988 and following years. Although the preliminary form apparently will be accepted for the 1987 reporting, the new form is strongly recommended, as it is designed to be compatible with the pertinent computer data banks.

Facility owners and operators may use readily available data collected under other provisions of the law to provide information on the releases and emissions needed for this form. Monitoring data may be included if available, but additional monitoring or measurements are not required beyond what may be necessary to meet other legal requirements. If data are not available to enable completion of the form, reasonable estimates must be made.

The EPA has provided assistance with this task in the form of a document, "Guidance for Determining Releases and Waste Treatment Efficiency for the Toxic Chemical Release Inventory Form." It is recommended that owners and operators of covered facilities obtain a copy of this document as well as the instructions for completing and submitting the inventory form. The telephone numbers of the Chemical Emergency Preparedness Program Hotline are (800) 535–0202 or (202) 479–2449.

Preventive Measures in the Workplace

7.1 THE WRITTEN HAZARD COMMUNICATION PROGRAM

Every facility that uses materials covered by the HCS must prepare a written hazard communication program. This written program must be available to all workers upon request, as well as to their designated representatives (such as union officials, attorneys, or physicians) and certain specified officials of the Department of Labor.

These requirements must be made known as a part of the worker information and training program, and procedures for requesting access to the written program should be disclosed at that time. Copies should be kept in the safety department and in the facility manager's office, as well as in other appropriate locations, consistent with company organization and policy.

As with any safety program, ultimate responsibility for the proper design and execution of the hazard communication program must rest with top management. Without such responsibility, the success of the plan cannot be assured, and it is top management personnel who risk a prison sentence in the ultimate worst-case scenario of plan failure. This responsibility should be acknowledged in the introductory section of the written plan.

Content

The written hazard communication program must include a list of hazardous materials by area within the facility, and it must also describe:

- how material safety data sheets are prepared or obtained, who has the responsibility of doing so, and how they are made available to workers
- how labels and other forms of warning are prepared or their correctness is checked, who is responsible for doing so, and how labels are to be used
- how workers will be trained and informed as required by the Standard, and who has the responsibility for the training program
- how workers will be informed of the hazards of nonroutine tasks such as cleaning and maintaining equipment and cleaning up spills
- how outside contractors' workers will be informed about the hazards present and appropriate protective measures to use while working in the facility

A sample generic written hazard communication program is given in Appendix G. This may be used as is by inserting appropriate facility-specific words and phrases, or it may be modified to fit specific needs as appropriate. It would also be appropriate to include the full text of the Standard in the written program so that workers may refer to it if they wish.

The Hazardous Materials List

As the inventory of hazardous materials is prepared, the list should be arranged alphabetically. The name used by the workers for each material must be included on the list, as well as the name on the label and any other names included on the MSDS or the label. These should be cross-referenced to the principal name on the MSDS so that it can be found as quickly as possible.

Often the name used by workers is a convenient one that has no obvious connection to the name on the MSDS, so cross-referencing is necessary. Following are a few examples of the types of nicknames that are sometimes used when speaking of particular chemicals:

- "Caustic" refers to sodium hydroxide, also known as caustic soda.
- "Monomer" refers to styrene monomer (or vinyl chloride monomer, or methyl methacrylate monomer). There are many different monomers, but in a particular facility the term may be used for a specific compound. In such cases the material should be referenced in that way on the hazardous materials list.
- "Plasticizer" refers to di-2-ethylhexyl phthalate.

Once the list is completed with cross-references, a sublist should be prepared for each area or department if a large operation is involved. The complete list must be kept up to date in the written program. It should include notations as to which sublists contain each material, and in which departments or areas each material is located. All chemicals need to be tracked from their initial introduction into the workplace so that the location of each material is known at all times.

This list is for those hazardous materials known to be present. If a reasonable effort is made to determine whether something is hazardous and no such properties are found, it isn't necessary to undertake a major research project to find some. However, the effort used must be reasonable and significant—not just a cursory inquiry—and it must be documented.

Also, a statement must be included describing how the hazards were determined for the listed materials. If the information supplied by the vendors was relied upon, a statement to that effect is needed. If, however, the employer or some other person made the determination of hazards, then the criteria used and the reference sources consulted need to be listed. This kind of statement will indicate whether a reasonable and significant effort was made to identify and characterize all the hazards involved.

The hazardous materials list contains much of the same information required on the lists to be submitted to local planning committees. Depending on how the list is organized, a single list may serve both purposes, or both may be generated from a single computer data base as described in Section 6.2. Thus, there is much overlapping information that doesn't need to be repeated as a separate project.

Material Safety Data Sheets

The supplier of a hazardous material is legally responsible for identifying the hazards present and for supplying a properly completed MSDS. Employers who buy and use hazardous materials are responsible for attempting to obtain the MSDSs for them. If an MSDS is not received for a particular substance that is suspected of being hazardous, a letter should be written to the vendor requesting one.

If a response is not received within a reasonable time, 15 days for example, a second letter should request either an MSDS or a disclaimer that the material is not hazardous under the Standard. Copies of these letters should be kept with the master hazardous materials list. If neither an MSDS nor a disclaimer is received, copies of the two letters will document diligence in the attempt to learn whether the material is hazardous.

The purchasing department should automatically request a copy of an MSDS, either new or updated if appropriate, as a part of each order for materials. This might involve using a rubber stamp or printed memo, with a notation of the date of preparation of the most recent version of the MSDS received from that vendor. Immediately on receipt of MSDSs, they should be photocopied as needed, copies forwarded to the appropriate locations in the facility, and notice of their receipt posted. A policy should be established to see that all materials purchased on petty cash, as well as sales samples of materials, also enter the system.

Labels and Other Forms of Warning

All containers should be checked to see that they are accurately labeled with the identity of their current contents, the hazards present, and the identity of a responsible party. If the supplier hasn't done this satisfactorily, someone in the receiving organization must do it on a routine basis. However, delivery of improperly labeled materials should not be accepted, even if there are plans to apply some kind of additional label.

The methods used to inform workers of the hazards associated with chemicals contained in unlabeled pipes, mixing tanks, reactors, etc., in their work areas must be included in the written program.

Worker Training and Information

A description of the methods used to inform the workers about the provisions of the Standard must be included, as well as the methods used to train them about material hazards in their workplace. If written objectives or outlines or both are used in conjunction with the training, it would be well to include them in the written program, probably as an appendix.

If packaged training materials from commercial sources are used, they should be specifically identified in the written program. If written materials such as pamphlets or handbooks are distributed to workers as a part of the training program, copies should be appended.

The written program must include a description of procedures used to inform occasionally exposed workers about the hazards in the various areas which they may visit in the course of their duties. Such personnel might include maintenance workers, security personnel, delivery crews, messengers, some office workers, and design engineers, for example. All workers may be given some abbreviated form of the training, or training may be selective.

Nonroutine Tasks

The written program must describe training for nonroutine maintenance and cleaning of equipment, and it must contain an outline of emergency procedures such as cleanup of spills and leaks. Emergencies such as fires, floods, or explosions can cause a release of hazardous substances into the environment, and are covered by the Standard.

Note that HCS training is not required for hazardous waste operations, but training of equivalent or greater extent is required for those working with RCRA hazardous waste (29 CFR 1910.120). This special training may effectively be coordinated with other safety training in the facility. If this is not to be done, reference to it would not need to be included in the written program. But language related to training of workers responsible for cleanup of spills and leaks does need to be included.

Outside Contractors' Workers

A policy must be developed to train and notify contractors and their workers operating in the facility. Written agreements should be prepared for repeat contractors such as cleaning or maintenance suppliers.

The contractor's foreman or superintendent may simply be briefed on the hazards, and provided with a copy of the written program. The hazardous materials list for the area within which the contractor's workers will be working should also be supplied, as well as access to MSDS files for the area. The contractor's representative should be advised to train his or her personnel as appropriate for their potential exposure.

Compliance Checklist

In order to help assure compliance with the HCS requirements, the following checklist of the necessary features of a written program is offered.

- Hazard evaluation
  - _____ person(s) or outside source(s) responsible
  - _____ sources consulted, if done in-house
  - _____ criteria used
  - _____ review plan for new information
- List of hazardous materials
  - _____ agreement of material names with those on MSDSs
  - _____ procedure for updating
  - _____ locations and availability
- MSDSs
  - _____ person(s) obtaining and maintaining them
  - _____ well-maintained and available
  - _____ procedure when MSDS not received
  - _____ procedure for updating
  - _____ descriptions of alternatives to MSDSs in the workplace, if used
- Labels
  - _____ person(s) responsible for special labeling at the facility
  - _____ person(s) responsible for shipping labels, both incoming and outgoing
  - _____ description of label system, if other than manufacturers' used
  - _____ description of alternatives to labels, if used
  - _____ review and update procedures

- Unlabeled pipe hazards
  _____ methods to inform workers
- Training
  _____ responsible person(s)
  _____ program format
  _____ program elements
  _____ initial training procedure
  _____ new-hazard training procedure
- Nonroutine task hazards
  _____ methods to inform workers
- On-site contractors
  _____ methods to inform them of hazards to which their workers may be exposed while working on the premises
  _____ methods to ascertain what hazards will be introduced by contractors' activities to which facility's own workers may be exposed

7.2 WORKER INFORMATION AND TRAINING

OSHA considers training to be one of "three vital components," along with labels and material safety data sheets, in a comprehensive hazard communication program. When OSHA inspectors visit a facility, they will interview some of the workers about their hazard communication training. The success of a worker training program will depend to a large extent on the quality of relations between employer and worker.

Information About the Standard

The provisions of the HCS, and the ways in which those provisions are met in the particular facility, must be explained to all workers. New workers must be given this information during their safety orientation.

The pertinent features should be outlined briefly, questions should be answered, and posters displaying the essentials of this information should be placed strategically throughout the facility. For example, posters should tell workers where the MSDSs are located, where the written hazard communication program is kept, when new MSDSs have been received, and how additional information can be requested.

Figures 17 and 18 are examples of posters that are recommended by the State of Michigan. Generally they will meet the needs of this part of the Standard, and something similar should be used. But some other states may have specific requirements of information to be included, or language to be used, on such informational posters.

Figure 17.   Poster for announcing locations of MSDSs in facility.

Rights of Workers

Workers must be told where the MSDSs are kept, and that they are available for inspection at any time during working hours. They should also have their rights explained regarding trade secrets. Workers and their designated representatives (e.g., union officials, attorneys, personal physicians, etc.) have a right to disclosure of the actual constituents of materials whose compositions are protected as trade secrets, but only upon written request and only under conditions that assure confidentiality.

The Hazard Communication Program

Workers must be informed about the nature of the written hazard communication program, including the list of hazardous materials in the work place. They should be told that the written program is available for inspection, how to use it, and where to find it.

As Required by the Michigan Right To Know Law

TO BE POSTED THROUGHOUT THE
WORKPLACE NEXT TO MSDS LOCATION POSTERS

New or Revised
MSDS

NEW OR REVISED TITLE	RECEIPT DATE	POSTING DATE	LOCATION OF NEW OR REVISED MSDS

MIOSHA
Michigan Occupational Safety and Health Act

SET #2106

Figure 18.   Poster for announcing availability of revised or new MSDSs.

Location and Use of Hazard Information

In addition to the general provisions of the Standard, workers must be told specifically where they can find hazard information when they want it. This is primarily a matter of telling them where to look for information on the labels, where to find the written program, what it includes, how to use the hazardous materials list, and where to find the MSDSs and how to read and understand them. Teaching the use and understanding of MSDSs will be a major instructional challenge, especially with the many workers who either are not technically oriented or are not academically motivated.

Content of Training

In general terms, the subjects to be covered in the training are mandated by the Standard, but exactly how to treat them is up to the individual employer. Following are some general guidelines.

All workers, old as well as new, must be told about the hazards posed by all the materials they work with that are covered under the Standard.

They must be told how to detect the presence of the materials, how to use personal protective equipment properly, what emergency procedures to use in case of leaks and spills, and what procedures are to be used in nonroutine tasks such as maintenance and cleaning of contaminated equipment. This training must be updated whenever a new hazardous material is introduced into the workplace. Critical aspects of the training should also be repeated periodically even if no new materials are introduced.

Recognizing Hazardous Materials

Workers must be told about the various forms of identity (synonyms) for the material. They may be accustomed to talking about a particular material with a common name that has little resemblance to its scientific name or proper shipping name. But they must be told about the various names and how to use the cross-indexed list of hazardous materials to find the MSDS for any material for which they want information.

Along with the name identity, they must be told how to recognize the material if it should be released in the workplace. This can include such things as its color, its state (solid, liquid, or gas), its general appearance and odor (if any), and where it is used in the process. Any monitoring instruments that are normally used or that are available for use as needed should be described. Those who actually use such monitoring instruments need to be instructed in their proper use.

Visual recognition is only one means of identification; odor is another. If the material is a gas, or a liquid or solid with a high enough vapor pressure to have an odor, workers need to know how to identify it from its odor. The workers should learn to identify the hazardous materials in their areas of the facility by actually smelling them. Each person should smell samples of the material. Then all workers should take a quiz to see how well they can identify by odor those materials to which they are likely to be exposed. When there are fairly toxic materials involved, they should be diluted to the point where the exercise will not pose a significant risk of undue exposure.

Understanding Hazards

All workers also must be told about the hazards themselves (what they are, and what they mean in terms of recognizable phenomena or symptoms) as well as the risks accompanying each hazard. The risk instruction will include some discussion of how serious the effect is that occurs if the hazard actually does result in harm to health, property, etc. It should also discuss how likely these things are to happen.

This is usually the most difficult part of any safety instruction program. It is very easy to overreact to a minimal hazard and become paranoid. It is even easier for foremen, supervisors, and long-term workers to ignore hazards that have been around for many years without a problem ever developing. This does not mean that a hazard does not exist, though, and the situation must be discussed openly, frankly, and honestly.

Are there materials used in your workplace that are so dangerous that accidental spills and releases may require evacuation and exposure may require CPR or other emergency measures? If so, workers certainly must be told about the potential need for such extreme measures, and should be instructed and drilled in the use of those measures. If there are other things that only have minor risks but that can still be problems, the workers need to be made aware of them. In other words, the workers need to know what they really have to watch out for, and what they simply have to be alert to in case things get out of hand. OSHA inspectors will expect workers to be familiar with these kinds of major differences.

Protecting Against Harm

A vital part of the instruction about risks and hazards is the discussion of precautionary procedures, equipment, and control measures that are designed to reduce the risk of harm from a particular hazard. This information should be stressed as an important aspect of reducing both the probability of exposure and the severity of harm if exposure should occur.

Workers should be told about various types of protection furnished in the workplace. These include both engineering controls and personal protective equipment.

Engineering controls. Engineering controls include all types of equipment that keep hazardous materials confined or otherwise controlled so that people are not unduly exposed to them.

One important type of control is adequate ventilation, either general or local. General ventilation normally removes air from the overall work area by mechanical blowers and exhausts the air outside the facility. In this way, any vapors, mists, fumes, or other materials that become dispersed in the air are removed before their concentrations can build up to worrisome levels. Local ventilation does the same thing, but by means of specially positioned exhaust equipment that keeps dispersed contaminants from moving very far from the source into the general workspace. Workers should be told about proper placement of work to maximize the effectiveness of the ventilation system and, if appropriate, how to use the controls of the system.

Another type of engineering control is automatic fire-extinguishing equipment such as sprinklers or halon release systems. Those workers who deal with such equipment should be told how to avoid setting it off inadvertently, and what procedures to follow when the equipment does operate (often evacuation of the area).

Some process equipment uses pressure relief valves to avoid dangerous pressure buildups, and some uses automatic level controls to prevent overfilling. These types of controls help to avoid explosions, spills, and leaks that could result in excessive amounts of hazardous materials getting out of control. Workers need to know what such controls do, how to set or reset them, and how to check their proper operation.

Still another important type of engineering control is the condition monitor, sometimes attached to a controlling device, sometimes to an alarm, and sometimes simply to a readout device. Properties monitored by such equipment can include pressure, temperature, liquid level, and concentration of specific contaminants in the air. Appropriate workers should be instructed in the proper use and interpretation of such monitors in order to maintain a safe environment.

Other types of controls and monitors are used in specific situations, but those mentioned here are probably the most widely used types. In all cases, workers should be told about such controls and monitors, what their functions are, how to tell when things are working correctly, and what to do when they are not.

Personal protective equipment. Types of equipment familiar to many include hard hats, safety-toe shoes, safety glasses, and heavy work gloves. However, other types of equipment are needed to protect against undue exposure to hazardous materials. And yet, many workers tend to resist using such protection, partly because the hazards are not so evident as a falling heavy object or flying metal chips.

Wearing of chemical splash goggles and face shields seems to be strongly resisted by people working with solvents, acids, bases, and other liquids. Yet, if such materials get into a person's eyes, they can sometimes cause blindness or permanently impaired vision. The type of eye protection to use under different conditions is not common knowledge. It must be taught.

It seems that many people resist wearing protective gloves when working around chemicals even more than they resist wearing proper eye protection. This apparently is at least partly because effects of chemicals absorbed through the skin are usually chronic, with few acute effects or symptoms, and there is no obviously apparent connection between exposure and effect. Workers need to be carefully instructed about such hazards, and they need to be informed about what types of gloves are appropriate and inappropriate for given hazardous materials, and why.

Some workers will need instruction in selection and proper use of respiratory protection, whether it is dust masks, canister gas masks, air supply masks, or self-contained breathing apparatus.

Nonroutine Operations

Equipment cleanup. Workers who are involved in cleanup of contaminated equipment, either on a routine periodic basis or when a process is shut down for a period of time, need to know about the special protective equipment and measures to be used. These usually are more stringent than during normal operations.

Spills and leaks. Specific training is needed for procedures used in the emergency cleanup of spills and leaks. Such emergencies often involve the presence of unusually large amounts of hazardous materials in an uncontrolled situation. Under such conditions, excessive exposure can occur easily, so special protective equipment may be in order. Also, quick response may be needed in order that a simple emergency does not become a disaster. Ideally, in any except the smallest facilities, specially trained teams should handle such situations and their training should be a part of the hazard communication program. This is also the point to introduce instruction about procedures for evacuation of personnel.

Personal exposure. Workers must be trained to recognize and to watch for symptoms of overexposure to materials with which they work. Appropriate first aid procedures need to be discussed, and sometimes even practiced.

Routine medical surveillance procedures, if used, should be carefully explained including the reason for using them. Frequently workers consider medical surveillance to be harassment. They need to recognize that it is designed to protect them by monitoring their systems for negligible effects that could become major problems if not detected and corrected.

If workers are represented by a union, there may well be contractual restrictions on medical testing that will have to be dealt with. In any case, such a program will need to be defined in writing with adequate protection against abuse. In the OSHA-proposed rules that are expected to require chemical hygiene plans for laboratories (see Section 2.2), provision is made for worker-initiated medical testing.

Fire emergencies. Another subject related to emergencies is that of fire control—whether to use extinguishers, what kind, how and at what stage to evacuate the area, and when and how to call the professionals. In some facilities everyone is instructed to deal with small fires until professional firefighters can arrive, and in others only special teams are to do so. In either case, all appropriate personnel should be given some fire training.

It is important to know what kinds of fire extinguishers are used for particular purposes, and why other extinguishers shouldn't be used. For example, you shouldn't use a stream of water on a flammable liquid—you will scatter the liquid around and spread the fire before you can put it out. Also, a flammable solid (one which will burn with a self-sustaining flame at a certain speed) generally cannot be extinguished by carbon dioxide or even some dry powder fire extinguishers. Some flammable solids will actually blaze more intensely in an atmosphere of pure carbon dioxide than they will in air.

Other exposed personnel. Workers who are in an area only part of the time, such as maintenance personnel, security personnel, delivery crews, messengers, and design engineers, must be informed about the hazards in those areas which they may visit in the course of their duties. They may receive the same training as those who regularly work in the areas or, if appropriate, they may receive some abbreviated form of the training.

Contractors' employees who will be working in areas where hazardous materials are present must receive training in the hazards and their avoidance. You may want to brief the contractor's foreman or superintendent on the hazards, furnishing him or her with a copy of the written program. The contractor should then train his or her personnel as appropriate for their potential exposure.

Note that if temporary personnel are hired through an agency such as Kelly Services or Manpower Incorporated, that agency technically is a contractor and the personnel are contractor's employees. However, such workers are to be trained under the HCS by the hiring facility just as if they were employees of the facility. In effect, the organization that assigns specific work duties is responsible for specific safety training.

The Training Program

OSHA suggests that training programs will be more effective if workers are involved in designing them. Some workers should be members of a safety committee, which should assist management not only in safety inspections but also in planning and carrying out the hazard communication instruction. Some workers may actually participate in delivery of the instruction, and others may simply assist in some other way. But if workers actually participate, the whole process will have their attention better than if it is just handed to them.

Planning

First of all, organize the materials in your facility according to types of hazards. List those that are flammable liquids, those that are flammable gases, those that are combustible liquids, those that are corrosive, those that are acute toxic hazards, etc. The first stage of training should be about each hazard type, and then specific applications and special situations should be taken up.

The starting point for any formal training program is to determine its goals and objectives. Under the requirements of this Standard, the primary goals are to impart to the workers an understanding of:

- the hazards presented by the materials in the work place
- the risks connected with the presence of those hazards
- the implications for precautions and emergency response

Objectives are specific, single-concept statements of what is to be accomplished by the training. Furthermore, the attainment (or failure) of each objective must be subject to verification through some type of evaluation. Some training programs may have dozens or even hundreds of individual objectives. An objective should state precisely what is to be learned, in terms of definite recognizable or testable features.

A possible set of objectives is as follows.
Upon completion of the training, for each hazardous material in the work area, each worker will be able to:

- know the chemical identity as well as the common name
- know where it is used in his or her area of the facility
- know how to identify it by odor and appearance when it is not mixed with other substances
- know what its physical hazards are
- know how to deal with an accidental fire in which it is involved
- know what its acute health hazards are
- know the symptoms to be expected for the acute health hazards
- know what first aid steps to use for a worker who has been exposed to it and is suffering its acute symptoms
- know what steps to take in the event a quantity of it is accidentally spilled
- know what its chronic health hazards are
- know what protective equipment to wear to minimize exposure
- know the proper way to wear the protective equipment
- know where to find any additional information needed about the hazardous properties of the material

In addition to these, there should be other objectives that are not specific to materials. These might include:

- the location of the written program
- the location of the list of hazardous materials
- the location of the MSDSs
- how to use the list to find the appropriate MSDSs
- evacuation procedures
- other emergency procedures, etc.

Individual orientation sessions, periodic safety training meetings, and evaluation of material learned should all revolve about these objectives.

Execution

One of the more difficult aspects of the whole hazard communication program is how to strike a good balance between frightening the worker and lulling him or her into a false sense of security. But it can be done, and workers can develop a realistic understanding of the situation.

The material must be presented as factually as possible, with explanations of the actual hazards and risks. It is very important to emphasize that there is very little risk in most cases when proper protective equipment is worn, when ventilation systems work properly, and when people treat materials with respect and don't slosh things around carelessly.

One of the first steps is a walk-through orientation of all workers who have not already been trained in these matters. This orientation is to show workers what hazardous materials are used in the facility, and where. At this time, material identification and the location of labels or batch tickets can be discussed.

Following the walk-through, the other aspects of the objectives that deal with the materials in the area should be covered in general safety meetings for the workers involved. The Standard is a complex subject. Depending on the nature and number of hazardous materials in a facility, it may not be learned well on the first attempt, and portions of it (particularly emergency response and first aid) that are not used often are easily forgotten.

It is suggested that such safety meetings be held weekly or biweekly, lasting for just 10–15 minutes each time. Training on each aspect of material hazards ideally should be repeated at least semiannually for all workers, and quarterly for those workers regularly working with a particular material. Appropriate portions of this training need to be repeated for all workers whenever new hazardous materials are introduced into the workplace.

The following approach is suggested: Don't just read through an MSDS and try to explain every item as you go. Training should be based on the various hazardous properties, what they mean, and how to guard against catastrophic problems with materials that have those properties. After a property has been discussed in general, the individual materials that do have that property and how they differ should be taken up. In other words, solvent A is a flammable liquid, so is lubricant B, and so is degreaser C, for example. All of these have similar properties even though they are quite different in their uses in the facility, and perhaps in terms of any other hazardous properties they might have. But at least those particular materials have a certain material hazard in common.

When people learn about a particular material hazard, they learn what that hazard really means. Then if that hazard is illustrated by materials they are familiar with, it will really have more meaning. It brings out the particular properties more forcefully than if an instructor just goes through an MSDS and talks briefly about each of the hazardous properties for the material.

These training sessions may take any form that will accomplish the goals and objectives. Some hands-on training is desirable, such as actual practice at putting on a self-contained breathing apparatus for emergency response. One only has to do this once to be convinced that it is not something that comes naturally, but has to be learned. Another example is training in the recognition of materials by odor and appearance. Small containers of materials used in the facility can be brought in and the workers asked to identify each. CAUTION: Smelling some of these substances can be very dangerous. It must be done properly, and particularly harmful materials should be diluted enough to prevent a dangerous dosage by careless sniffing.

In the training sessions, the workers on the safety committee can participate in setting up and carrying out some of the demonstrations, if any are used. Demonstrations can be a real flop, though, particularly when they don't work well, and industrial workers are less likely to be patient with inept demonstrators than are college students. An instructor should only use those demonstrations with which he or she is very, very familiar, and then only after adequate rehearsal.

In the case of flammable liquids, perhaps a professional firefighter can put on a demonstration of liquid fires, some of their properties, and how they are extinguished. Such demonstrations often are designed to involve the trainees actively, leaving them with a more permanent impression of what they were taught.

Training aids are available to help with many aspects of this training. Movies, tape-slide programs, slides with narrative scripts, posters, and other aids are available from safety supply companies, national, local, or regional safety councils, manufacturers of safety equipment, and other sources. Some of these are not as good as others, but nearly all can be useful within their intended range of purposes. A list of companies that have advertised the availability of such training aids is given in Appendix B.4.

Evaluation

At the end of each session, trainees should be given a quiz over the material covered. Every aspect of the training should be followed up by a quiz. It doesn't need to be a very difficult one. Only if the workers get involved directly through quizzes will they be made aware that they really need to learn the material. They should be told before the instruction that they are going to be given a quiz on the material to be covered. This can be an important part of worker feedback. The exact nature and use of the quizzes is a matter for local determination.

One purpose of the quizzes is to determine how well the workers learned the material. Quizzes should be graded and problem areas discussed with the individuals. Things that are not learned well can then be scheduled for additional instruction. Some workers will need to have repeated training on certain subjects before they really understand and are able to deal with things in the way the Standard implies they should. Subjects that are learned very well may need only an occasional refresher session. Testing the workers is the only way to tell how successful the training is, short of waiting for an emergency to occur and observing how they react.

Many facilities will not use materials that can pose a significant threat to life. In such cases, quiz results will most likely be used just to decide whether a worker needs more instruction on a particular topic. However, those workers who deal with any of the materials in three specific classes should be required to pass tests on proper handling of such material or they should be transferred to another job where they will not endanger their own and other people's lives. (Sometimes this may mean discharge if there is no other place to transfer them.) One of the specific classes is the list of materials with spill reporting requirements under CERCLA and SARA Title III; the second is the list of extremely hazardous materials for emergency planning purposes under SARA Title III; and the third is the group of carcinogens indicated in the table in Appendix D.1.

Some unions may object to giving quizzes, though they usually are very supportive of safety programs in general. If objections are raised, it may be necessary to negotiate such testing policies and procedures into the next contract. However, note that safety training and testing are already required before people can do certain jobs such as operate automobiles, trucks, forklifts, and cranes. This is a well-established procedure, and the testing is usually done before a person is considered to be a permanent employee. That is, the employee is probationary until he or she has successfully completed the training.

The problem with establishing such procedures in the case of the HCS is that sudden imposition of a new job requirement for workers who have been around a long time, with no change in the actual nature of the job, is not likely to be accepted easily. But it is a serious problem if some workers endanger others because of carelessness, ignorance, or failure to learn how to deal safely with critically hazardous materials. They constitute a real danger to themselves and others and shouldn't be on the job.

One form of quiz that very definitely should be used is to supply the workers with a copy of one or more MSDSs that are actually used in the facility. Ask them to locate certain information on an MSDS, and to answer one or more questions about the information they have located. For example, if they find that a certain liquid has a flash point of 137°F, is it a flammable liquid or not? That is, is it something that is easily ignited in the workplace? It is important that they learn to interpret flash point data.

They should also be quizzed on identification of materials, both by visual means and by odor, to make sure they are thoroughly trained. They should be asked to identify what is in each of several bottles containing various substances, labeled A, B, C, D, etc.

Documentation

All training and information sessions of any kind need to be documented as to subjects covered, times, dates, instructors and/or packaged programs used, and names of workers attending. An up-to-date summary of this information should be maintained in each worker's personal file for all training received.

Ideally, the worker's signature or initials should appear on an attendance sheet to verify that he or she actually received that training. A photocopy of the signed sheet, or an entry on an individual training log, should be placed in the personnel file of each worker who attends, and the original attendance sheet should be kept in a file of training accomplishments.

A copy of each quiz should also be kept on file. If such quizzes are given frequently, individual files could become very bulky. Thus, it may be preferable to keep one copy of each quiz in a master training file. An entry on the training log in each individual file should include the average and range of scores of all the workers taking the quiz in addition to the individual worker's score. An OSHA inspector can then see that the training had been provided, and how well the workers learned the material, at least for short-term retention.

8

Emergency Planning and Preparedness

Section 303 of SARA Title III requires that each local emergency planning committee prepare an emergency response plan for dealing with releases of extremely hazardous substances into the community. Such plans must have been completed by October 17, 1988. They must be reviewed at least once a year, or more frequently as may be needed because of changes in circumstances.

The amount of additional emergency planning actually required to comply with this section, and the resources and expertise available to carry out that planning with this section, will vary widely by community. For communities which already have emergency response plans for dealing with hazardous materials or nuclear power accidents, complying with this requirement may be a relatively simple matter of updating the existing plan and submitting it for review. Other communities which have plans in place for natural disasters such as floods and hurricanes may need to step back a bit and decide whether it is best to have a single, integrated plan or develop separate plans for separate situations. Still other communities may not have any plans already developed. They will have to start from scratch.

This chapter outlines the major steps in the planning process, and the factors to be considered in fulfilling the specific requirements of SARA Title III. It is intended to serve as a general guide to members of planning committees and other interested persons, not as a comprehensive primer on emergency planning and how to go about it. Further advice on emergency planning and preparedness, as well as resources to assist in the task, may be obtained from sources discussed in Section 8.4.

8.1 THE PLANNING PROCESS

Congress required that members of local emergency planning committees be appointed by each state emergency response commission no later than August 17, 1987. Each committee must include, at a minimum, representatives from each of the following groups or organizations:

- elected state and local officials
- law enforcement, civil defense, fire fighting, first aid, health, local environmental, hospital, and transportation personnel
- broadcast and print media
- community groups
- owners and operators of facilities subject to notification and recordkeeping requirements

Membership appointments may be revised by the state commission, either on its own initiative or in response to petitions from interested persons.

Each planning committee must appoint a chairman and establish rules under which it will operate. These rules must include provisions for:

- public notification of committee activities
- public meetings to discuss the emergency plan
- receiving public comments on the plan
- providing responses to comments received
- distribution of the emergency plan

Congress intended the local planning process to be a truly community-based activity, and not simply an exercise carried out by a few representatives of industry and the government bureaucracy in a back room at city hall. Achieving this goal of public participation while also completing the substantive requirements for emergency response plans by the October 17, 1988 deadline presents a significant challenge, especially for those communities without any mechanism already in place. It may be that initial plans prepared to meet the statutory deadline will be subject to considerable refinement in the future, as more people become aware of the full range of issues involved.

Developing an emergency response plan is an iterative feedback process, wherein consideration of issues raised with respect to one part of the plan may well lead to changes elsewhere. Although the planning elements discussed below are presented more or less in logical order, they cannot practically be addressed in isolation from one another. Figure 19 presents a flow chart of the planning process which is based on materials developed by the National Response Team (the NRT).

The NRT is an interagency group consisting of members from 14 federal agencies having responsibilities in environmental, transportation, emergency management, worker safety, and public health agencies. It is

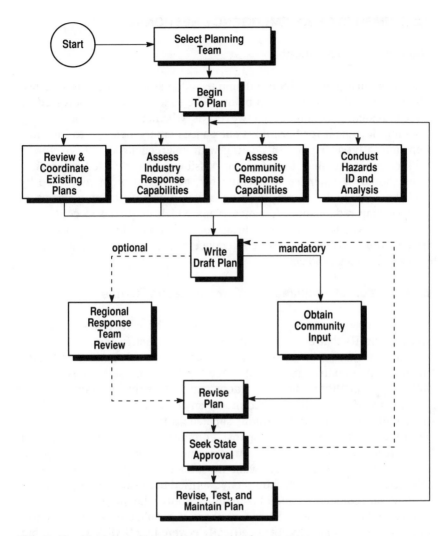

Figure 19. Flow chart of local emergency planning process.

responsible for overall planning and coordination of the National Oil and Hazardous Substances Pollution Contingency Plan described in Section 8.4 below. The NRT has developed a *Hazardous Materials Emergency Planning Guide* in response to Congress' direction that it publish guidance materials to assist local committees in developing emergency response teams. Much of the information in the following section is taken from this document.

8.2 ELEMENTS OF AN EMERGENCY RESPONSE PLAN

The Planning Document

Each local emergency response plan must address a specific list of issues, which will be discussed in turn below. The first issue to be addressed by a local planning committee, however, is whether to create a separate planning document for hazardous substances emergencies or to integrate them into a broader plan for dealing with emergencies of all types. The NRT recommends that, to the extent possible, currently used emergency plans should be amended to account for the special problems posed by hazardous materials, thereby making efficient use of planning resources and avoiding redundant emergency plans which may lead to confusion. Even plans which are no longer in use or which have been developed by private groups to deal with particular concerns may serve as useful starting points or sources of ideas.

Identification of Facilities and Transportation Routes

Each emergency response plan must identify those facilities subject to SARA Title III that are within the planning district, routes likely to be used for the transportation of extremely hazardous substances, and additional facilities contributing or subjected to additional risk because of their proximity to facilities where hazardous materials incidents may occur.

The bulk of the facilities listed in any emergency response plan will be those required to submit hazardous substances lists or MSDSs, or identified by the state emergency response commission. These facilities must have already notified the local committee within 30 days after the committee was created, and in any event no later than September 17, 1987. Additional facilities which should be identified because of their proximity to potential accidents include hospitals, schools, shopping centers, theaters, and other places where large numbers of people are likely to congregate. Facilities which are not specifically covered by SARA Title III, but which might contribute to the danger created by a hazardous materials accident, such as railroad yards and natural gas facilities, should also be identified.

Transportation routes to be identified will include all railroad lines and main highways used by traffic passing through the area, as well as local access routes to the facilities where extremely hazardous substances are located. In addition, it would be prudent to identify those transportation routes located near potential accident sites and the traffic patterns (e.g., typical rush hours) on each.

The list of facilities and transportation routes included in the plan will need to be updated from time to time, as the community develops and businesses begin, move, change, fold, etc. The need for this should be assessed as part of the annual review of the plan.

Designation of Community and Facility Coordinators

Each emergency response plan must also include a designation of a community emergency coordinator and facility emergency coordinators, who shall make determinations necessary to implement the plan. It would also be a good idea to designate alternate coordinators, at least for the community and larger facilities.

The NRT recommends that a separate person or organization be given primary responsibility for coordinating each of three separate phases of a hazardous materials incident: pre-response (planning and prevention), response (implementing the plan during an incident), and post-response (cleanup and restoration).

Facilities required to provide hazardous substance lists or MSDSs, and other facilities identified by the state emergency response commission, must have already appointed their own representative to participate in the emergency planning process no later than September 17, 1987. Additional coordinators should be designated for those facilities identified because of their proximity to potential accident sites, as well as for transportation routes included in the plan.

Description of Emergency Equipment and Facilities

Plans must also include a description of emergency equipment and facilities in the community and an identification of persons responsible for them. This essentially requires a listing of police and firefighting facilities, hospitals, and other more specialized facilities, along with the particular equipment or capabilities available at each. Planning committees should contact each such facility to determine what capabilities they provide and who is responsible at all times. Each planning coordinator for a facility subject to recordkeeping and notification requirements should also be asked to provide information on the specific emergency equipment available at the facility itself.

This section of the plan should also include identification of potential evacuation shelters, and places where injured persons may be taken for treatment in the event that the number of injured exceeds available hospital space.

Methods for Identifying Releases and Affected Populations

Many local committees will probably find one of the most difficult tasks in the planning process to be the development of methods for determining the occurrence of a release, and the area or population likely to be affected. The plan itself will necessarily include only a summary of methods and a preliminary assessment of where the effects of a release are likely to be felt. This will be of little use without actual monitoring systems for detecting releases, some means of tracking a release after it occurs, and possibly even computer modeling of plume development under assumed conditions to determine what areas and populations are likely to be affected.

The NRT recommends that planning committees begin tackling this difficult task by making a three-part hazards analysis, consisting of *hazard identification;* analysis of different types of *potential harm* to life, property, or the environment; and assessment of the *probability* that each type of harm will occur. Hazard identification will be greatly aided by the lists of extremely hazardous chemicals or MSDSs which facility owners or operators are required to submit, as well as (eventually) the toxic release forms. Planning committees should carefully review the annual updates which facilities must supply to determine whether new hazards exist that are to be incorporated into the plan, or whether the nature or extent of existing hazards has changed significantly. Additional information about possible hazards may be supplied by concerned employees or community members, and committees are authorized to obtain the information necessary to determine whether a potential hazard other than one reported by the facility owner or operator exists and should be included in the plan.

The EPA has prepared a guide to assist planners in assessing exposure of the community to health hazards. Copies of the guide, titled *Technical Guidance for Hazards Analysis,* may be obtained by contacting the EPA regional office preparedness staffs or the EPA Chemical Emergency Preparedness Program Hotline at (800) 535–0202 or (202) 479–2449. The guide is said to outline the complex process as simply as possible while ensuring the requisite accuracy for good planning. Another guide is being developed to deal with analysis of physical hazards.

Analysis of the type and extent of damage which could occur in the event of a release should focus on the facilities identified because of their proximity to potential releases, as well as the transportation routes along which hazardous materials travel. It should include an examination of the extent of the vulnerable zone; the size and types of population that can be expected to be within the zone; the private and public property that may be damaged, including essential services such as water, food, power, and medical services; and the natural environment that may be affected, including the impact on sensitive natural areas and endangered species.

Assessment of the probability of various types of harm to life, property, or the environment should be based on the probability that different types of releases will occur and the surrounding environmental conditions. These may include unusual geographic features such as flood plains; the possibility of simultaneous emergency incidents; the type of harm to people (acute, chronic, or delayed) and the associated high-risk groups; the type of damage to property (temporary, repairable, or permanent); and the type of damage to the environment (recoverable or permanent).

Once hazards have been identified and each potential harm and the probability of its occurrence have been assessed, decisions can be made regarding where monitoring for releases is most critical, and whether monitoring at individual sites may be relatively informal, or whether highly sophisticated techniques and equipment are required. Decisions can also be made regarding the need for technical tools used to track releases or predict plume development and the likely effects of an actual release. Sources of information on monitoring equipment and analytical tools are discussed in Section 8.4.

Procedures for Public Notification

Procedures providing reliable, effective, and timely notification of a release to the public and specific persons designated in the plan are obviously a crucial element of any response plan. The plan should include 24-hour emergency response telephone numbers for key response personnel, a number to notify state authorities and the National Response Center ((800) 424–8802; (202) 426–2675), and other agencies such as hospitals and the Red Cross. The plan should also indicate how volunteer and off-duty personnel will be summoned, and a method for notifying facilities such as schools, nursing homes, and shopping centers which may be threatened, depending on the severity of the incident.

The plan should also include the title and phone number of persons responsible for alerting the public *as soon as word of the incident is received,* as well as a list of essential data to be passed on. Since speed of release of this information is crucial, there should be one organization specifically assigned the task of public notification. Most communities already have an emergency alert system developed in cooperation with radio stations for civil defense and other disasters. This may provide an appropriate means of initial notification.

On the one hand, notification should be carried out as widely and as quickly as possible, e.g., through the broadcast media, including as many radio and television stations as possible in the affected area. On the other hand, public panic is always a possibility to be avoided. The person or persons responsible for notifying the media should therefore be well versed in the type of information which must be released, as well as the need to provide an accurate, balanced summary of the situation which avoids blowing things out of proportion.

Notification procedures should be consistent with section 304(b) of SARA Title III, which requires that facility owners or operators give immediate notification to local emergency coordinators of hazardous substance releases. (See Section 6.4.) All persons likely to be on the receiving end of such notification should be provided with checklists on which they can record the following information:

- the chemical name or identity of any substance involved in the release
- an indication of whether the substance is on the list of extremely hazardous substances for that facility
- an estimate of the quantity of the substance released into the environment
- the time and duration of the release
- the medium or media into which the release occurred
- any known or anticipated acute or chronic health risks associated with the emergency and, where appropriate, advice regarding medical attention necessary for exposed individuals
- proper precautions to take as a result of the release, including evacuation
- the name and telephone number of the person or persons to be contacted for further information

The public and persons designated in the plan should be alerted on the basis of this information, without waiting for the required written follow-up notice. Public updates should then be broadcast at regular intervals as more information becomes available.

Methods and Procedures for Responding to a Release

Although SARA Title III requires that each plan include the methods and procedures to be followed in response to any releases of an extremely hazardous substance, it is obviously impossible to do so with any level of detail, as the appropriate response will depend on dozens of variables which cannot fully be predicted. What can be done, however, is to make an assessment of which agencies and facilities need to be included in response efforts for different levels of hazardous substance incidents, and to be sure that each agency or facility is aware in advance of those tasks it might be called upon to perform.

The NRT suggests dividing up potential hazardous substance incidents according to potential emergency conditions, limited emergency conditions, and full emergency conditions. A *potential emergency condition* is an incident or threat of a release which can be controlled by the first response agencies and does not require evacuation of other than the immediate area. The incident is confined to a small area and does not pose an immediate threat to life or property. The appropriate response may be carried out by local fire, police, and emergency medical services personnel.

A *limited emergency condition* is an incident involving a greater hazard or larger area which poses a potential threat to life or property and which may require a limited evacuation of the surrounding area. Agencies such as the local public works department, health department, Red Cross, county emergency management agency, state police, and public utilities may need to become involved.

A *full emergency condition* is an incident involving a severe hazard or a large area which poses an extreme threat to life and property. It will probably require a large-scale evacuation, or at least the expertise or resources of county, state, federal, or private agencies and organizations. Additional agencies to be contacted may include state departments of health and environmental resources, the state emergency response commission, the EPA, the U.S. Coast Guard, the Federal Emergency Management Agency (FEMA), and the NRT.

The emergency response plan should also include an assessment of the equipment and expertise available in the immediate area for containing different types of releases and disposing of the released substances. If local resources do not appear adequate, additional resources should be obtained or arrangements made with nearby communities or state agencies for assistance, and these arrangements should be clearly spelled out in the plan.

Evacuation Plans

Evacuation is the most sweeping response to an accidental release, and the one most likely to generate public panic. It is therefore a response which should be used only when necessary, as defined by conditions which have been carefully considered and spelled out in advance. In many situations, it may actually be safer to tell people to stay inside rather than try to evacuate the area, at least until the initial release has been controlled or given a chance to disperse.

Evacuation procedures spelled out in the plan should include the title of the person who can order or recommend an evacuation; identification of vulnerable zones where evacuation could be necessary and a method for notifying persons in those zones; provisions for a precautionary evacuation; methods for controlling traffic flow and providing alternative traffic routes; shelter locations and other provisions (for example, special assistance of hospitals); agreements with nearby jurisdictions to receive evacuees; agreements with hospitals outside the local jurisdictions; protective shelter for relocated populations; reception and care of evacuees; and reentry procedures.

In addition to evacuation procedures, the plan should include some provisions for protection of drinking water supplies and the sewage system. If these essential services become contaminated by a release, it may prove necessary to evacuate the area even though the immediate threat to health and property has long since disappeared. Protection of essential services is thus a less onerous alternative to potential evacuation of the area.

Training Programs and Simulation Exercises

The emergency response plan must also include a description of training programs, including schedules for training of local emergency response and medical personnel. Training programs, exercises, and drills are necessary to insure that each person potentially involved in plan implementation is aware of his or her duties and capable of carrying them out. They also provide a useful source of feedback which can be used to identify difficulties in the plan and areas which require updating.

Training is available from a variety of sources in the public and private sectors. At the federal level, EPA, OSHA, DOT, FEMA, and the Coast Guard all offer some sort of hazardous materials training. State and county emergency management agencies, as well as local police and fire departments, may also provide training which is geared to the particular problems of the local community. Local universities and community colleges, industry associations, special interest groups, and safety consultants may also provide training. The Chemical Manufacturers Association has a lending library of audio-visual training aids which may be borrowed at no charge by emergency response personnel and the public sector. See Appendix A.4 for a list of companies providing "packaged" training courses.

Exercises for testing and updating the plan may include tabletop, functional, and full-scale exercises. These are all simulation exercises of various levels of realism. This section of the plan should specify the organization in charge of the exercise; the types of exercises; the frequency of exercises; and a procedure for evaluating performance. Planners should work with local industry and the private medical community when conducting simulation exercises, and should be sure to comply with state and local requirements concerning the content and frequency of emergency drills. After each exercise, the plan should be revised and retested as needed until the planning team is confident that it is ready.

There will be a temptation, once the plan is developed, to leave it on the shelf and not engage in regular and rigorous training exercises. This temptation should be avoided. No matter how good a plan looks on paper, realistic exercises will likely draw attention to unforeseen problems. It is not unheard of for a significant portion of the participants in a training exercise to be declared "dead" by referees for various reasons, such as approaching from the wrong direction or being located in the path of a sudden wind shift, or using water on chemicals which react violently when wet, or exposing vulnerable equipment to disabling conditions. Although such results may be discouraging at the time, it is far better to experience them in a drill and learn from that mistake than to suffer a real catastrophe.

8.3 COORDINATION AND REVIEW OF PLANS

Each plan developed by a local emergency response committee must be reviewed by the state emergency response commission, which shall make recommendations on revisions of the plan as may be necessary to ensure coordination with other emergency planning districts. It is not clear whether review and incorporation of comments is required in order to satisfy the October 17, 1988 deadline for completion of local plans, but Congress did specify that, to the maximum extent practicable, state review shall not delay implementation of local plans.

Local committees may also request that their plan be reviewed by the regional response team established under the National Contingency Plan for responding to oil and hazardous substances pollution emergencies on a national scale. The National Contingency Plan, which is published at 10 CFR Part 300, covers many of the same topics as must be addressed under SARA Title III, but with emphasis on coordination of different levels of government. It is actually administered by 13 regional response teams (RRTs). Each RRT is responsible for planning and preparedness activities before an oil spill or hazardous substances incident occurs, and for coordination and advice during response actions within its region. RRTs may provide comments on the plan itself and any other issues related to preparation, implementation, or exercise of the plan, but will only do so upon request by the local committee. This optional review must not be allowed to delay implementation of the plan by the statutory deadline.

A list of RRTs and their telephone numbers is included in Appendix B.3. It is suggested that local committees request that the appropriate RRT review and comment on their plan, even if this process cannot be completed until after the plan has technically been implemented.

Community response plans should also be subject to peer review and community input on an ongoing basis. Peer review involves objective comments provided by individuals similar in qualifications to those who drafted the plan. Examples of appropriate individuals may include private industry safety or environmental engineers, responsible authorities from other political jurisdictions, local college professors familiar with chemical safety procedures and hazardous materials response operations, and concerned citizens' groups which are both aware of the environmental issues and reasonably objective.

Community involvement is vital to success throughout the planning process. Community input at the plan appraisal stage does more than encourage consensus building, although the chances of a successful plan are greatly increased if all or nearly all of the community understands it and agrees that it provides for viable and appropriate response measures. Public input also improves the soundness of the plan by exposing it to scrutiny from persons whose chief concerns are nontechnical and who may be able to identify areas which were simply overlooked by the experts.

Congress has in fact specified that local planning committees must provide for community input through public notification of committee activities and meetings to receive and respond to comments. Other approaches for obtaining community input include community workshops with short presentations followed by questions and answers; publication of notices in local newspapers outlining issues and requesting written comments; invited reviews by key interest groups which provide for direct participation in the planning process; and advisory councils composed of several interested parties that can independently review and comment on the plan. One or more of these opportunities must be provided each time the plan is reviewed, as well as when it is initially developed.

8.4 ADDITIONAL SOURCES OF INFORMATION AND ASSISTANCE

This chapter has, of necessity, treated the subject of emergency response planning in a highly general fashion which seeks to track the specific provisions of SARA Title III. Actual development of emergency response plans will require considerably more detailed expertise.

Every local emergency planning committee should have a copy of the *Hazardous Materials Emergency Planning Guide* developed the National Response Team. The Guide may be obtained free of charge by writing:

HAZMAT Planning Guide (WH-562A)
401 M Street, S.W.
Washington, D.C. 20460

The EPA has also established a Chemical Emergency Preparedness Program Hotline, which can be reached at (800) 535–0202 or (202) 479–2449.

Various other books, data bases, government agencies, and private groups from which additional information may be obtained are listed in Appendix B. These lists are not intended to be comprehensive, nor should they be interpreted as recommendations of the sources listed. They should, however, provide reasonable starting points for local committees, who may then proceed to develop their own sources and expertise.

Two sources which local planning committees may find particularly helpful are the regional response teams discussed above and FEMA.

Every local plan should include a contact name and telephone number for the appropriate RRT, and procedures for coordination during a response.

FEMA was created in 1978 for the purpose of consolidating and coordinating federal emergency preparedness, mitigation, and response activities. It is responsible for the national flood insurance program, federal crime insurance program, and federal disaster assistance, in addition to emergency preparedness. FEMA provides disaster preparedness assistance to the states, reviews state and local emergency plans for nuclear power plants, and conducts state assistance programs for training and education in comprehensive emergency management. FEMA has ten regional offices, which are listed in Appendix B.3.

In addition to information, local emergency planning committees will obviously need access to financial resources or in-kind assistance in order to implement the planning requirements of SARA Title III. Congress has directed each local emergency planning committee to evaluate the need for resources necessary to develop, implement, and exercise the emergency plan, and to make recommendations with respect to additional resources that may be required and the means for providing such additional resources. These recommendations should presumably be made to the state emergency response commission or the EPA (or both). Congress has also authorized officials carrying out existing federal programs for emergency training to provide training to federal, state, and local personnel in hazard mitigation and emergency preparedness. FEMA will receive up to $5 million each year through 1990 to support programs designed to improve emergency planning, preparedness, mitigation, response, and recovery capabilities.

Finally, the EPA was directed to report to Congress no later than April 17, 1988 on the status and feasibility of various capabilities and techniques for monitoring, detecting, and preventing releases of extremely hazardous substances, and for alerting the public. The report was also to include recommendations for further development and improvement of technologies, devices, and systems. Although the report will no doubt be made public, no commitments have been made to provide financial or other support for recommended actions.

Part C

Policy Issues and Related Laws

Although the acts required by the HCS and SARA Title III must largely be performed by employers, facility owners or operators, and local planning committees, federal Right-to-Know requirements raise a number of other legal and policy issues involving other interests, other persons, and other laws. Part C of this book looks at these types of issues.

Chapter 9 deals with the conflict between protection of commercially sensitive trade secrets and maximum availability of information on potential hazards. Chapter 10 describes briefly the relationship of federal Right-to-Know requirements to other federal, state, and local laws. It includes a discussion of the legal question of preemption, raised when more than one law appears to apply to a given situation. Chapter 11 describes options available to workers and community members who are dissatisfied with compliance by employers, owners or operators, and local committees. It also emphasizes the fact that workers and community members also have an active role to play if the goals of Right-to-Know are to be realized. Finally, Chapter 12 provides a "summing up" of federal Right-to-Know requirements. It discusses attempts to measure the costs and benefits associated with such requirements, the reactions of various persons to existing statutes and regulations, and further developments which may occur in the near future.

9

Trade Secrets

One issue which has received considerable attention from Congress, OSHA, the EPA, and the courts is the conflict between providing access to information required to avert or respond to medical emergencies and protecting commercially sensitive information which might be vital to maintaining a successful business. Both the HCS and SARA Title III attempt to strike a reasonable balance by allowing chemical manufacturers, importers, and employers to withhold the identity of specific chemicals when complying with reporting requirements, provided it can be demonstrated that the identity constitutes a legitimate trade secret. Even this information, however, must be disclosed under certain conditions to qualified health professionals, workers, and worker representatives.

9.1 AUTHORITY TO WITHHOLD SPECIFIC CHEMICAL IDENTITIES

Section (i) of the HCS states that chemical manufacturers, importers, or employers may omit the specific chemical identity of a hazardous substance from MSDSs, provided that (1) it can be demonstrated the information being withheld is a trade secret, (2) information concerning the properties and effects of the hazardous chemical is disclosed, and (3) the specific chemical identity is made available to health professionals, workers, and designated representatives under appropriate circumstances.

Similarly, section 322(a) of SARA Title III authorizes the withholding of the specific identity of any hazardous, extremely hazardous, or toxic chemical from planning documents, MSDSs, hazardous substance lists, hazardous chemical inventories, and toxic release forms, provided that (1) a proper claim of trade secret protection has been made, and (2) the document from which the information is omitted includes information for the generic class or category to which the chemical belongs. Section 323 requires the release of specific identities to qualified health professionals under appropriate circumstances.

9.2 WHAT IS A TRADE SECRET?

"Trade secret" is defined in the HCS as "any confidential formula, pattern, process, device, information or compilation of information that is used in an employer's business, and that gives the employer an opportunity to obtain an advantage over competitors who do not know or use it." Following the decision of the Third Circuit Court of Appeals in *United Steelworkers of America v. Auchter*, 763 F.2d 728 (1985), OSHA has clarified this definition by stating that chemical identities which may be discovered though reverse engineering are *not* considered trade secrets. OSHA has also added Appendix D to the HCS, which sets forth specific factors to be used in determining whether a trade secret claim is legitimate.

Reverse engineering is a technique whereby information that would allow one to replicate a product is discovered by working backwards from the finished product and breaking it into its constituent parts, without the aid of formulas or other information about its composition. A practical consequence of the reverse engineering clause is that only certain types of complex mixtures can be expected to qualify as legitimate trade secrets. Identities of pure compounds can almost always be determined by reverse engineering, and with that information they can almost always be duplicated by competent chemists.

SARA Title III states that a specific chemical identity may not be entitled to protection as a trade secret unless it is shown that:

- The information has not been disclosed to any other person, other than a member of a local emergency planning committee, a government officer or employee, an employee of the person claiming the trade secret, or a person bound by a confidentiality agreement, and reasonable measures have been taken to preserve confidentiality.
- The information is not required to be disclosed, or otherwise made available, to the public under any other federal or state law.
- Disclosure of the information is likely to cause substantial harm to the competitive position of the person making the claim.
- The chemical identity is not readily discoverable through reverse engineering.

Trade secret claims for information required under SARA Title III may be made for specific chemical identities *only*—other "patterns, processes, devices or compilations of information" may not be claimed as trade secrets.

The two fundamental elements of a trade secret, then, are secrecy and commercial value. Consequently, any information provided under the HCS to persons who are not legally bound by law or agreement to keep it confidential and not use it for commercial purposes may not be claimed as a trade secret under SARA Title III, and vice versa.

9.3 PROCEDURES FOR MAKING AND CHALLENGING TRADE SECRET CLAIMS

Both the HCS and SARA Title III require more than a bald assertion of trade secret protection: the claimant must be able to demonstrate that the definition of trade secret is satisfied for each specific chemical identity claimed as a secret. Under the HCS, any denial of a written request for a specific chemical identity must include evidence to support the claim that the specific identity sought is a trade secret. Furthermore, the person seeking the information may refer the denial to OSHA for consideration, following which OSHA will make a determination whether the information sought is a bona fide trade secret, and, if so, whether adequate means exist to protect confidentiality.

Since all MSDSs prepared under the HCS now must either be submitted under SARA Title III, or must be made available upon request from the local planning committee, each trade secret claim must, for all practical purposes, comply with the procedural requirements established by EPA as well. The most important difference is that trade secret claims under SARA Title III must be submitted and documented *simultaneously* with the withholding of information. For all practical purposes, therefore, firms claiming trade secret protection for specific chemical identities must be prepared to substantiate their claim at the same time the information is withheld.

The EPA has proposed a detailed and somewhat convoluted set of regulations for making, challenging, and evaluating trade secret claims (52 FR 38312 [Oct. 15, 1987]). The centerpiece of the proposed regulations is standard Form 9510-1, which requests specific information describing the nature of the trade secret and the steps taken to maintain confidentiality. This form is reprinted as Appendix E.4.

Under the proposed regulations, each time a chemical identity is withheld, the claimant must submit a complete copy of the required form or document (including the chemical identity) to the EPA, but with the specific chemical identity marked as "TRADE SECRET" and an indication in parentheses of the generic class or category of the chemical claimed as trade secret. The claimant must also submit a "sanitized" version of the same form of the same document to the EPA, and to the appropriate state or local body. This sanitized version lists the generic class or category in place of the specific chemical identity each time it is requested. For example, if a firm wishes to claim trade secrecy for a specific chemical identity on a Tier I inventory form required under section 312(a), it must submit a completed copy of the form which *includes* the specific chemical identity to the EPA, and sanitized versions to the EPA, local emergency planning committee, state emergency response commission, and local fire department.

The claimant must also submit a completed Form 9510-1 to EPA that includes the following information:

- the specific measures taken to safeguard the confidentiality of the chemical identity claimed as trade secret
- any disclosure to persons other than company or government employees who have not signed a confidentiality agreement
- all government entities to which the specific chemical identity has been disclosed, and whether any confidentiality claim was asserted and denied
- the specific use of the chemical substance
- whether the company has been linked to the specific substance in publications or other information available to the public
- how competitors could deduce the use from disclosure of the chemical identity plus other information available through SARA Title III forms
- why information regarding use of the substance would be valuable to competitors
- the nature of the competitive harm that would likely result from disclosure, including an estimate of lost sales or profitability
- the extent to which the substance is available to the public or competitors in products, articles, or environmental releases
- whether the claimed use of the substance is subject to any United States patent

Once this procedure is completed, the specific chemical identity need not be released except under the circumstances described in Section 9.4 below, or if the claim is challenged and found to be invalid.

Any person may challenge a trade secret claim by submitting a written petition to the EPA. Once a petition is received, the EPA will make an initial determination whether answers provided on Form 9510-1 appear to be sufficient to support a trade secret claim, assuming all the facts stated are correct. If the initial finding is that the answers are sufficient, the claimant will be allowed 30 days to submit additional information demonstrating the truth of the facts stated. Trade secret claims may also be asserted for this supplemental information, and need not be limited to specific chemical identities. If no supplemental information is submitted, a final determination will be made based only on Form 9510-1.

If the answers provided on Form 9510-1 appear insufficient to support the claim, the claimant will be notified of this, and of his or her opportunity to appeal the determination. EPA has emphasized that the answers originally submitted on the Form 9510-1 must include facts sufficient to support a trade secret claim, and that penalties of up to $25,000 for each frivolous claim may be assessed for failure to satisfy this standard.

EPA is required to make a final determination on the validity of a challenged trade secret claim within nine months after a petition is filed. If EPA finds that the claim is valid, the petitioner may appeal that determination to a federal district court within 30 days. If the claim is determined to be invalid, the claimant may appeal to a district court within 20 days. If no appeal is made, notice of intent to release the information must be provided within ten more days.

If a document other than Form 9510-1 is submitted under any requirement of SARA Title III which contains the specific chemical identity, then that identity cannot be claimed as a trade secret under any other provision of SARA Title III. For example, a facility owner or operator who includes the specific identity on his list of hazardous chemicals submitted to the local planning committee in lieu of MSDSs cannot later claim trade secret protection for that same chemical identity on a toxic release form or Tier I inventory form.

Given the detail required to support trade secret claims and EPA's stated intent to assess strict fines for frivolous claims, chemical manufacturers, importers, employers, and facility owners or operators contemplating making trade secret claims under either HCS or SARA Title III should review their existing records to determine whether they will be able to supply all of the required information. If all of the information does not exist or cannot be located, then any trade secret claim will likely be found frivolous, and should not be filed. Similarly, each time a new product is developed which might be the subject of a trade secret claim, a file should be kept documenting each of the elements required to substantiate a claim.

9.4 RELEASE OF PROTECTED INFORMATION

Even specific chemical identities which are protected as bona fide trade secrets must be released under certain specified circumstances.

Medical Emergencies

Where a treating physician or nurse determines that a medical emergency exists and the specific chemical identity of a hazardous chemical is necessary for diagnosis or treatment, the identity must be disclosed immediately. The treating physician or nurse may be required to supply a written statement of need and a confidentiality agreement as soon as circumstances permit.

Nonemergency Situations

In nonemergency situations, the specific chemical identity must be supplied following a written request from a qualified person, provided that the request contains an adequate statement of need and provided the requester (as well as anyone contracting for his or her services) signs a written confidentiality agreement. Qualified persons under the HCS include health professionals providing medical or other occupational health services to exposed workers, as well as exposed workers and their representatives. No definition of the term "employee representative" is given in the Standard. Presumably, it includes appropriate union officials, but it is uncertain whether other persons such as attorneys also qualify. Qualified persons under SARA Title III must be health professionals employed by or under contract with the local government.

The written request must describe in reasonable detail one or more of the following needs for the requested information:

- to assess the hazards to workers or community members exposed to the chemical concerned
- to conduct or assess sampling of exposure levels of workers or community population groups
- to conduct preassignment or periodic medical surveillance of exposed workers or population groups
- to provide nonemergency medical treatment to exposed workers, individuals, or population groups
- to conduct studies to determine the health effects of exposure
- to select or assess appropriate personal protective equipment for exposed workers (HCS only)
- to design or assess engineering controls or other protective measures for exposed workers (HCS only)
- to conduct studies to aid in the identification of a chemical that may reasonably be anticipated to cause an observed health effect (SARA Title III only)

The person filing the request must also certify that the facts stated in the written request are true.

Under the HCS, the written request must also explain in detail why other information in lieu of the specific chemical identity would not be sufficient. It is not clear whether this requirement is effectively removed by SARA Title III, or whether the "population groups" referred to in SARA Title III are different from "employees" receiving treatment under the HCS.

Denial of a written request under the HCS must be provided in writing within 30 days of the request, must provide evidence to support the trade secrecy claim, must state the specific reasons why the claim is being denied, and must explain how alternative information may satisfy the specific medical or occupational health need without revealing the chemical identity. For example, if the request is for the purpose of monitoring workplace air contamination, and if the claimant can tell the requester how to monitor precisely without knowing the specific chemical identity, this would be a satisfactory substitute for providing the identity itself.

Denial of a written request made under the HCS may be appealed to OSHA. Qualified health professionals whose written request is denied under SARA Title III may bring an action in federal district court to obtain the requested information.

Confidentiality Agreements

Persons filing written confidentiality agreements in exchange for specific chemical identities must describe the procedures to be used to maintain the confidentiality of the disclosed information, and agree not to use the information for any purpose other than the health reasons contained in the statement of need, unless specifically authorized by the person providing the agreement. Confidentiality agreements are treated as legally binding contracts, and failure to abide by the terms of the agreement may result in a claim for damages. Persons seeking trade secret information may not be required to post a penalty bond, however.

Under the HCS, confidentiality agreements may also be required from the employer or contractor of the services of the person making the request. This would include downstream employers, labor organizations, and individual workers who employ or contract with persons making requests.

A sample confidentiality agreement is provided in Figure 20. It is suggested that suitably modified versions of this or a similar agreement be prepared and kept on file so that they will be available on short notice.

Release of Chemical Identities to Government Officials

Finally, specific chemical identities must be provided to OSHA upon request, and the EPA may give information supplied in support of a trade secret claim, including the specific chemical identity, to state government officials. Health professionals, workers, or worker representatives bound by confidentiality agreements who provide specific identities to OSHA must notify the claimant of this disclosure.

Unauthorized disclosure or use of trade secret information by government officials is prohibited and may result in criminal sanctions.

(SAMPLE) CONFIDENTIALITY AGREEMENT

The undersigned hereby recognizes that the specific chemical identity provided in response to { **describe written request** } is a protected trade secret under { **insert one: 29 CFR Sec. 1910.1200(i) or 40 CFR Part 350** }. He/she certifies that the specific chemical identity will not be used for any purpose other than that stated in the written request or disclosed to any person other than the following without specific prior authorization by { **name of claimant** }: { **list other allowed purposes or persons** }.

The undersigned agrees further that any violation of the terms of this agreement will result in liability for damages. It is agreed further that such damages may reasonably be expected to be at least: { **insert estimated minimum damages** }.

Name: _____

Position: _____

Date: _____

Figure 20. Sample confidentiality agreement.

10

Related Laws

Although they provide the principal federal standards for Right-to-Know, neither the HCS nor SARA Title III stands in isolation. Each must be considered in combination with various other federal statutes and regulations, as well as state and local laws.

10.1 FEDERAL LAWS

The following discussion summarizes in very brief terms several related federal statutes. Citations for these statutes and important regulations are listed in Appendix C.3.

Comprehensive Environmental Response, Compensation and Liability Act (Superfund)

The federal Emergency Planning and Community Right-to-Know Act was passed by Congress as Title III of the Superfund Amendments and Reauthorization Act of 1986. Titles I and II of SARA amended the Comprehensive Environmental Response, Compensation and Liability Act (42 USC § 9601 *et seq.*), also known as "Superfund."

Most of CERCLA is concerned with cleaning up the various hazardous substance waste sites that have been created over the past decades. In order to track the release of wastes from disposal sites, however, EPA has established RQs for a list of hazardous substances. As noted in section III.A.4, emergency notification requirements under SARA Title III are explicitly tied to whether the substance released is also covered by Superfund. These substances are listed in Appendix D.2. That portion of the National Contingency Plan (discussed in Section 8.4) which deals with hazardous substances is also required by Superfund.

Although hazardous wastes are generally exempt from the HCS, Congress included in SARA Title I the requirement that OSHA adopt regulations similar to the HCS for workers engaged in hazardous waste operations. These regulations are discussed in Section 2.3.

Finally, Congress has also authorized applications by local communities for reimbursement for money spent to respond to a release or threatened release of any hazardous substance or pollutant or contaminant. This reimbursement, which is limited to $25,000 for each response, must not supplant local funds normally provided for emergency response. The EPA published an interim final rule establishing application procedures for these funds on October 21, 1987 (52 FR 39386).

Consumer Product Safety Act

The Consumer Product Safety Act (15 USC § 2051 *et seq.*) establishes the Consumer Product Safety Commission for the purpose of regulating "any article, or component part thereof, produced or distributed" for sale to or personal use, consumption, or enjoyment by a consumer. The Commission may require the use and dictate the content of labels for consumer products for which it adopts safety standards. These products are then exempt from the labeling requirements of the HCS.

To date, the Commission has adopted standards for only a handful of consumer products. These include architectural glazing materials, matchbooks, omnidirectional citizens' band base station antennas, walk-behind power lawn mowers, swimming pool slides, and cellulose insulation. This act currently has little practical effect on Right-to-Know requirements.

Energy Reorganization Act

The Energy Reorganization Act of 1974 (42 USC § 5801 *et seq.*) amended the Atomic Energy Act of 1954 and created the Nuclear Regulatory Commission for the purpose of licensing and otherwise regulating nuclear power reactors and waste storage facilities. The NRC has adopted regulations stating it will not issue an operating license for a nuclear power reactor unless there is "reasonable assurance that adequate protective measures can and will be taken in the event of a radiological emergency" [10 CFR § 50.47]. The NRC has established a list of standards to be met by both on-site and off-site emergency response plans. It relies for evaluation of the off-site plans on the Federal Emergency Management Agency discussed in Section 8.4.

The NRC has also adopted an appendix to its regulations which is a discussion of emergency planning and preparedness for production and utilization facilities. This appendix discusses in some detail each of the following elements of emergency plan content:

- organization
- assessment actions
- activation of emergency organization
- notification procedures
- emergency facilities and equipment
- training
- maintaining emergency preparedness
- recovery

(10 CFR Part 50, Appendix E)

Federal Food, Drug, and Cosmetic Act

Food, food additives, drugs, and cosmetics subject to labeling requirements issued by the Food and Drug Administration (FDA) under the Food, Drug, and Cosmetic Act (21 USC § 301 *et seq.*) are also exempt from the labeling requirements of the HCS. This Act prohibits the adulteration or "misbranding" of "any food, drug, device, or cosmetic" that is in interstate commerce. Food, drugs, devices, and cosmetics are generally deemed to be misbranded unless they bear a label which accurately describes the product and its ingredients and contains any necessary warnings.

The FDA has adopted separate regulations governing labeling of each type of item covered by this Act (21 CFR Parts 101 [food], 201 [drugs], 501 [animal food], 701 [cosmetics], 801 [medical devices]).

Federal Hazardous Substances Act

The Consumer Products Safety Commission also has authority under the Federal Hazardous Substances Act (15 USC § 1261 *et seq.*) to regulate labeling of hazardous substances. A "hazardous substance" is defined for purposes of this Act as

Any substance or mixture of substances which (i) is toxic, (ii) is corrosive, (iii) is an irritant, (iv) is a strong sensitizer, (v) is flammable or combustible, or (vi) generates pressure through decomposition, heat, or other means, if such substance or mixture of substances may cause substantial personal injury or substantial illness during or as a proximate result of any customary or reasonably foreseeable handling or use, including reasonably foreseeable ingestion by children;

as well as certain radioactive substances and toys or other articles for use by children which present electrical, mechanical, or thermal hazards.

Hazardous substances which are subject to a consumer product safety standard or labeling requirement under this Act are exempt from the labeling requirements of the HCS. The Commission has again issued labeling regulations for only a limited number of specific substances under this Act (16 CFR Part 1500).

Federal Insecticide, Fungicide, and Rodenticide Act

Pesticides defined in the Federal Insecticide, Fungicide, and Rodenticide Act (7 USC § 136 *et seq.*) which are subject to labeling regulations issued by EPA are also exempt from the labeling requirements of the HCS. FIFRA requires that all new pesticide products used in the United States, with minor exceptions, be registered with EPA. Registration requires, among other things, submittal of a proposed label which must be approved before the pesticide may be used. The EPA has adopted extensive labeling requirements, which are published in 40 CFR § 162.10.

Hazardous Materials Transportation Act

In exempting transportation from most requirements of SARA Title III, Congress recognized that transportation is already regulated under various statutes, including the Hazardous Materials Transportation Act (49 USC App. § 1801 *et seq.*). Under the HMTA, the Department of Transportation has designated a list of hazardous materials, and has issued extensive regulations governing their safe transportation. These regulations cover the packing, handling, labeling or placards, marking, and routing of hazardous materials. (See 49 CFR Parts 171–177.)

The DOT is also required under the HMTA to establish and maintain facilities and technical staff for evaluating risks connected with the transportation of hazardous materials, and a central reporting system and data center to provide information and advice on meeting emergencies. FEMA is also required by the 1984 amendments to evaluate training programs conducted by federal, state, and local government and private agencies for incident prevention and response.

Note that notice of a release which occurs during transportation of an extremely hazardous substance is still required under SARA Title III. This may be accomplished by dialing 911.

Resource Conservation and Recovery Act

Just as Superfund was created by Congress to deal with the problem of environmental pollution from inactive hazardous waste sites, the Resource

Conservation and Recovery Act (42 USC § 6901 *et seq.*) requires management of hazardous waste by generators and transporters, and by owners and operators of treatment, storage, and disposal (TSD) facilities.

The exemption from the HCS of hazardous wastes relies on the definition included in RCRA. A "hazardous waste" is defined in RCRA as a *solid waste,* or combination of solid wastes, which, because of its quantity, concentration, or physical, chemical, or infectious characteristics, may

(A) cause, or significantly contribute to an increase in mortality or an increase in serious irreversible, or incapacitating reversible illness; or

(B) pose a substantial present or potential hazard to human health or the environment when improperly treated, stored, transported, or disposed of, or otherwise managed.

RCRA requires that persons who generate, produce, or transport hazardous wastes must comply with standards established by the EPA and DOT. The EPA must also designate and issue permits governing the operation of certain TSD facilities. RCRA also requires hazardous waste site inventories, and authorizes site inspections, monitoring, and enforcement actions. The regulations adopted by EPA to implement these programs are published at 40 CFR Parts 240–280.

Toxic Substances Control Act

The Toxic Substances Control Act of 1976 (15 USC § 2601 *et seq.*) requires manufacturers and importers to notify the EPA 90 days before producing a new chemical substance, or whenever a significant new use is discovered for an existing toxic chemical which increases human or environmental exposure. Regulations adopted by the EPA require annual reports containing production, release, and exposure data on approximately 250 chemicals chosen because of toxicity or exposure levels. Firms which manufacture or import at least 10,000 pounds per year of certain toxic substances at a single site must also report current inventory data on production, volume, plant site, and site-limited status of the substances to the EPA. Small manufacturers and importers are exempt from these reports, which are required every four years. (See 40 CFR Part 702 *et seq.*)

Although many of the chemicals covered by TSCA will also be subject to the inventory reporting requirements of SARA Title III, TSCA does not require that reports be filed with state and local government bodies.

10.2 STATE AND LOCAL LAWS

Preemption Issues

One legal issue created whenever there are overlapping federal and state or local laws is what is known as the preemption question—which of two laws which seem to apply to the same situation controls? The answer in the case of Right-to-Know laws is not always clear, and depends on whether it is a worker or community Right-to-Know provision which is at issue.

It now appears settled that the HCS preempts both state and local laws dealing with worker Right-to-Know in those sectors of the economy subject to HCS requirements. The Court of Appeals for the Third Circuit ruled in *United Steelworkers of America v. Auchter,* 763 F.2d 728 (1985), that the original HCS preempted state worker Right-to-Know laws in the manufacturing sector only. The Court held that the HCS did not preempt community Right-to-Know provisions, or worker Right-to-Know provisions in nonmanufacturing sectors. In *Ohio Manufacturers Association v. City of Akron,* 801 F.2d 824 (1986), the Sixth Circuit applied the same rule to local regulations. (The Court reversed a district court opinion holding that the HCS preempted state laws, but did not preempt local laws covering the same subject matter!) Now that OSHA has expanded the standard to include both manufacturing and nonmanufacturing sectors, however, it appears that all state worker Right-to-Know laws are preempted. OSHA in fact specifically revised the preemption paragraph of the Standard at the time it expanded coverage to make clear its intent to preempt "all state or local laws which . . . would conflict with, complement, or supplement the Federal standard . . ." [52 FR 31860 (Aug. 24, 1987)].

Even though a state or local worker Right-to-Know law is preempted, it may still take effect if it is part of a federally approved state plan to implement OSHA standards and regulations. In general, state plans will be approved if they are consistent with the federal standard and are at least as effective in protecting worker safety and do not restrict interstate commerce unreasonably. A total of 25 states have their own regulations as part of a federally approved state plan as of this writing, most of which either adopt the HCS verbatim, or follow it with minor variations including expanded coverage of chemicals or workers. These states are indicated in Table 11 and Figure 21.

Congress has specifically said that SARA Title III is *not* intended to preempt any state or local law (42 USC § 11041(a)(1)). This is consistent with the intent to establish a federal floor for community Right-to-Know, and essentially means that whichever is the stricter of two consistent requirements will control. In the event of inconsistent provisions, how-

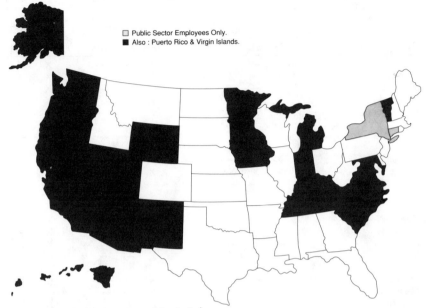

Figure 21. State plan states (shaded).

ever, the result is less clear. It seems reasonable that the law which requires that more information be made available or a greater degree of preparedness be achieved will control, but such issues will probably have to be resolved on a case-by-case basis.

In summary, the federal HCS preempts state and local laws, unless there is a federally approved state plan which includes stricter requirements. State and local community Right-to-Know laws, on the other hand, will generally preempt the requirements of SARA Title III if they are stricter or more comprehensive. There remains the question of whether the primary intent behind a particular provision of a state or local law or regulation is to protect employees from workplace hazards, or the surrounding community from more general environmental hazards. Two court decisions, both by the Third Circuit Court of Appeals, which provide examples of how this question may be resolved are *New Jersey State Chamber of Commerce v. Hughey*, 774 F.2d 587 (1985), and *Manufacturers Association of Tri-County v. Knepper*, 801 F.2d 130 (1986). In both cases, the court engaged in section-by-section analysis of the state laws (New Jersey and Pennsylvania, respectively), holding that some sections were preempted and others were not.

The "primary purpose" test applied by the Third Circuit is not particularly helpful in predicting the outcome of other cases challenging different state statutes. First, different courts may have different notions of what

Table 11. Provisions of State Right-to-Know Statutes

State or Terr.	Worker				Community			Trade Secrets	
	SP	SL	PS	WR	IER	PA	ERP	TSP	TSR
Alabama	—	—	—	—	X	—	—	X	X
Alaska	X	—	X	—	X	X	—	X	X
Arizona	X	—	—	—	—	—	—	X	X
California	X	—	—	—	X	X	e	X	X
Connecticut	b	X	X	c,d	X	—	—	f	—
Delaware	—	—	X	—	X	—	—	—	—
Florida	—	X	X	d	X	—	—	X	X
Hawaii	X	—	—	—	—	—	—	X	X
Illinois	—	X	X	X	X	—	e	X	—
Indiana[a]	X	—	—	—	—	—	—	X	X
Iowa	X	—	X	X	X	X	—	X	X
Kansas[a]	—	—	—	—	X	X	X	X	X
Kentucky	X	—	—	—	—	—	—	X	X
Louisiana[a]	—	X	—	—	X	X	—	X	X
Maine[a]	—	—	X	X	X	—	—	X	—
Maryland	X	—	X	c	X	X	—	g	X
Massachusetts	—	—	—	—	X	—	—	—	—
Michigan	X	—	X	c	X	X	—	X	X
Minnesota	X	—	X	—	X	—	—	X	X
Montana	—	—	X	—	X	—	—	X	—
Nevada	X	—	—	—	—	—	—	X	X
New Hampshire	—	X	X	c	X	X	—	X	—
New Jersey[a]	—	X	—	—	X	X	—	f	—
New Mexico	X	—	—	—	—	—	—	X	X
New York	b	X	X	—	—	—	—	X	—
North Carolina	X	—	X	—	X	X	e	X	X
North Dakota	—	—	X	—	X	—	—	X	—
Oklahoma	—	—	—	—	X	—	—	f	—
Oregon[a]	X	—	—	—	X	X	e	X	X
Pennsylvania	—	—	X	d	X	—	—	—	—
Puerto Rico	X	—	—	—	—	—	—	X	X
Rhode Island[a]	—	—	X	—	X	—	—	X	—
South Carolina	X	—	—	—	—	—	—	X	X
Tennessee	X	—	X	—	X	X	—	X	X
Texas[a]	—	—	—	—	X	—	—	X	X
Utah	X	—	—	—	—	—	—	X	X
Vermont	X	—	X	—	X	—	—	g	X
Virginia	X	—	—	—	—	—	—	X	X
Virgin Islands	X	—	—	—	—	—	—	X	X
Washington[a]	X	—	—	—	X	X	—	X	—
West Virginia	—	X	—	—	X	X	e	X	—
Wisconsin[a]	—	X	X	—	—	—	—	—	—
Wyoming	X	—	—	—	—	—	—	X	X

Source: Bureau of National Affairs, Inc.

Key to column headings:
SP   State plan state.
SL   Maintains separate state list of hazardous substances.
PS   Coverage includes state and local public sector employers.
WR   Statute includes workers' rights provision.
IER   Information must be provided to emergency response units.
PA   Includes provisions for public access to information.
ERP   Emergency response planning required.
TSP   Trade secrets are protected.
TSR   Trade secret information must be released in emergencies.

Footnotes:
a.   Employer fees collected to cover costs.
b.   State plan applies to public sector employers only.
c.   Workers may refuse to work or request a transfer in the event of noncompliance.
d.   Provides for worker self-enforcement suits.
e.   Emergency response plans required for covered facilities only.
f.   Trade secrets must be registered.
g.   No trade secret protection for carcinogens, teratogens, mutagens.

constitutes a "primary" purpose. Second, the primary purpose of each individual statute must be found in the legislative history of that particular statute, meaning that the conclusion in one case is not necessarily applicable to laws in other states. Third, states which wish to adopt their own worker Right-to-Know requirements may simply intertwine them with community Right-to-Know provisions, in hopes that a court will hold that they are not preempted under SARA Title III.

Persons who may be confused by all of this (and who isn't?) are advised to do two things: first, make sure you know whether there are any state or local laws in your jurisdiction with which you should be concerned; second, when in doubt, assume that the stricter standard applies. If an error is made, it will therefore be on the side of maximum information and preparedness.

Comparative State Law Provisions

Table 11 summarizes some of the important features of worker and community Right-to-Know laws adopted by various states and territories. Over 40 states and territories have at least some form of worker or community Right-to-Know law, including those which administer the HCS through a federally approved state plan. All states and territories, including those which have no Right-to-Know laws of their own, are of course subject at a minimum to the requirements of the HCS and SARA Title III.

Once the expanded HCS coverage took effect on May 23, 1988, the specific provisions of state worker Right-to-Know laws became less important, except to the extent that they are stricter than the HCS or that they cover public sector workers. Several states maintain their own list of hazardous substances, which may be more extensive than the federal list, while others include physical agents such as sound, radiation, etc., as well as the hazardous substances themselves within the scope of their statute. Some states also have requirements for employer compliance which are not included in the HCS, such as labeling of pipelines. Another significant difference is that several states provide coverage for state and local government workers, which are not covered by the HCS. Some states have also included a provision for some form of workers' rights to self-help measures such as withholding work or filing citizen suits. Finally, a few states fund implementation of their requirements by collecting fees from covered employers.

Although many states provide for community Right-to-Know in the form of providing MSDSs or the equivalent to local police, fire, and emergency response personnel, only the federal SARA Title III includes comprehensive community planning and notification requirements. Many states, however, have emergency planning requirements for natural disasters and other types of accidents which may overlap with hazardous chemical planning requirements and require notification of public safety officials in the event of a hazardous chemical release.

Trade secret provisions are included in virtually all state statutes. Several states require (similarly to SARA Title III) that trade secret claims must be registered with the state authority. Some states also exclude certain types of chemicals, such as carcinogens, mutagens, and teratogens, from eligibility for trade secret protection.

Although the information provided herein is believed to be accurate as of January 1988, it is necessarily general, and state Right-to-Know requirements can change rapidly. Appendix B.3 lists state safety and health agencies which administer the HCS, and which may be contacted for further information on current requirements. Further information on state and local community Right-to-Know requirements may be obtained from the local emergency planning committee or state emergency response commission. The full text of state or territory Right-to-Know laws can also be obtained by contacting BNA PLUS at (202) 452-4323.

11

Making the "Right" to Know a Reality

Although it may seem to anyone who has read this far that the HCS and SARA Title III require a tremendous amount of information and cover nearly every possible contingency in exhaustive detail, there will still be instances where workers or community members are dissatisfied with the best efforts of employers, facility owners, or operators and local planning committees, or where individual facility owners or operators or employers will intentionally or unintentionally fail to come into compliance. Additionally, OSHA and the EPA both have only a limited amount of resources to administer and enforce a large number of programs. Some persons have also expressed the view that neither OSHA nor EPA are always as eager to make use of the available resources as they should be.

This chapter discusses some of the options available for "self-help" by workers and community members interested in obtaining more information regarding the hazardous chemicals to which they may be exposed. Many of the specific provisions have already been discussed in the context of the duties they impose on employers and facility owners or operators or government officials. Still, it may be useful to collect them once again and look at them from the flip-side perspective of the "right" to know or, more accurately, the right to be able to find out. First, this chapter describes the ability of workers to engage in self-help enforcement of the HCS worker Right-to-Know requirements. Second, it discusses the procedures for releasing information collected and supplied to emergency planning committees and the EPA under SARA Title III to the public, and for citizen enforcement suits brought to ensure that the community Right-to-Know provisions are carried out. Third, it emphasizes the fact that both workers and community members must take upon themselves the responsibility of becoming informed on the potential hazards affecting them and of using that information constructively.

11.1 WORKER SELF-HELP UNDER THE HCS

OSHA's HCS requires that hazardous substance containers be labeled, that some means be provided of telling workers what is in pipelines, that a file of MSDSs be maintained in a location accessible to all workers, and that each employer have a written hazard communication program, which includes worker training. But what recourse do workers have if they are not satisfied with the amount of information provided?

Access to Specific Chemical Identities

If the problem is simply a matter of obtaining a specific chemical identity which has been withheld as a trade secret, workers and their representatives are specifically authorized to make written requests for specific chemical identities needed to:

- assess the hazards of the chemicals to which workers will be exposed
- conduct or assess sampling of the workplace atmosphere to determine exposure levels
- conduct preassignment or periodic medical surveillance
- provide medical treatment
- select or assess appropriate personal protective equipment
- design or assess engineering controls or other protective measures
- conduct studies to determine the health effects of exposure

The specific procedures for making such requests are explained in Section 9.4.

Filing Complaints With OSHA

Should the problem or inadequacy go beyond information covered by trade secret claims, workers or their representatives who believe that the requirements of the HCS are not being satisfied may file a complaint which requests that OSHA conduct an inspection under the procedures published in 29 CFR § 1903.11. Requests must be in writing to the Area Director or to a Compliance Safety and Health Officer (an inspector), must specify the grounds for the suspected violation, and must be signed by the worker or representative. A copy shall be provided to the employer at the time of the inspection, except that the name of the complainant and any other workers referred to by name will be deleted upon request from the employer's copy.

Filing a complaint does not guarantee that an inspection will take place. The Area Director will review complaints and determine whether there are reasonable grounds to believe that a violation or danger exists. If the Area Director determines that no reasonable grounds exist, he will notify the complainant, who may appeal the determination to the Assistant Regional Director. If the Area Director determines that reasonable grounds do exist, OSHA regulations state that an inspection will take place subject to the procedures discussed in Section 2.4.

Even so, whether any inspection will actually take place within a reasonable time will depend on the availability of inspectors. OSHA has been subject to frequent criticism for not having enough inspectors to enforce all of its standards, and has indicated that it intends to conserve resources by focusing enforcement activities for the expanded HCS on only the largest establishments. Workers in smaller establishments may therefore have to file multiple complaints or bring some other form of pressure to bear in order to get relief.

Workers who file complaints or discuss hazard communication issues with an OSHA inspector during the inspection (assuming one occurs) are protected from being discharged or disciplined for their actions. The Occupational Safety and Health Act under which the HCS was adopted specifically provides: "No person shall discharge or in any manner discriminate against any employee because such employee has filed any complaint or instituted or caused to be instituted any proceeding under or related to this Act or has testified or is about to testify in any such proceeding or because of the exercise by such employee on behalf of himself or others of any right afforded by this Act" [29 USC § 660(c)(1)].

Going Outside the Standard

If OSHA ultimately determines not to conduct an inspection, workers are left with no direct recourse under the HCS. As indicated in Section 10.2, however, several states have Right-to-Know laws which provide that a worker who submits a valid written request for information which is not provided within a specified time period (ten days, for example) may refuse to work in that area of the facility until the request is satisfied.

Finally, workers may resort to less direct means of obtaining information, such as filing a grievance under their collective bargaining agreement or employment contract (if a grievance procedure is available), contacting the parent company if their employer is a subsidiary, and resorting to outside pressure by contacting the local press or community groups.

Workers and their representatives may, of course, also seek information as members of the public under the community Right-to-Know provisions discussed in the next section.

11.2 SARA TITLE III AND THE COMMUNITY

Participation in the Planning Process

Congress stipulated that local emergency planning committees include representatives of various groups or organizations, and that each committee adopt rules providing for public notification, meetings, comments, and receipt of the emergency plan. Although there is no minimum level of actual public participation required by Congress, local committees will have to be able to demonstrate that they have made a "good faith" effort to provide reasonable opportunity for input from all affected segments of the community. (See Section 8.1.)

Availability of Local Facility Records

Section 324 of SARA Title III (42 USC § 11044) states that each emergency response plan, MSDS, extremely hazardous substance list, inventory form, and follow-up emergency notice shall be made available to the public during normal working hours at designated locations. Local emergency planning committees must publish an annual notice in local newspapers stating that the emergency response plan, MSDSs, and inventory forms are available for inspection. Information which is subject to trade secret protections need not be released to the public, except that any member of the public may file a petition challenging the validity of a trade secret claim, as discussed above in Section 9.3.

Any person may submit a written request for MSDSs from specific facilities to the local planning committee, which must then obtain the MSDS from the subject facility. *All* MSDSs are subject to this requirement, and not just those for chemicals which are present in quantities exceeding the threshold reporting quantities. Requests for Tier II inventory information may be submitted in writing to either the local planning committee or state emergency response commission, which must request the form from the facility if the request is limited to hazardous chemicals stored at the facility in excess of 10,000 pounds. Requests for Tier II information on chemicals which are stored in amounts less than 10,000 pounds may still be granted if the request includes a general statement of need. The EPA has decided to leave the development of more specific criteria for release of Tier II information up to each local committee. The location of specific chemicals identified in the Tier II form may be withheld upon request of the facility owner or operator, regardless of whether the particular chemical is a protected trade secret.

Local committees in communities with more than a minimal number of covered facilities will need to establish a filing system, and perhaps employ staff for the purpose of keeping records in order and making sure they are available to the public. Alternatively, records may be made available in public libraries or at the offices of political subdivisions within the region.

National Toxic Release Data Base

The purpose of the national computerized toxic release data base required by section 323 is to "inform persons about releases of toxic chemicals to the environment; to assist governmental agencies, researchers and other persons in the conduct of research and data gathering; and to aid in the development of appropriate regulations, guidelines and standards; and for other similar purposes."

EPA is currently considering four different options for making the data submitted on toxic release forms available to the public: creation of an EPA-funded clearinghouse; use of the National Technical Information Service as a broker to deal with commercial vendors; making the information available through the National Library of Medicine's TOXNET data base; and a cooperative agreement with Purdue University, which currently maintains the National Pesticides Information Retrieval System for the U.S. Department of Agriculture (52 FR 10135 [March 30, 1987]). Once the data base has been established and the data for the first reporting period have been entered, EPA will issue a *Federal Register* notice instructing potential users regarding access to the data base and procedures for its use.

As with records available at the local level, trade secret claims may be asserted for specific chemical identities on the toxic release forms which supply the raw data for the national computerized data base. Such claims may again be challenged using the procedure described in Section 9.3.

Citizen Self-Help Enforcement Suits

Section 323 creates the possibility of community self-help enforcement of SARA Title III by allowing suits against the facility owners or operators, state and local governments, or the EPA. Because no suits have yet been filed under this provision, it is difficult to determine what standards will be applied or how effective such suits are likely to be.

Similar provisions for citizen enforcement suits exist, however, in at least 11 other federal environmental statutes: the Clean Air Act, the Federal Water Pollution Control Act, the Toxic Substances Control Act, the Resource Conservation and Recovery Act, the Noise Control Act, the Safe Drinking Water Act, the Endangered Species Act, the Deep Water Port Act, the Outer Continental Shelf Lands Act, the Surface Mine Control and Reclamation Act, and the Marine Protection, Research, and Sanctuaries Act. Suits filed under these provisions have become increasingly popular, as funds for enforcement activities continue to diminish, and many environmental interest groups turn their attention from passing new laws to enforcing the laws which already are on the books.

Experience under these related statutes has shown that the following questions are likely to be asked. The answers suggested are educated guesses and should not be taken as definitive answers, which can only come with experience under this particular statute.

Who can file a suit?

Although section 326(a)(1) states that "any person" may bring a citizen self-enforcement suit, the question is whether this language is to be taken literally. Courts generally have been liberal in allowing persons who have not suffered actual, physical damage to file environmental suit. It generally is enough that the plaintiff be subject to potential harm, particularly where the statute states that "any person" may file a suit. Still, it is likely that the plaintiff will have to show a "direct interest" in potential harm, i.e., that he or his property is actually threatened. It may not be possible for a group consisting entirely of persons outside the potential zone of exposure to bring suit.

Where must a suit be filed?

Section 326(b) states that suits against a facility owner or operator must be filed in the local federal district court. Suits against the EPA may also be filed in the federal district court for the District of Columbia.

What form of advance notice is required?

Citizen suits must be preceded by 60 days' notice to the facility owner or operator, the state in which the violation occurs, and the EPA. The notice should be in writing and a copy retained. It should be delivered by certified mail or other means of obtaining proof of delivery. The notice will also need to be sufficiently detailed so that the owner or operator will have an opportunity to correct the violation. Although it is probably not necessary to spell out the exact sections of that statute which have been violated, the notice should contain as much information as possible regarding what information has not been provided or what actions have not been taken.

What must be proven?

Section 326(a) states that suits may be filed for "failure" to perform acts required under the statute. The obvious question is whether suits may also be filed for performance which is deficient. For example, suppose MSDSs are filed as required, but it is believed that some information is inaccurate or incomplete. This question can only be answered by actual court decisions. It may be, however, that simply providing notice of intent to file a suit will be sufficient to get the owner or operator to correct any deficiencies.

Of course, it will probably prove more efficient to simply bring the deficiency to the firm's attention and request that it be fixed. There is no need to file a formal notice of intent to sue unless such a request is unreasonably denied.

What form of relief is available?

Assuming that a suit is brought and won, the court will issue an order called an injunction, which directs the defendant to perform the actions necessary to comply with the statute within a certain time period. If the defendant still does not comply, it will be necessary to return to court and seek a contempt citation. Fines may be levied for failure to comply with the statute, but they will be paid to the federal treasury.

What about the "diligent prosecution" exception?

Citizen suits may not be brought under SARA Title III if the EPA "has commenced and is diligently pursuing" its own enforcement action for the same violation. Congress emphasized in the Conference Committee report that "diligent pursuit" requires a vigorous and active enforcement involving the expenditure and commitment of resources—mere plans and good intentions will not do. Moreover, it is the EPA which bears the burden of proving that it has commenced and is diligently pursuing enforcement activities.

Who pays legal fees?

Section 326(f) states that the court may award "costs of litigation (including reasonable attorney and expert witness fees) to the prevailing or the substantially prevailing party whenever the court determines such an award is appropriate." This means that a plaintiff who loses a citizen suit may end up paying the litigation costs of the other side, but only if the court determines that this would be "appropriate."

What about government intervention?

There are really three questions here. First, the EPA or state government may intervene as a party in any citizen suit. It is likely that the government would want to do so, because (second) it may be that a citizen suit will bar the government from bringing a later suit based on the same violation. By intervening in the original suit, the government hopes to avoid the possibility of an unfavorable precedent resulting from inadequate or underfunded litigation by a citizen plaintiff.

Finally, citizens may also intervene in suits originally brought by the government, but only if (1) the citizen can show that he or she would be directly affected by the outcome of the case, and (2) the government is unable to show that it adequately represents the interests of the intervening citizen.

The Bureau of National Affairs has recently published a special report titled *Environmental Citizen Suits: Confronting the Corporation*. Although the authors have not yet obtained a copy of this report, it is being promoted as a comprehensive survey of requirements for and experience with citizen suits written for nonlawyers. Copies may be obtained from The Bureau of National Affairs, Inc., Circulation Department, P.O. Box 40947, Washington, D.C. 20077–4928; (800) 372–1033.

11.3 RIGHTS AND RESPONSIBILITIES

As noted above, the term "right to know" is something of a misnomer, except perhaps as applied to worker training programs under the HCS. In all other respects, the so-called right to know might better be thought of as the right to find out without having to expend unreasonable time, energy, and money. It can also be thought of as the duty of the owner or operator to make information available. In order for the goals of the HCS to be fully satisfied, each worker must take training programs seriously, read the materials provided, and make an effort to be particularly aware of the hazards to which he or she is potentially exposed and the proper responses in the event of a release. Similarly, members of the community should take it upon themselves to make use of the opportunity required under SARA Title III for public input and review in the emergency planning process, and to cooperate with practice drills and other preparedness exercises.

Conversely, once a conscientious effort is made by all concerned to fulfill their respective duties and take advantage of the opportunities for obtaining necessary knowledge, there is little or nothing to be gained by alarmist tactics or random "fishing expeditions" in search of information which may be blown out of proportion to score political or public relations points. This type of overreaction will no doubt occur in some locations, but it is hoped that it will be the exception rather than the rule.

12

Experience and Consequences

Any new government requirements can cause problems among those who are regulated, however well-intentioned the requirements may be. On the average, the net overall effect of Right-to-Know requirements is more likely to be positive than negative. However, there are and will continue to be many specific negative effects. Many of the problems that do arise will be temporary results of changing requirements rather than lasting effects.

In any case, how the facility management personnel handle particular situations will often have more to do with the overall consequences of compliance than will the requirements themselves. The best way to deal with this new "higher profile" is to maintain a conscientious program of compliance with safety and environmental requirements.

12.1 MEASURING COSTS AND BENEFITS

As with any new legislation or regulation, there are both costs and benefits to all parties concerned. In the area of Right-to-Know, it is almost certain that some of the benefits will easily be worth the attendant costs. It is also likely that some of the costs will outweigh their attendant benefits by a large margin.

Undoubtedly the combined effect of the HCS and SARA Title III (if they are not mismanaged) can contribute significantly to an improvement in the quality of life in this country. Whether the amount of improvement will justify the cost remains to be seen.

Costs

Employers

For employers, compliance with the HCS will have many real, nontrivial costs on which dollar figures can be placed. Very likely, those costs will come largely from profits. Some might be passed through to customers by way of price increases, but the ability to do so will vary widely. Some marginal enterprises could be forced out of business by the cost of compliance.

Some of the economic costs to employers include mounds of newly required paperwork: filing of reports, maintaining inventories and other records, and the written program and MSDS files themselves. These all take significant amounts of time and effort by people who are not at the bottom of the pay scale, and in larger organizations additional personnel, or even entire departments, may be required to do the job.

Other costs include the required training, either the cost of "nonproductive" time or the cost of training materials or both. There may sometimes be a need to retain technical consultants or attorneys to deal with specific problems that arise.

Finally, if compliance is inadequate there may be expensive fines, penalties, and lawsuits. Even if litigation against an employer is unsuccessful, defending against it can be very expensive. The accompanying unfavorable publicity can also have a negative impact on the business.

Retail trade concerns have some specific problems under the HCS, but they must deal with them. Under the expanded HCS, they must carry MSDSs for all hazardous materials they sell, and supply copies of the MSDSs to those purchasers who request them. Moreover, since there is always the possibility that packages of hazardous materials may be broken in handling and hence people may be exposed to hazards, training is in order for the entire inventory of such materials.

In many cases, the help in retail establishments consists largely of part-time workers, often high school or college students. Such workers may work night hours, with little supervision, and it can be difficult to operate a good safety training system under such circumstances.

Workers

There are not likely to be many costs to workers as a result of the Right-to-Know movement, but there will be some. Among the intangible costs is the fact that the complexity of knowledge required for many jobs will increase significantly. A large fraction of workers in many of the more heavily affected industries are not trained in chemistry, physics, and health science. But now they will have to learn new facts about materials that may not make sense to them. In some cases, failure to learn those facts could well result in job reassignment.

Tangible costs may accrue to workers who have the misfortune of being employed by firms that will fail, partly because of the cost of these new requirements. It is unlikely that this will be a large portion of the work force, but surely it will happen to some workers.

Government and the Public

Compliance with Right-to-Know requirements will very likely contribute to inflation. Not only will some costs be passed through to consumers, but also some marginal, low-cost firms will go out of business, thereby relieving some anti-inflationary pressures. In addition, taxes may be increased in some locations to pay the added costs to state and local governments, or else other government services may be curtailed. It may be a considerable time before the effects of these factors can be analyzed quantitatively, but they will almost certainly not be trivial.

Costs to the governmental bodies will include the immense amounts of extra information management, which in many cases will require additional personnel as well as additional filing and computer facilities. Other costs will include large numbers of person-hours newly required to be spent in meetings, conferences, training sessions, test exercises, etc.

Benefits

Employers

Although new federal regulations are frequently viewed as unwanted intrusions, the HCS should provide some real benefits by virtue of being primarily a single standard applied uniformly. States that wish to impose their own special standards must secure the approval of the Secretary of Labor. Not only will firms that have facilities in several different states benefit from more uniform standards, but also "downstream" employers will be able to depend on receiving MSDSs and labels for chemicals they purchase from out-of-state suppliers.

When workers know of the potential material hazards and the risks involved, they may be more careful when handling such materials. This in turn can lead to better cooperation in the use of protective equipment and procedures, as well as less frequent and less serious accidental spills and releases. All these factors should lead to fewer accidents and less time lost, as well as generally better productivity.

As a result of having fewer and less serious injuries and illnesses, first aid costs might be reduced. Workers' compensation and health insurance claims could even decline, thereby perhaps lowering premiums the company pays, or at least resisting a tendency for increasing premiums with inflation.

In a well-managed hazard communication program, worker morale might be expected to improve. As a consequence, a decreased turnover rate in the work force should be experienced, as well as fewer labor relations problems. The net savings in overhead costs can be significant.

One very realistic beneficial prospect of this Standard is that management at all levels, not just the workers, will become increasingly aware of the nature of material hazards. As a result, changes in equipment or procedures may be motivated, resulting in fewer spills and accidents. Sometimes such awareness provides motivation to replace one material with another that is less hazardous. These factors mean less downtime, less expense of emergency response, and less lost worker time—in short, improved productivity.

Workers

Under the HCS, nearly all workers will benefit somewhat, at least in intangible ways. If training is well done, they will learn the characteristic hazardous properties of the materials they work with. Thus, they will know better how and why to protect their health on the job. This knowledge can provide a measure of peace of mind, in addition to a lower likelihood of chemical-source illnesses and injuries.

The existence of the HCS and SARA Title III will undoubtedly force many employers to reevaluate their reasons for using some of the more hazardous materials in their facilities. In particular, the informing and reporting requirements of SARA Title III will focus attention on the relative hazards of materials. As a result, decisions may often be made to replace materials with low TPQs with other materials having higher TPQs, or with materials not on the list of substances with TPQs. The result of such changes will be intrinsically safer work environments.

Neighboring Communities

Much bad publicity, often ill-informed, has been given to various problems related to hazardous materials in recent years. The Right-to-Know movement should help to counter the misinformation. In addition, when the public realizes that positive steps are being taken to minimize such problems, some of the concerns about individual materials may subside.

In addition, some of these facility managers may well be motivated to adopt the use of less hazardous materials or more effective engineering controls in their operations. One result of such changes will be intrinsically safer environments near such facilities because of the reduced risk of hazardous releases and spills.

The Environment

It is to be hoped and expected that the immense amount of attention focused on hazardous materials will indirectly result in environmental benefits. Most of these have been mentioned above in other connections, but they also are environmentally beneficial as well. Those results that will sometimes, but not always, occur include:

- more conscious attentiveness, tending to decrease incidents of spills and releases
- replacement of some materials by other, less hazardous materials
- improved engineering controls in equipment and procedures involving handling and processing materials

These factors reduce the probability or severity, or both, of environmental contamination from hazardous materials.

12.2 REACTION TO EXISTING STATUTES AND REGULATIONS

The HCS

Suppliers

Generally, most manufacturers, importers, and distributors of chemical products seem to be making a good faith effort to meet the spirit as well as the letter of this Standard. Some may not believe that their products could be hazardous, though, and will simply not do an adequate job of characterizing them.

Some chemical suppliers will not provide MSDSs free of charge to customers for materials that had been ordered in the past, unless another order is placed. The charges for supplying MSDSs in such cases often seem exorbitant—certainly far greater than the cost of copying and mailing them. Some suppliers have been unwilling to furnish MSDSs to customers who are not in the manufacturing industry (prior to the date of extension to all employers), even though those customers may be covered by state laws that require their participation.

There are a great many problems with MSDSs that are being supplied. For example, there is much variation in format used by different manufacturers, making it difficult to locate wanted information quickly. This problem could have been avoided if OSHA had decided to mandate the use of a single MSDS form, e.g., Form 174, but it chose not to do so. To many, the result of this decision is a seemingly unnecessary confusion. Some manufacturers and suppliers use the obsolete OSHA Form 20 or some other form that does not contain all the information required by the

Standard. Other manufacturers use computer-generated MSDSs that can run as long as eight to ten pages.

Even when MSDSs meet all legal requirements, they often are inadequate to employers and workers who are not technically oriented. Very often the information given in the sections on protective equipment and spill cleanup is so general as to be of little value. Other times, protective equipment specified is of the type needed for an emergency cleanup operation of a major spill and does not apply to normal use, but that fact is not mentioned.

Employers

The HCS is written in rather vague language, leaving to employers the responsibility to develop the specific policies, MSDSs, labels, and training programs. Most employers are not trained in chemistry, nor do they employ any chemists. Most chemists are not specifically trained to recognize and deal with all of the types of hazards covered anyway, so conforming with the Standard may be a serious problem for many employers.

Whenever an OSHA inspector visits a facility for any purpose at all, HCS compliance is almost certain to be scrutinized. He or she will ask to see the written hazard communication program and the MSDS file. A few workers will also be interviewed to determine how well they have been trained and informed under the HCS.

Many small employers do not even know how to determine which of the materials they use might be hazardous. Therefore they don't know just what needs to be done. The usual procedure seems to be to rely on the MSDSs sent by suppliers and to ignore materials for which MSDSs are not received automatically. Some employers make a determined effort to obtain MSDSs for other materials, but many do not.

A particular problem seems to be the preparation of a written hazard communication program. That requirement seems to bewilder or intimidate many people charged with administering the HCS in their facility— they aren't sure how to go about writing such a document. In nearly one-third of their HCS inspections, OSHA inspectors have found that written programs are completely nonexistent. To assist in meeting this requirement, a generic written program is given in Appendix G of this book. It is easily modified as needed to fit almost any facility.

Training is another serious problem. OSHA inspectors report that when they interview workers, they find few who seem to have the understanding they should have about the material hazards they use on the job. It appears that training is often absent or inadequate, or it needs more frequent refreshing if it is to be truly useful. Numbers of workers have commented to OSHA interviewers that they felt a need for repeated training in these matters.

How will the existence of Right-to-Know requirements affect the likelihood of lawsuits? No one can predict such things with any degree of certainty. A facility that is in reasonable compliance may actually be less likely to attract suits for negligence than it would have under previous conditions where workers had not been informed about materials they worked with. But the fact that a worker knows more about potential harm from particular substances in the work place may increase the chances of a suit in case of illness years later.

Actually, some of the liability suits that have been filed in the past might not have occurred if the workers had been informed in the manner prescribed by the HCS. That is, sometimes a former employee "presumes" that his health problems were caused by something he had been exposed to in the workplace when in fact the particular kind of problem is not known to be associated with the particular chemical. If full disclosure had been made much earlier, it is possible some such suits would never have been filed.

Workers

Some workers may overreact to the disclosure of hazards and effectively refuse to work with such materials. The appropriate response is to persuade such individuals that the danger is less than they fear, provided they follow the precautions that are recommended for each material.

Some managers fear that the existence of this Standard might increase the number of workers who "call in sick." Thus, if workers are told that overexposure to vapors of a certain material can cause headaches or nausea, some may use that fact as an excuse to collect sick leave pay, or they may become honestly convinced that they suffer the symptoms described. Indeed, this may happen in some cases, particularly if labor relations are poor, but the frequency of such incidents probably will not vary much from what it has been all along.

A related problem is that sometimes it may be difficult to hire secretarial and clerical personnel to work in an office where such hazards are known or rumored to exist in an adjacent facility. Education is probably the only appropriate answer to this problem. If the company's safety record is a good one, the task should be somewhat easier.

Another problem is worker apathy. Some workers apparently do not believe that significant hazards exist with materials they have worked with for years without adverse effects. Such individuals may not pay enough attention during safety instruction to learn the important information. Workers must be impressed with the fact that learning this material is in their own self-interest as well as being a very important part of their overall job responsibilities.

SARA Title III

There is a fundamental problem underlying the concepts behind SARA Title III—one that is likely to be overlooked often. The underlying assumption is that if emergency response personnel are provided information about existing hazards, they will utilize that information to plan in such a way as to prevent accidents from becoming crises. However, most of the personnel who will serve on community planning committees are not trained scientists and may not be able to use the information adequately without additional expert help.

Facility Owners and Operators

Many small business spokesmen feel that compliance with Right-to-Know laws is too large an economic burden, particularly since many such businesses operate with marginal profits as it is. In addition, there is the problem of dealing with high-tech issues in low-tech business by people who have no significant education or background in science and engineering. As a result of such considerations, one might expect a large fraction of the theoretically affected business facilities to simply "slip through the net" and not comply. It is questionable whether government officials will have the personnel and budgets needed to force compliance by such facilities. However, Congress provided for citizen suits when necessary to force compliance.

Schools and colleges could also be faced with an extraordinary compliance burden. They frequently have a very large number of hazardous materials within their facilities, and their budgets are inadequate for meeting compliance requirements. Little hope exists of improving those budgets just to comply. Whether this actually will become a major problem in academic institutions depends somewhat on future threshold quantities for reporting the presence of materials.

The need to document claims of trade secrets is seen by many in industry to be an unreasonable burden. Much of American industry depends on trade secrets rather than patents to give them the economic advantage they view as necessary to thrive in their industry.

Justification of a trade secret under SARA Title III will require much more documentation than many companies currently have, and they may lose the protection of secrecy simply because they cannot support the claim. In the process, some may be subject to fines for filing frivolous trade secret claims.

Some business people do not trust the government to maintain trade secrets that are disclosed. Many fear that disclosure of secret composition information to health personnel and workers and their representatives will likely compromise the secrecy. Undoubtedly this will occasionally happen, in spite of the penalties for disclosure of bona fide trade secrets in this system. It is to be hoped that this is largely a groundless fear, but some small companies could lose their economic edge and be forced out of business.

Local Government

Costs. The cost of compliance with SARA Title III will be a significant item in the budgets of states and local communities. The paperwork alone will require much handling, storage, and indexing to facilitate retrieval. In addition, there will be the cost of meetings, of mailings to local citizens and industry, and of emergency response, both in training sessions and in actual disasters. Many volunteer fire departments are not in a position to deal with the administrative paperwork that could be required of them, and in many cases they probably cannot get the extra personnel needed to do the job.

In general, SARA Title III is an example of Congress mandating large expenditures by state and local governments without providing an adequate means of funding those expenditures. Federal support may be available for some of the costs, such as training, but it remains to be seen whether citizens will be willing to pay the price for the remainder in terms of increased taxes. When citizen suits are brought to force officials to act, they may or may not be successful, but in any case they will be costly.

Committee makeup. In addition to the outright costs to states and communities, it is reportedly difficult to persuade appropriate people to serve on emergency planning committees. There are major questions about legal liability of committee members in the event of litigation related to emergency planning and disaster response. Some communities cover such liability with general insurance arrangements, and others do not.

Another problem has arisen from the requirement that such committees include representatives of the broadcast and print media. Many news people resist such involvement because they believe it would compromise their objectivity. They feel they would become captive publicity agents and would not be free to criticize the committee's actions or inaction.

There have been complaints regarding inadequate representation of environmental, community, and labor representatives on community emergency planning committees. Some people have alleged that this is the result of a pro-industry bias, but it is also because of difficulties in finding someone from each of the specified categories to serve on such committees, especially in rural areas.

There is considerable reason to be concerned about whether planning personnel will receive adequate advice and training to generate the kind of understanding that is really needed. MSDSs and the types of reporting required provide primarily data, rather than useful information. Whether the ability exists to process the data into the needed kinds of information is a major concern.

Response personnel training. Essentially the same problem regarding technical expertise applies to response personnel. Almost no police and fire personnel have received a significant amount of education in chemistry and the hazardous properties of materials. Of the estimated 33,000 fire departments in the United States, only about 3,000 are staffed exclusively with full-time professional firefighters. The remaining 30,000 are staffed entirely or in part by volunteers whose training, even in firefighting, runs the gamut from none to fully professional.

Many of the volunteers have little or no hazardous material training. Many professional firefighters, at least at the command level, have received some training about hazardous materials, but such training is very sketchy and terribly inadequate at best. Unfortunately, many anecdotal reports have been circulating about how emergency response officials refuse offers of technical help from knowledgeable chemists, both professors and consultants. Officials reportedly have responded that they know what they are doing and don't need help from citizens who are "not familiar with disaster emergency procedures." It seems that an uncooperative attitude is one consequence of misguided pride.

Responses from facilities. Based on scant information and evidence available at the time of this writing, it appears that the emergency planning system is not working exactly as Congress had hoped. In many cases, facility owners or operators have simply not yet submitted required reports, sometimes because they do not think they are obligated to do so. This is often because they think their facilities do not use any hazardous materials, and sometimes because they think their operations probably are not covered by the law.

The response to the initial reporting requirement that is simplest for facility owners or operators is the one that tends to frustrate the system the most. They simply make photocopies of their MSDSs and send them in rather than the preferred lists of materials. This poses a real danger of flooding the governmental agencies and fire departments with mounds of paper that they cannot realistically assimilate. Warehouses full of completed forms constitute data, but not useful information.

The Public

Provision of information on potential disasters to the general public can be viewed as a two-edged sword. On the one hand, if a program is well implemented and publicized, reduced fear of the unknown among neighbors can lead to improved relations with the nearby community. In fact, for several years some major chemical companies have been voluntarily disclosing material hazard information to the communities in which their facilities are located, just for this reason.

On the other hand, the mere fact that more information is available to the public regarding the nature of the chemicals manufactured or used in their community may lead to an increase in civil suits based on personal injury claims from persons who otherwise would not have identified a hazardous chemical as a possible cause. Similarly, it is possible that suits to enjoin certain industrial activities based on public nuisance claims or other environmental statutes will increase for much the same reason. Of course, Congress encouraged some measure of litigation by including the citizen suit provisions in the statute.

One possible consequence of public fears that can be aroused by knowledge of the presence of hazardous materials is that official boards and commissions may be motivated to enact much more restrictive zoning laws and ordinances than in the past. Thus, facilities using more than a small amount of hazardous materials might be forced to locate in an isolated area. In fact, it is almost certain that the influence of SARA Title III will increase the amount of attention paid to hazardous operations when zoning matters are considered.

12.3 FUTURE DEVELOPMENTS

Modifications

Industries using construction workers, transient workers, and mobile work sites (e.g. repair services) have special problems with the location of MSDS files, training of short-term employees and those moving frequently from one work site to another, etc. OSHA felt these problems did not justify either exempting such workers from protection by the HCS or writing separate standards for such industries. Rather, it was decided that their special problems could be addressed more effectively by modifying provisions of the current HCS where appropriate. It thus appears that the construction industry and other segments with special problems will be prime candidates for further rulemaking in the relatively near future. There are currently at least two lawsuits pending that deal with provisions of the HCS relative to the construction industry.

Current reporting requirements for spills and releases are somewhat different under SARA Title III Section 304 and CERCLA Section 103. The EPA is aware that reporting requirements are so confusing that compliance may be difficult. At the time of this writing, efforts are under way to simplify the lists and remove at least some of the confusion.

In addition, reportable quantities probably will be changed as a result of further, more careful study than was possible initially under the Congressionally mandated time constraints. Threshold planning quantities are also under review, and some may be changed as a result. Finally, it is likely (though not certain) that EPA will change threshold quantities requiring submission of inventories.

Extensions

The fact that a large portion of disaster emergencies involving hazardous materials occurs within the transportation industry would seem to indicate a need for drastic action. And yet, in the aftermath of the deregulation of the trucking industry there are increasing numbers of poorly trained operators violating existing laws and operating defective equipment in a dangerous fashion.

Most states do not enforce existing transportation laws adequately; even so, there perhaps should be still more restrictions placed on operators hauling hazardous materials. Congress has been loath to deal effectively with this problem, as have most state legislatures. However, at the time of this writing, the DOT and several states are considering changes to licensing requirements for truckers hauling hazardous materials. It remains to be seen how effective those efforts will be.

The petroleum and agriculture industries are involved in the vast majority of deaths, injuries, evacuations, and incidents involving hazardous materials. Experience with SARA Title III may focus attention on this problem more sharply, and perhaps lead to some effective action.

Perhaps the largest negative factor in dealing with such problems in the transportation, petroleum, and agriculture industries is the fact that effective statutes and regulations undoubtedly would add to the cost of operations in those industries. Either some segments of industry would be affected more than others, or costs would be driven up for all, or both would happen. The decisions to be made will depend largely on political priorities, and it may be many years before the basic problems will be dealt with effectively.

Laboratories are currently exempted from portions of the HCS requirements in both the original and the expanded versions of the Standard. However, at the time of this writing, OSHA is putting the "finishing touches" on a new laboratory standard or regulation for laboratory workers that would deal with the same matters as the HCS, but recognizing the peculiarities of laboratories. Primarily these involve the presence of many small packages of a wide variety of materials, some of whose hazardous properties are not yet known, as well as the fact that the workers in laboratories generally are more knowledgeable about matters of importance within the overall subject of hazardous material safety than are workers in general.

RCRA-regulated hazardous wastes are exempted from the HCS because such wastes are regulated by EPA, and OSHA has not had the jurisdiction to regulate activities involving them. However, Congress has mandated that OSHA develop safety programs including more specific training requirements for workers dealing with such materials. In response, OSHA has published an interim final rule dealing with the subject (51 FR 45654 [Dec. 19, 1986]).

Related Areas

One matter that is related to the thrust of the Right-to-Know movement is that of risk notification. Bills have been considered in Congress to require that employers notify any workers who are, or ever have been, exposed to high-risk materials such as carcinogens, and to recommend medical surveillance. Some approaches would require employers to pay the cost of such surveillance, as well as the medical costs of treatment if the feared disease develops.

This type of risk notification legislation is a political "hot potato," but it is likely to be passed in some form in the near future. This will be seen as a disaster by many small business people who might have exposed workers to carcinogens in the past without even knowing about it, and who cannot afford the costs of such a risk notification process without facing possible bankruptcy. Consequently, some people favor governmental funding of such a program. Even if this is done, there still might be major lawsuits as a response to the required disclosures.

SARA Title III addresses the symptoms and results, rather than the causes, of unfortunate incidents associated with the use of hazardous materials in our economy. If hazardous materials were properly controlled in their manufacturing, transportation, and end-use environments, there would rarely, if ever, be any need for emergency response to hazardous material disasters. It is entirely possible that the attention focused on the problem by this statute will have a secondary effect of forcing the introduction of better engineering controls.

Possibly OSHA and EPA could accomplish the same end by a more direct, individual approach in facilities with a record of chemically-related accidents and environmental releases. They have not yet done so, but to many observers, this would seem a more appropriate and much less expensive approach. In fact, some states are already considering legislation which would force businesses to take steps to prevent accidents from occurring.

APPENDIX A

Glossary of Terms, Abbreviations, and Acronyms

NOTE: The following terms, abbreviations, and acronyms are commonly encountered when dealing with various aspects of Right-to-Know statutes, rules, and regulations. Not included below are the precise definitions of the various material hazards defined in the HCS. Those definitions, together with examples and some explanatory material, are given in detail in Chapter 4.

absorption   the taking in of a foreign substance by a solid, a liquid, or an organism, whereby the foreign substance becomes distributed throughout the host substance or organism, as through the blood stream.

ACGIH   American Conference of Governmental Industrial Hygienists.

acid   a substance which dissociates in water to form hydrogen ions; acids have sour taste, turn litmus paper red, and have pH values below 7.

acute effect   a health effect which occurs soon after a brief exposure to the offending agent.

acute exposure   an exposure, often intense, over a relatively short time.

acute toxicity   the property of a substance that causes an acute effect.

adsorption   the attachment of molecules of a foreign substance to the surface of another substance, usually a solid.

aerosol   a fine suspension of particles in a gas or in air, in which the particles are so small that they do not settle out rapidly. The particles may be either liquid (as in fog) or solid (as in smoke).

alkali   a chemical substance known as a base.

allergic reaction   an abnormal physiological response to a stimulus by a person who is sensitive to that particular stimulus.

anesthetic   a substance that causes a loss of sensation, totally or partially.

ANSI   American National Standards Institute.

antidote   a substance that will counteract the effects of a poison.

API   American Petroleum Institute.

asbestosis   a lung disease caused by inhaled asbestos fibers.

asphyxiant   a vapor or gas that can result in an insufficient oxygen supply, either by lowering the oxygen content of air or by preventing the blood from transporting oxygen efficiently. The result can be unconsciousness, or even death.

ASTM   American Society for Testing and Materials.

asymptomatic   a condition of lack of identifiable symptoms.

atm   atmosphere, a unit of pressure equal to 14.7 psi, or 101.2 kPa, or 760 mm Hg.

ATF   U. S. Treasury Department's Bureau of Alcohol, Tobacco, and Firearms.

auto-ignition temperature   the lowest temperature at which a liquid will ignite spontaneously or burn when heated in the presence of air.

base   a substance which produces hydroxide ions in water solution; bases have bitter taste, turn litmus paper blue, and have pH values above 7.

Black Lung Disease   a lung disease caused by inhaled coal dust.

boiling point   the temperature at which a liquid vaporizes at a given atmospheric pressure. Normal boiling point is the boiling point at a pressure of 1 atm.

bona fide   in or with good faith, without deceit or fraud.

BP   boiling point.

Brown Lung Disease   a lung disease caused by inhaled cotton fibers.

C   centigrade, or Celsius—a unit of temperature.

'C'   see ceiling.

CAA   Clean Air Act.

CAER   Community Awareness and Emergency Response program developed by the CMA.

carcinogenicity   the ability of a substance or condition that can cause a cancer to be formed in the exposed organism.

carcinoma   a form of cancer.

CAS   Chemical Abstracts Service.

caustic   generally considered synonymous with base, though sometimes used to refer to a material that is corrosive to skin and flesh.

cc   cubic centimeter, a unit of volume in the metric, or SI, system, equal to one-thousandth of a liter.

ceiling   a descriptive indication used in connection with TLVs or PELs to indicate a concentration that should never be exceeded, even briefly.

central nervous system   the brain and spinal cord, which together control the entire nervous system of the body.

CEPP   Chemical Emergency Preparedness Program (EPA), developed to deal with accidental releases of acutely toxic chemicals.

CERCLA   Comprehensive Environmental Response, Compensation, and Liability Act of 1980, commonly referred to as "Superfund."

CFR   *Code of Federal Regulations.*   An annual compilation of administrative regulations issued by OSHA, the EPA, and other federal agencies in carrying out their statutory duties. Regulations appearing in CFR have the force of law. CFR is organized by titles, which are divided into chapters, subchapters, parts, and then sections. The citation "29 CFR Part 1900" means title 29, part 1900. "29 CFR 1910.1200" means title 29, section 1910.1200.

chemical name   a name given to a chemical substance that is descriptive enough to allow a chemist to deduce its identity, composition, and structure.

chemical family   a group of elements or compounds with a common structural feature and with similar behavior in chemical reactions.

chemical reaction   a change in the structure of atoms or molecules to make new substances of entirely different composition and properties.

CHEMNET   a mutual aid network of more than fifty chemical shippers and contractors with emergency teams.

CHEMTREC   Chemical Transportation Emergency Center, operated by CMA.

CHRIS/HACS   Chemical Hazards Response Information System/Hazard Assessment Computer System developed by the USCG.

chronic effect   a health effect which occurs over a long period of time as a result of continued or periodic exposure to the offending agent.

chronic exposure   an exposure, often mild, over a long time.

chronic toxicity   the property of a substance that causes a chronic effect.

CMA   Chemical Manufacturers Association.

CNS   central nervous system.

COC   Cleveland Open Cup method of determining flash point.

common name   any identification other than a chemical name that is used to identify a chemical.

concentration   the relative amount of a given substance in a mixture with other substances.

conjunctivitis   inflammation of the membrane (conjunctiva) that lines the eyelids.

consumer product   a product that is packaged and distributed for sale through retail sales agencies for consumption by individuals for personal care or household use; includes drugs and medicines.

contingency plan   a document containing an organized course of action to use in case of fire, explosion, or release of hazardous materials that could threaten human health or environment.

CPSA   Consumer Product Safety Act.

CPSC   Consumer Product Safety Commission.

cubic meter   m^3, a measure of volume in the metric or SI system.

cutaneous   having to do with the skin.

cylinder   a pressure vessel designed to contain gases at pressures significantly above ambient atmospheric pressure; cylinders are circular in cross section, but do not include portable tanks, tank cars, tank trucks, or cargo tanks.

CWA   Clean Water Act.

cyst   a fluid-filled sac.

cytology   the scientific study and knowledge of cells of organisms.

decomposition   breakdown of a material or substance into smaller parts or compounds or elements.

de minimis   a trifling amount below which a law does not apply.

density   the mass of a substance contained in a given volume of the substance.

depressant   a substance that reduces a physiological activity or drive.

dermal   having to do with the skin.

dermatitis   an inflammation of the skin.

DHHS   U. S. Department of Health and Human Services.

dike   a barrier to control or confine a liquid to a given area and thus to prevent its spreading.

DOL   U. S. Department of Labor.

DOT   U. S. Department of Transportation.

dry chemical   a powdered fire extinguishing agent.

dyspnea   shortness of breath, difficulty in breathing.

edema   an abnormal accumulation of fluid in body tissues.

EPA   U. S. Environmental Protection Agency.

EPCRA   Emergency Planning and Community Right-to-Know Act, also known as SARA Title III.

ERA   Energy Reorganization Act of 1974.

epidemiology   the science and study of disease in general population through application of statistical techniques to collections of data relating the disease to various factors characteristic of some or all of the population.

erythema   redness or inflammation of the skin or mucous membranes.

etiologic   disease-causing, infectious.

evaporation rate   the rate at which a material will vaporize when compared to the rate at which a reference substance (often ether or normal butyl acetate) vaporizes under specified conditions.

extinguishing medium   the material used to control a fire.

Extremely Hazardous Substance   a substance determined by EPA to be so potentially hazardous to life and health if released accidentally that it requires special notification to planning committees if present in a facility.

eye protection   device worn to protect the eyes from exposure to splash or impact; includes safety glasses, goggles, and face shields.

F   Fahrenheit—a unit of temperature.

facility   1. (SARA Title III) all buildings, equipment, structures, and other stationary items which are located on a single site or on contiguous or adjacent sites and which are owned or operated by the same person (or by any person which controls, is controlled by, or under common control with, such person). For purposes of section 304, the term includes motor vehicles, rolling stock, and aircraft. **2.** (CERCLA) any building, structure, installation, equipment, pipe or pipeline (including any pipe into a sewer or publicly owned treatment works), well, pit, pond, lagoon, impoundment, ditch, landfill, storage container, motor vehicle, rolling stock, or aircraft; or any site or area where a hazardous substance has been deposited, stored, disposed of, or placed, or otherwise come to be located; but does not include any consumer product in consumer use or any vessel.

f/cc   number of fibers suspended or distributed in each cubic centimeter of air.

FDA   U. S. Food and Drug Administration.

FDCA   Federal Food, Drug, and Cosmetic Act.

Fed. Reg.   also FR. *Federal Register*.

FEMA   Federal Emergency Management Agency.

FHSA   Federal Hazardous Substances Act.

fibrosis   an abnormal thickening of fibrous connective tissue, often referring to a condition in the lungs.

FIFRA   Federal Insecticide, Fungicide, and Rodenticide Act.

first aid   immediate emergency measures taken when an individual suffers symptoms of exposure to a hazardous material, while awaiting the availability of professional medical care.

flash point   the lowest temperature at which the vapors of a liquid near its surface are concentrated enough to ignite when tested by Tagliaube, Pensky-Martens, or Setaflash test methods.

formula   the chemical expression of the composition of a substance, which indicates the number of atoms of each element in a single structural unit (molecule, ion, or formula unit).

FP   Flash Point.

FR   *Federal Register*.   A daily publication of notices, proposed regulations, and final regulations issued by federal agencies. Documents published in the *Federal Register* are cited by volume (corresponding to year), page, and date. Citations are usually given in the form "52

FR 10135, Mar. 30, 1987," meaning "page 10135 of Volume 52, Federal Register." Sometimes the volume number is not quoted, but merely the date.

frivolous   in legal usage, of little weight or importance, i.e., clearly insufficient on its face, and presumably interposed for mere purposes of delay or to embarrass the opponent.

fume   an air suspension (aerosol) of tiny, condensed solid particles commonly generated during evaporation of a liquid such as molten metal.

g   gram, a unit of mass in the metric, or SI, system of units.

general exhaust   a system for removing contaminated air from a large space.

general ventilation   see general exhaust.

genetic   an inherited characteristic, carried by genes from one generation to the next through biological reproduction.

GI   gastro-intestinal.

g/kg   grams dose per kilogram of body weight.

hand protection   device worn to protect the hands, and sometimes the forearms, from exposure to hazardous materials.

hazard warning   words or symbols used on labels, placards, or other communicating media to warn of the presence of potentially harmful materials or conditions.

Hazmat   hazardous material, usually referring to a material defined as such by DOT, but can refer to any material known to be hazardous.

HCS   Hazard Communication Standard (OSHA: 29 CFR 1910.1200).

health hazard   a material for which there is significant scientific evidence that acute or chronic health effects may occur to exposed persons or animals.

hematology   the science and study of the blood.

hematopoietic system   the mechanism of the body by which blood is formed.

HMIS   Hazardous Material Identification System.

HMTA   Hazardous Materials Transportation Act.

HSWA   Hazardous and Solid Waste Amendments to RCRA.

IARC   International Agency for Research on Cancer.

IDLH   Immediately Dangerous to Life and Health; maximum level of exposure without irreparable effects within about one-half hour.

impermeable   impervious.

impervious   the property of a material that prevents another substance from passing through or penetrating it.

incompatible   a descriptive term for materials that can react together in a dangerous fashion if they should contact or be mixed with each other.

ingestion   taking in by the mouth, usually referring to swallowing.

inhalation   breathing in.

inhibitor   a chemical added to another substance to prevent an unwanted chemical reaction or decomposition.

insoluble   incapable of being dissolved in a particular liquid to form a clear (not necessarily colorless) solution.

ips   inches per second.

kg   kilogram, a unit of mass in the metric, or SI, system of units.

kPa   kilopascal, a unit of measure of pressure in the metric, or SI, system of units.

L   abbreviation for liter.

lachrymation   formation of tears, usually referring to an excessive amount.

LC_{50}   median lethal concentration.

LD_{50}   median lethal dose.

LEL   lower explosion limit—identical to LFL.

LFL   lower flammability limit—the lowest concentration of substance in air that will burn.

lesion   any damage to a living tissue.

liter   a unit of measure of volume in the metric, or SI, system, also spelled litre; a volume slightly larger than a quart.

local exhaust   a system for removing contaminated air from the immediate vicinity of the point of generation of contamination.

local ventilation   see local exhaust.

m   meter.

m^3, cubic meter, a unit of measure of volume in the metric, or SI, system.

malaise   a general feeling of discomfort, uneasiness or lack of well-being.

malignant   tending to become progressively worse and potentially to result in death.

melting point   the temperature at which a solid substance melts, or fuses, to become a liquid.

MEK   methyl ethyl ketone.

metabolism   chemical processes taking place within the body, normally related to the use of food for sustenance and repair of the body.

metastasis   the spread, or transfer, of disease from one organ or part of the body to another that is not directly connected with it.

meter   a unit of measure of length in the metric, or SI, system, also spelled metre; a length slightly more than a yard.

mg   milligram, a unit of measure of mass in the metric, or SI, system; one-thousandth the mass of a gram.

mg/kg   milligrams dose per kilogram of body weight.

mg/m^3, concentration unit indicating milligrams of a substance in each cubic meter of air.

micron   a unit of measure of length equal to one millionth of a meter.

mist   a suspension of tiny liquid droplets in the air.

mixture   any combination of two or more substances that is not the result of a chemical reaction.

mL   milliliter, one thousandth of a liter, equal to a cubic centimeter; a unit of volume in the metric, or SI, system.

mm Hg   millimeters of mercury, a unit of measurement of pressure of air or any other gas.

molecular weight   the mass of a molecule stated on a relative scale of the masses of individual atoms and molecules.

mpcf   million particles of suspended material per cubic foot of air.

mp   melting point.

MSDS   Material Safety Data Sheet.

MSHA   Mine Safety and Health Administration, U. S. DOL.

mutagen   a substance that is capable of changing the genetic material in a living cell.

MW   molecular weight.

NA   North America (shipping identification number).

n/a   not available, or not applicable.

narcosis   a state of arrested activity, including stupor or unconsciousness, brought on by exposure to a narcotic or other material.

nausea   a sickness of the stomach, or a tendency to vomit.

NCI   National Cancer Institute, of the National Institutes of Health.

NCP   National Contingency Plan (40 CFR 300).

neoplasia   a condition characterized by the presence of new growths, such as tumors.

neutralize   to eliminate potential hazards by inactivating the materials that exhibit the hazards; a reaction of acids and bases together to form compounds that are normally nonhazardous or much less hazardous than the original materials.

NFPA   National Fire Protection Association.

ng   nanogram, one billionth of a gram.

NIOSH   National Institute of Occupational Safety and Health, a part of DHHS.

NPIRS   National Pesticide Information Retrieval System.

NRC   National Response Center, or Nuclear Regulatory Commission.

NRT   National Response Team.

NTP   National Toxicology Program.

odor threshold   the lowest concentration at which the presence of a vapor can be detected in air by smelling.

olfactory   pertaining to the sense of smell.

OMB   (federal) Office of Management and Budget.

oral   having to do with the mouth, such as oral administration of toxin.

OSH Act   Occupational Safety and Health Act.

OSHA   Occupational Safety and Health Administration.

oxidation   the process in which an atom loses electrons, often in a chemical reaction of a substance with oxygen, though also in a reaction of a substance with a peroxide or any other oxidizer.

oxygen deficiency   an atmosphere having less than the normal concentration of oxygen. Normal air contains 21% oxygen at sea level; air containing less than 19.5% oxygen is considered oxygen deficient; air containing less than about 16% oxygen will not sustain human life.

overexposure   exposure to a hazardous material beyond the PELs, or beyond a level at which health effects can occur.

pathologic   pertaining to disease.

pathology   science or study of changes produced by disease.

PBB   polybrominated biphenyl.

PCB   polychlorinated biphenyl.

PEL   Permissible Exposure Limit (OSHA).

percent volatile   the percentage of a mixture that can evaporate.

person   (SARA Title III) any individual, trust, firm, joint stock company, corporation (including a government corporation), partnership, association, State, municipality, commission, political subdivision of a State, or interstate body.

personal protective equipment   articles worn by a worker to protect against environmental hazards; examples include gloves, chemical splash goggles, respirators, full face shields, SCBA.

pH   a measure of the concentration of hydrogen ions or hydroxide ions in a water system, hence a measure of how acidic or basic the system is.

PHS   Public Health Service, a part of DHHS.

PMCC   Pensky-Martens Closed Cup flash point test.

pneumoconiosis   a condition of permanent deposition of particulate matter in lung tissue, and the tissue reaction to it.

polymerization   a chemical reaction in which very large molecules are formed by many small molecules joining together, usually with the release of large amounts of heat energy.

ppb   parts of contaminating material per billion parts of air.

ppm   parts of contaminating material per million parts of air.

psi   pounds per square inch, a unit of pressure measurement.

pulmonary   related to the lungs and their functioning.

pulmonary edema   fluid in the lung tissues.

RCRA   Resource Conservation and Recovery Act of 1968.

reaction   a chemical transformation, in which substances react to form new substances, not merely a mixture of the otherwise unchanged original substances.

reactive   unstable, will react easily alone or with other materials.

reactivity   ability to undergo chemical reaction, often with hazardous consequence.

relative vapor density   the mass of a given volume of vapor or gas divided by the mass of the same volume of air at the same temperature and pressure.

release   (SARA Title III) any spilling, leaking, pumping, pouring, emitting, emptying, discharging, injecting, escaping, leaching, dumping, or disposing into the environment (including the abandonment or discarding of barrels, containers, and other closed receptacles) of any hazardous chemical, extremely hazardous substance, or toxic chemical.

respirator   a device worn by a worker to protect against inhaling harmful contaminants.

respiratory protection   a device that protects against the inhalation of harmful materials; may include dust masks, respirators, gas masks, air-supply masks, and SCBAs.

respiratory system   the breathing system of the body including the nose and mouth, larynx, trachea, bronchia, and lungs.

routes of entry   the means by which foreign material gains access to the body.

RQ   Reportable Quantity, under CERCLA or SARA.

RRT   Regional Response Team.

SARA   Superfund Amendments and Reauthorization Act of 1986.

sarcoma   a tumor which is often malignant.

SCBA   self-contained breathing apparatus.

self-contained breathing apparatus   respiratory protection that includes its own supply of breathable air in a tank that is carried by the wearer.

Seta   Setaflash Closed Tester for flash point.

SIC   Standard Industrial Classification, a code used to classify the area of business engaged in by an organization.

silicosis   a lung disease caused by inhaled silica dust.

'skin'   a designation sometimes used with a TLV or PEL value to indicate the possibility of skin absorption.

skin absorption   the passage of substances, whether solid, liquid, or gas, through the skin into the blood stream.

solubility in water   the amount of a substance that will dissolve in a given amount of water to form a clear (not necessarily colorless) mixture at a particular temperature which is often taken to be "room temperature."

solvent   a liquid, water or otherwise, in which other substances are dissolved.

species   type of test animals used in toxicity tests.

specific chemical identity   the chemical name, CAS Registry Number, or any other precise identifier of a chemical substance.

specific gravity   the density of a substance divided by the density of water.

spill   accidental spilling, leaking, pumping, emitting, emptying, or dumping of hazardous wastes or materials which, when spilled, become hazardous wastes on land or in water.

spill or leak procedures   the overall procedures and equipment to be used to contain and clean up a leak or spill of the material in question.

spontaneously combustible   the property of a material that can ignite as a result of the buildup of heat from a slow reaction, such as with air or moisture, until it reaches its ignition temperature.

squamous   scaly or platelike, as in squamous cell tumors.

stability   the ability of a material to resist chemical change or to remain unchanged in the absence of overt procedures to react it.

STEL   short-term exposure limit.

stenosis   narrowing of an opening or passage in the body.

subcutaneous   beneath the layers of the skin.

substance   any chemical entity, normally used in connection with a pure compound rather than a mixture.

Superfund   see CERCLA.

systemic poison   a poison which spreads throughout the body rather than producing a purely localized effect.

systemic toxicity   adverse effect to the body generally rather than lo-cally.

synonyms   different names by which the same material is known.

target organ effects   toxic effects primarily confined to specific indi-vidual organs of the body.

target organ toxin   a material which produces a harmful target organ effect.

TCC   Tagliaube Closed Cup test for flash point.

TCLo   toxic concentration low, the lowest concentration of a vapor shown to produce a toxic effect under specified conditions.

TDLo   toxic dose low, the lowest dose of an agent shown to produce a toxic effect under specified conditions.

teratogen   a material that can cause malformations in a fetus carried by a female who is exposed to the material.

tfx   toxic effects.

Threshold Quantity   see TPQ.

Title III   (of SARA), see EPCRA.

TLV   Threshold Limit Value (ACGIH).

TOC   Tagliaube Open Cup test for flash point.

torr   a unit of gas pressure equal to one millimeter of mercury (1 mm Hg).

tort   a private or civil wrong or injury, for which the court may provide a remedy in the form of an action for damages.

toxicity   the adverse health effects caused by exposure to a material.

TPQ   threshold planning quantity under SARA Title III.

trade name   the commercial trade name or trademark name by which a commercial product is known.

transplacental   a description of a material that causes physical defects in a developing embryo.

TSCA   Toxic Substances Control Act of 1976.

TSDF   Treatment, Storage, or Disposal Facility (under RCRA).

TWA   time-weighted average.

UEL   upper explosion limit—identical to UFL.

UFL   upper flammability limit—the highest concentration of a substance in air that will burn.

UN   United Nations (shipping identification number).

unstable   tending toward decomposition or other unwanted chemical change even in the absence of any overt attempt to institute such change.

USC   United States Code. A compilation of all statutes enacted by Congress which are currently in effect. USC is organized by titles, which are divided into chapters, subchapters, and sections. The notation "*et seq.*" following a section number indicates that the act being cited includes the next several sections. Citations are usually given in the form "42 USC 501, *et seq.*," meaning "section 501 and following sections of Title 42, United States Code."

USCG   United States Coast Guard.

vapor   the gaseous form of a material as it evaporates from the liquid form or sublimes from the solid form.

vapor density   the mass of a given volume of vapor or gas, often recorded as the relative vapor density.

vapor pressure   the pressure exerted by a saturated vapor in contact with its own liquid in a closed container at a given temperature.

vermiculite   an expanded mica sometimes used to adsorb liquids in spill cleanup procedures.

volatility   the tendency of a substance to evaporate, usually stated in terms of its evaporation rate.

viscosity   the tendency of a fluid to resist flow.

vp   vapor pressure.

APPENDIX B.1

Books and Compendia

Anderson, K., and R. Scott. *Fundamentals of Industrial Toxicology* (Ann Arbor, MI: Ann Arbor Science Publishers, Inc., 1981).

Asfahl, C. R. *Industrial Safety and Health Management* (Englewood Cliffs, NJ: Prentice-Hall, 1984).

Atherley, G. R. C. *Occupational Health and Safety Concepts* (London: Applied Science Publishers, 1978).

Bennett, G., F. S. Feates, and I. Wilder, Eds. *Hazardous Materials Spill Handbook* (New York: McGraw-Hill, 1984).

Black, H. C. *Black's Law Dictionary* (St. Paul, MN: West Publishing Co., 1979).

Braker, W., and A. L. Mossman. *Matheson Gas Data Book, 5th Ed.* (East Rutherford, NJ: Matheson Gas Products, 1971).

Braker, W., A. L. Mossman, and D. Siegel. *Effects of Exposure to Toxic Gases—First Aid and Medical Treatment* (Lyndhurst, NJ: Matheson, 1977).

Bretherick, L. *Handbook of Reactive Chemical Hazards, 2nd Ed.* (Cleveland, OH: CRC Press, Inc., 1979).

Browning, E. *Toxicity of Industrial Metals* (London: Butterworth & Co., 1969).

Browning, E. *Toxicity of Industrial Organic Solvents, Rev. Ed.* (London: H. M. Stationery Office, 1953).

Browning, E. *Toxicity and Metabolism of Industrial Solvents* (New York: Elsevier Publishing Co., 1965).

Browning, R. L. *The Loss Rate Concept in Safety Engineering* (New York: Marcel Dekker, Inc., 1980).

Cheremisinoff, P. N. *Management of Hazardous Occupational Environments* (Lancaster, PA: Technomic Publishing Co., 1984).

Chissick, S. S., and R. Derricott, Eds. *Occupational Health and Safety Management* (Chichester, U.K.: Wiley, 1984).

Clayson, D. B., D. Krewski, and I. Munro, Eds. *Toxicological Risk Assessment* (Boca Raton, FL: CRC Press, Inc., 1985).

Clayton, G. D., and F. E. Clayton, Eds. *Patty's Industrial Hygiene and Toxicology, 3rd Rev. Ed., Vol. I, General Practices* (New York: John Wiley & Sons, 1978).

Clayton, G. D., and F. E. Clayton, Eds. *Patty's Industrial Hygiene and Toxicology, 3rd Rev. Ed., Vol. II, Toxicology* (New York: John Wiley & Sons, 1982).

Cralley, L. J., and L. V. Cralley, Eds. *Patty's Industrial Hygiene and Toxicology, 3rd Rev. Ed., Vol. III, Theory and Rationale of Industrial Hygiene Practice* (New York: John Wiley & Sons, 1978).

DeBruin, A. *Biochemical Toxicology of Environmental Agents* (Amsterdam: Elsevier, 1976).

Documentation of the Threshold Limit Values, 5th Ed. (Cincinnati, OH: ACGIH, 1986).

Elkins, H. B. *The Chemistry of Industrial Toxicology, 2nd Ed.* (New York: John Wiley & Sons, 1959).

Emergency Response Guidebook (Washington, DC: U.S. Department of Transportation).

Encyclopedia of Chemical Technology, 12 vols. (New York: Wiley Interscience, 1978–80).

Encyclopedia of Occupational Health and Safety, 3rd Rev. Ed. (Geneva, Switzerland: International Labor Office, 1983).

Fairhall, L. T. *Industrial Toxicology, 2nd Ed.* (Baltimore, MD: The Williams and Wilkins Co., 1957).

Fawcett, H. H., and W. S. Wood. *Safety and Accident Prevention in Chemical Operations* (New York: Wiley Interscience, 1965).

Fawcett, H. H. *Hazardous and Toxic Materials: Safe Handling and Disposal* (New York: Wiley Interscience, 1984).

Fire Protection Guide on Hazardous Materials, 7th Ed. (Boston, MA: National Fire Protection Association, 1978).

Glanze, W. D., Ed. *Mosby's Medical Dictionary, Rev. 2nd Ed.* (St. Louis, MO: The C. V. Mosby Co., 1986).

Gleason, M. N. *Clinical Toxicology of Commercial Products, 3rd Ed.* (Baltimore, MD: Williams & Wilkins Co., 1969).

Hallenbeck, W. H., and K. M. Cunningham. *Quantitative Risk Assessment for Environmental and Occupational Health* (Chelsea, MI: Lewis Publishers, Inc., 1986).

Hazardous Materials Emergency Planning Guide (Washington, DC: National Response Team).

Hazardous Materials: Emergency Response Guidebook (Washington, DC: U.S. Department of Transportation, Materials Transport Bureau).

Hunter, D. *The Disease of Occupations, 4th Ed.* (Boston, MA: Little, Brown and Co., 1969).

Joseph, E. Z., Ed. *Chemical Safety Data Guide* (Washington, DC: The Bureau of National Affairs, 1985).

Kamrin, M. A. *Toxicology* (Chelsea, MI: Lewis Publishers, Inc., 1988).

Kopfler, F., and G. Craun. *Environmental Epidemiology* (Chelsea, MI: Lewis Publishers, Inc., 1986).

Leemann, J. E. *Spill Reporting Procedures Guide* (Washington, DC: The Bureau of National Affairs, 1987).

Levy, B. S., and D. H. Wegman, Eds. *Occupational Health: Recognizing and Preventing Work-Related Diseases,* (Boston, MA: Little, Brown and Co., 1984).

Lowrance, W. W. *Of Acceptable Risk: Science and the Determination of Safety* (Los Altos, CA: William Kaufmann Inc., 1976).

Manual of Hazardous Chemical Reactions, 5th Ed., 491-M (Boston, MA: National Fire Protection Association).

Meidl, J. H. *Flammable Hazardous Materials, 2nd Ed.* (New York: Macmillan Publishing Co., 1978).

Meyer, E. *Chemistry of Hazardous Materials* (Englewood Cliffs, NJ: Prentice-Hall, 1977).

Muir, G. D. *Hazards in the Chemical Laboratory* (London: The Chemical Society, 1977).

National Research Council. *Prudent Practices for Handling Hazardous Chemicals in Laboratories* (Washington, DC: National Academy Press, 1981).

Nothstien, G. Z. *Law of Occupational Safety and Health* (New York: Macmillan Publishing Co., 1984).

O'Reilly, J. T. *Emergency Response to Chemical Accidents: Planning and Coordinating Solutions* (New York: McGraw-Hill, 1987).

OSHA. *Training Guidelines* (48 FR 39317–23, August 30, 1983).

Ottoboni, M. A. *The Dose Makes the Poison* (Berkeley, CA: Vincente Books, 1984).

L. Parmeggiani, Ed. *Encyclopedia of Occupational Health and Safety* (Geneva: International Labor Office, 1983).

Pipitone, D. *Safe Storage of Laboratory Chemicals* (New York: Wiley Interscience, 1984).

Proctor, N. H., and J. P. Hughes. *Chemical Hazards of the Workplace,* (Philadelphia, PA: Lippincott, 1978).

Registry of Toxic Effects of Chemical Substances (Rockville, MD: U.S. Department of Health and Human Services, Public Health Service, NIOSH).

Risk Assessment and Risk Control (Washington, DC: The Conservation Foundation, 1985).

Rodricks, J. V., and R. G. Tardiff, Eds. *Assessment and Management of Chemical Risks* (Washington, DC: American Chemical Society, 1984).

Rom, W., Ed. *Environmental and Occupational Medicine,* (Boston, MA: Little, Brown and Co., 1983).

Ruch, W. E., and B. J. Held. *Respiratory Protection: OSHA and the Small Businessman* (Ann Arbor, MI: Ann Arbor Science Publishers, 1975).

Safe Handling of Chemical Carcinogens, Mutagens, Teratogens, and Highly Toxic Substances (Woburn, MA: Butterworths, 1980).

Sax, N. I. *Cancer Causing Chemicals* (New York: Van Nostrand Reinhold Co., 1981).

Sax, N. I., and B. Feiner. *Dangerous Properties of Industrial Materials, 6th Ed.* (New York: Van Nostrand Reinhold Co., 1984).

Shrader-Frechette, K. S. *Risk Analysis and Scientific Method* (Boston, MA: D. Reidel Publishing Co., 1985).

Shrivastava, P. *Bhopal: Anatomy of a Crisis* (Cambridge, MA: Ballinger Publishing Co., 1987).

Sittig, M. *Handbook of Toxic and Hazardous Chemicals* (Park Ridge, NJ: Noyes Publications, 1981).

Sittig, M., Ed. *Hazardous and Toxic Effects of Industrial Chemicals* (Park Ridge, NJ: Noyes Data Corp., 1979).

Sittig, M., Ed. *Priority Toxic Pollutants—Health Effects and Allowable Limits* (Park Ridge, NJ: Noyes Data Corp., 1979).

Sittig, M. *Toxic Metals, Pollution Control and Waste Protection* (Park Ridge, NJ: Noyes Data Corp., 1976).

Smith, A. J. *Managing Hazardous Substances Accidents* (New York: McGraw-Hill, 1981).

Steere, N. V., Ed. *Handbook of Laboratory Safety, 2nd Ed.* (Cleveland, OH: CRC Press, Inc., 1971).

Technical Guidance for Hazards Analysis (Washington, DC: Environmental Protection Agency).

Threshold Limit Values and Biological Exposure Indices (Cincinnati, OH: ACGIH, 1987 *et seq.*).

Toxic and Hazardous Industrial Chemicals Safety Manual (Tokyo, Japan: The International Technical Information Institute, 1977).

Tuve, R. L. *Principles of Fire Protection Chemistry* (Boston, MA: National Fire Protection Association, 1976).

van Oettingen, W. F. *Poisoning—a Guide to Clinical Diagnosis and Treatment, 2nd Ed.* (Philadelphia, PA: W. B. Saunders Co., 1963).

Verschueren, K. *Handbook of Environmental Data on Organic Chemicals* (New York: Van Nostrand Reinhold Co., 1977).

Weinstein, A. S., A. D. Twerski, H. R. Piehler, and W. A. Donaher. *Products Liability and the Reasonably Safe Product* (New York: John Wiley & Sons, 1978).

Wexley, P. *Information Resources in Toxicology* (Amsterdam: Elsevier, 1975).

Windholz, M., S. Budavari, L. Y. Stroumtsos, and M. N. Fertig, Eds. *The Merck Index: An Encyclopedia of Chemicals and Drugs* (Rahway, NJ: Merck & Co. Inc., 1976).

APPENDIX B.2

Computer Data Banks

The following is a list of public data banks which contain information of use in various aspects of managing HCS and SARA programs. Some data banks include technical data on the properties of various materials. In some cases, the data are available in the form of an MSDS, and in others specific items of data can be obtained interactively. Some of the data banks contain research information that is useful to industrial hygienists and other health professionals in understanding the materials more thoroughly than can be done with an MSDS. Some include information such as statutes, regulations, court decisions, and various other citations useful in researching legal cases.

This list has been compiled from several sources, but it probably is not complete. It is for information only. No endorsement of any of these data banks is expressed or implied by the authors.

Some individual data bases are available from more than one data bank. It is suggested that several vendors be contacted to obtain more detailed information in order to determine the most suitable data bank for a particular application.

BRS: Bibliographic Retrieval
   Services
1200 Route 7
Latham, NY
(800) 833-4707

CAS: Chemical Abstracts Service
P. O. Box 3012
Columbus, OH 43210
(800) 848-6538, (614) 421-3600

CIS: Chemical Information Systems, Inc.
7215 York Road
Baltimore, MD 21212
(301) 821-5980

DIALOG: Dialog Information Systems, Inc.
3460 Hillview Avenue
Palo Alto, CA 94304
(800) 2-DIALOG

ICIS: ICI Consultants
1133 15th Street, N.W.
Suite 300
Washington, DC 20005
(202) 822-5200

LAB-LINK: Mallinckrodt, Inc.
Science Products Division
P. O. Box 5840
St. Louis, MO 63134
(800) 354-2050

LEXIS: Mead Data Central, Inc.
P. O. Box 933
Dayton, OH 45401
(513) 865-6800

NLM: National Library of Medicine
Specialized Information Services
Building 38A
8600 Rockville Pike
Bethesda, MD 20309
(301) 496-1131

NPIRS: National Pesticides Information Retrieval System
Entomology Building
Purdue University
West Lafayette, IN 47907
(317) 494-6616

OHS: Occupational Health Services, Inc.
Suite 2407
457th Avenue
New York, NY 10123
(212) 967-1100

PERGAMON INFOLINE: Pergamon Infoline, Inc.
1340 Old Chain Bridge Road
McLean, VA 22101
(703) 442-0900

QUESTEL: Questel
5201 Leesburg Pike
Suite 603
Falls Church, VA 22041
(703) 845-1133

SDC: SDC Information Services
2500 Colorado Avenue
Santa Monica, CA 90406
(213) 820-4111

WESTLAW: West Publishing Co.
P. O. Box 64526
St. Paul, MN 56164
(800) 328-0109, (612) 228-2450

APPENDIX B.3

Federal and State Agencies

AGENCIES RELATED TO THE HAZARD COMMUNICATION STANDARD

Occupational Safety and Health Administration

200 Constitution Avenue, N.W.
Washington, DC 20210
(202) 523-6666

Region I: CT, ME, MA, NH, RI, VT
16-18 North Street
1 Dock Square Building
4th Floor
Boston, MA 02109
(617) 223-6710

Region II: NJ, NY, PR, VI
1515 Broadway (1 Astor Plaza)
Room 3445
New York, NY 10036
(212) 944-3432

Region III: DC, DE, MD, PA, VA, WV
Suite 2100, Gateway Building
3535 Market Street
Philadelphia, PA 19104
(215) 596-1201

Region IV: AL, FL, GA, KY, MS, NC, SC, TN
Suite 587
1375 Peachtree Street, N.E.
Atlanta, GA 30367
(404) 347-4495

Region V: IL, IN, MI, MN, OH, WI
Room 3244
230 South Dearborn Street
Chicago, IL 60604
(312) 353-2220

Region VI: AR, LA, NM, OK, TX
Room 602
555 Griffin Square Building
Dallas, TX 75202
(214) 767-4731

Region VII: IA, KS, MO, NB
Room 406
911 Walnut Street
Kansas City, MO 64106
(816) 758-5861

Region VIII: CO, MT, ND, SD, UT, WY
Room 1576, Federal Building
1961 Stout Street
Denver, CO 80294
(303) 844-3061

Region IX: AZ, CA, HI, NV,
American Samoa, Guam, Northern

Mariana Islands
11349 Federal Building
450 Golden Gate Avenue
San Francisco, CA 94102
(415) 556-7260

Region X: AK, ID, OR, WA
Room 6003
Federal Office Building
909 First Avenue
Seattle, WA 98174
(206) 442-5930

State Safety and Health Agencies

Listed below are the administration agencies of the various states under the OSH Act. In most states, the administration of the HCS is under a single state agency, but in the states of Louisiana, Michigan, New Hampshire, and West Virginia, responsibility is divided between two state agencies. Generally one of these is concerned with labor safety matters and the other is concerned with public health matters.

ALABAMA   Region IV
Alabama Department of Labor
600 Administration Building
64 North Union Street
Montgomery, AL 36130
(205) 261-3460

ALASKA   Region X
Alaska Department of Labor
P. O. Box 1149
Juneau, AK 99802
(907) 465-2700

AMERICAN SAMOA   Region IX
Department of Manpower Resources
Pago Pago, American Samoa
   96799
(681) 633-4485

ARIZONA   Region IX
Occupational Safety & Health Division
Industrial Commission of Arizona
800 West Washington
Phoenix, AZ 85007
(602) 255-5795

ARKANSAS   Region VI
Department of Labor
1022 High Street
Little Rock, AR 72202
(501) 375-8442

CALIFORNIA   Region IX
Department of Industrial Relations
525 Golden Gate Avenue

San Francisco, CA 94102
(415) 557-3356

COLORADO   Region VIII
Division of Labor
Department of Labor and Employ-
    ment
1313 Sherman Street
Room 314
Denver, CO 80203
(303)866-2782

CONNECTICUT   Region I
Connecticut Department of Labor
200 Folly Brook Boulevard
Wethersfield, CT 06109
(203) 566-5123

DELAWARE   Region III
Department of Labor
820 North French Street
Albert N. Corvel Bldg.
6th Floor
Wilmington, DE 19801
(302) 571-2710

D.C.   Region III
Office of Occupational Safety and
    Health
950 Upshur Street, N.W.
Washington, DC 20011
(202) 576-6651

FLORIDA   Region IV
Dept. of Labor and Employment
    Security
2590 Executive Center Circle, East
Suite 206, Berkeley Building
Tallahassee, FL 32301
(904) 488-4398

GEORGIA   Region IV
Department of Labor
288 State Labor Building

Atlanta, GA 30334
(404) 656-3011

GUAM   Region IX
Department of Labor
Government of Guam
P. O. Box 23548, GMF
Agana, Guam 96910
011-671-477-9821

HAWAII   Region IX
Hawaii Department of Labor and
    Industrial Relations
825 Mililani Street
Honolulu, HI 96813
(808) 548-3150

IDAHO   Region X
Dept of Labor and Industrial Ser-
    vices
277 North Sixth Street
State House Mall
Boise, ID 83720
(208) 334-3950

ILLINOIS   Region V
Department of Labor
310 South Michigan Avenue
10th Floor, Room 1803
Chicago, IL 60604
(312) 793-2800

INDIANA   Region V
Indiana Division of Labor
1013 State Office Building
100 North Senate Avenue
Indianapolis, IN 46204
(317) 232-2663

IOWA   Region VII
Bureau of Labor, State House
307 East Seventh Street
Des Moines, IA 50319
(515) 281-3447

KANSAS   Region VII
Department of Human Resources
401 Topeka Avenue
Topeka, KS 66603
(913) 296-7474

KENTUCKY   Region IV
Kentucky Labor Cabinet
U S Highway 127 South
Frankfort, KY 40601
(502) 564-3070

LOUISIANA   Region VI
Department of Labor
Office of Labor
5360 Florida Blvd.
Baton Rouge, LA 70806
(504) 925-4246

Office of Health Services and Envi-
    ronmental Quality (OHSEQ)
P. O. Box 60630
New Orleans, LA 70160
(504) 568-5050

MAINE   Region I
Department of Manpower Affairs
20 Union Street
Augusta, ME 04330
(207) 289-3788

MARYLAND   Region III
Maryland Division of Labor and In-
    dustry
Department of Licensing and Regu-
    lation
501 St. Paul Place
Baltimore, MD 21202
(301) 659-4176

MASSACHUSETTS   Region I
Department of Labor and Industries
Leverett Saltonstall Building
100 Cambridge Street

Room 1100
Boston, MA 02202
(617) 727-3454

MICHIGAN   Region V
Michigan Department of Labor
7150 Harris Drive
Box 30015
Lansing, MI 48909
(517) 373-9600

Michigan Department of Public
    Health
3500 North Logan Street
Box 30035
Lansing, MI 48909
(517) 373-1320

MINNESOTA   Region V
Department of Labor and Industry
444 Lafayette Road
St. Paul, MN 55101
(612) 296-2342

MISSISSIPPI   Region IV
State Board of Health
P. O. Box 1700
Jackson, MS 39215
(601) 354-6646

MISSOURI   Region VII
Division of Labor and Industrial
    Relations
1904 Missouri Blvd.
Jefferson City, MO 65101
(314) 751-2461

MONTANA   Region VIII
Bureau of Safety and Health
Division of Worker's Compensa-
    tion
P. O. Box 4759
Helena, MT 59604
(406) 499-2047

NEBRASKA   Region VII
Department of Labor
P. O. Box 95024
Lincoln, NB 68509
(402) 471-2239

NEVADA   Region IX
Department of Industrial Relations
Division of Occupational Safety
  and Health
1370 South Curry Street
Carson City, NV 89710
(702) 885-5240

NEW HAMPSHIRE   Region I
Department of Labor
19 Pillsbury Street
Concord, NH 03301
(603) 271-3176

Bureau of Environmental Health
Division of Public Health Services
Health and Welfare Building
Hazen Drive
Concord, NH 03301
(603) 271-4587

NEW JERSEY   Region II
Department of Labor
John Fitch Plaza, CN 110
Trenton, NJ 08625
(609) 292-2323

NEW MEXICO   Region VI
Environmental Improvement Divi-
  sion
Health and Environment Depart-
  ment
P. O. Box 968
Santa Fe, NM 87504
(505) 827-2850

NEW YORK   Region II
New York Department of Labor

Two World Trade Center
Room 7308
New York, NY 10047
(212) 488-6300

NORTH CAROLINA   Region IV
North Carolina Department of La-
  bor
P. O. Box 27407
Raleigh, NC 27603
(919) 733-7166

NORTH DAKOTA   Region VIII
North Dakota Workmen's Compen-
  sation Bureau
Russell Building
Highway 83 North
4007 North State Street
Bismarck, ND 58501
(701) 224-2700

OHIO   Region V
Division of OSHA On-Site Consul-
  tation
Department of Industrial Relations
P. O. Box 825
Columbus, OH 43216
(614) 466-7485

OKLAHOMA   Region VI
Department of Labor
1315 Broadway Place
Oklahoma City, OK 73103
(405) 521-2461

OREGON   Region X
Workers' Compensation Depart-
  ment
Labor and Industries Building
Salem, OR 97310
(503) 378-3304

PENNSYLVANIA   Region III
Department of Labor and Industry

1700 Labor and Industry Building
7th and Foster Streets
Harrisburg, PA 17120
(717) 787-3756

PUERTO RICO   Region II
Puerto Rico Department of Labor
  and Human Resources
Prudencio Rivera Martinez Building
505 Munoz Rivera Avenue
Hato Rey, PR 00918
(809) 754-2119

RHODE ISLAND   Region I
Rhode Island Department of Labor
220 Elmwood Avenue
Providence, RI 02907
(401) 277-2741

SOUTH CAROLINA   Region IV
South Carolina Department of La-
  bor
P. O. Box 11329
Columbia, SC 29211
(803) 758-2851

SOUTH DAKOTA   Region VIII
South Dakota Department of Labor
700 Illinois North
Pierre, SD 57501
(605) 733-3101

TENNESSEE   Region IV
Tennessee Department of Labor
501 Union Building
Suite "A"—Second Floor
Nashville, TN 37219
(615) 741-2582

TEXAS   Region VI
Bureau of Environmental Health
Texas Department of Health

1100 West 49th Street
Austin, TX 78756
(512) 458-7542

UTAH   Region VIII
Utah Industrial Commission
Utah Occupational Safety and
  Health
P.O. Box 5800
Salt Lake City, UT 84110
(801) 530-6900

VERMONT   Region I
Department of Labor and Industry
120 State Street
Montpelier, VT 05602
(802) 828-2765

VIRGIN ISLANDS   Region II
Department of Labor
Government of Virgin Islands
Box 890
Christiansted, St. Croix, VI 00820
(809) 773-1994

VIRGINIA   Region III
Department of Labor and Industry
P. O. Box 12064
Richmond, VA 23241
(804) 786-2376

WASHINGTON   Region X
Department of Labor and Industries
General Administration Building
308 East Fourth Avenue
Room 334—AX 31
Olympia, WA 98504
(206) 753-6307

WEST VIRGINIA   Region III
West Virginia Department of Labor
State Office Building No. 3
Room 319, Capitol Complex

1800 Washington Street, East
Charleston, WV 25305
(304) 348-7890

West Virginia Department of
  Health
1800 Washington Street, East
Charleston, WV 25305
(304) 348-2971

WISCONSIN   Region V
Division of Safety and Buildings
Department of Labor, Industry,
  and Human Relations

P. O. Box 7969
Madison, WI 53707
(608) 266-3151

WYOMING   Region VIII
Department of Occupational Safety
  and Health
604 East 25th Street
Cheyenne, WY 82002
(307) 777-7786

AGENCIES RELATED TO SARA TITLE III

**Federal Emergency Management
  Agency**

500 C Street S.W.
Washington, DC 20472
(202) 646-4600

The Federal Emergency Management Agency (FEMA) coordinates emergency preparedness and response resources at the federal, state, and local levels in preparing for and responding to natural, manmade, and nuclear emergencies. Listed below are the FEMA Regional Training and Education Office addresses and telephone numbers.

Region I: CT, ME, MA, NH, RI,
VT
John W. McCormack POCH
Room 422
Boston, MA 02109
(617) 223-9532

Region II: NJ, NY, PR, VI
26 Federal Plaza
Room 1337
New York, NY 10278
(212) 264-8980

Region III: DC, DE, MD, PA, VA,
WV
Liberty Square Building
2nd Floor
105 South Seventh Street
Philadelphia, PA 19106
(215) 597-5154

Region IV: AL, FL, GA, KY, MS,
NC, SC, TN
1371 Peachtree Street, N.E.
Suite 700
Atlanta, GA 30309
(404) 347-2410

Region V: IL, IN, MI, MN, OH, WI
300 South Wacker Drive
24th Floor
Chicago, IL 60606
(312) 353-3040

Region VI: AR, LA, NM, OK, TX
Federal Regional Center 206
800 North Loop 288
Denton, TX 76201
(817) 898-5262

Region VII: IA, KS, MO, NB
911 Walnut Street
Room 300
Kansas City, MO 64106
(816) 374-6912

Region VIII: CO, MT, ND, SD, UT, WY
Denver Federal Center
Building 710
Box 25267
Denver, CO 80225
(303) 235-4929

Region IX: AZ, CA, HI, NV, American Samoa, Guam
Building 105
Presidio of San Francisco, CA 94129
(415) 963-7108

Region X: AK, ID, OR, WA
Federal Regional Center
130 228th Street, SW
Bothell, WA 98021
(206) 487-4603

Regional Response Teams

Each Regional Response Team (RRT) is cochaired by a representative of the Environmental Protection Agency and an officer of the Coast Guard. Addresses and telephone numbers of the cochairs of the RRTs are given below. Note that there are two RRTs each for EPA Regions II, IX, and X.

Region I: CT, ME, MA, NH, RI, VT
Environmental Services Division
Environmental Protection Agency
Room 2203
JFK Federal Building
Boston, MA 02203
(617) 861-6700

First Coast Guard District
408 Atlantic Avenue
Boston, MA 02110
(617) 223-8444

Region II: NJ, NY
Emergency and Remedial Response
Environmental Protection Agency
Room 900
26 Federal Plaza
New York, NY 10278
(212) 264-2647

First Coast Guard District
408 Atlantic Avenue
Boston, MA 02110
(617) 223-8444

Region II: PR, VI
Office of Emergency and Remedial
  Response
Environmental Protection Agency
Room 900
26 Federal Plaza
New York, NY 10278
(212) 264-2647

Seventh Coast Guard District
Federal Building
51 S.W., 1st Avenue
Miami, FL 33130
(305) 350-5651

Region III: DC, DE, MD, PA, VA,
WV
Superfund Branch
Environmental Protection Agency
6th and Walnut Streets
Philadelphia, PA 19106
(215) 597-8132

Fifth Coast Guard District
Federal Building
431 Crawford Street
Portsmouth, VA 23705
(804) 398-6372

Region IV: AL, FL, GA, KY, MS,
NC, SC, TN
Emergency Response and Control
  Section
Environmental Protection Agency
345 Courtland Street, N.E.
Atlanta, GA 30365
(404) 347-3931

Seventh Coast Guard District
Federal Building
51 S.W. 1st Avenue
Miami, FL 33130
(305) 350-5651

Region V: IL, IN, MI, MN, OH,
WI
Remedial Response Branch
Environmental Protection Agency
230 South Dearborn Street
Chicago, IL 60604
(312) 353-9773

Ninth Coast Guard District
1240 East 9th Street
Cleveland, OH 44199
(216) 522-3944

Region VI: AR, LA, NM, OK, TX
Environmental Services Division
Environmental Protection Agency
1201 Elm Street
Dallas, TX 75270
(214) 767-2697

Eighth Coast Guard District
Hale Boggs Federal Building
500 Camp Street
New Orleans, LA 70130
(504) 589-6271

Region VII: IA, KS, MO, NB
Environmental Services Division
Environmental Protection Agency
25 Funston Road
Kansas City, KS 66115
(913) 236-3720

Second Coast Guard District
1430 Olive Street
St. Louis, MO 63103
(314) 425-4655

Region VIII: CO, MT, ND, SD,
UT, WY
Emergency Response Branch
Environmental Protection Agency
999 18th Street, Suite 500

Denver, CO 80202
(303) 293-1723

Second Coast Guard District
1430 Olive Street
St. Louis, MO 63103
(314) 425-4655

Region IX: AZ, CA, NV
Toxics and Waste Management
  Division
Environmental Protection Agency
215 Fremont Street
San Francisco, CA 94105
(415) 974-7460

Eleventh Coast Guard District
Union Bank Building
400 Oceangate
Long Beach, CA 90822
(213) 590-2301

Region IX: HI, American Samoa,
Guam, Northern Mariana Islands
Toxic and Waste Management
  Division
Environmental Protection Agency
215 Fremont Street
San Francisco, CA 94105
(415) 974-7460

Fourteenth Coast Guard District
Prince Kalanianacle Fed. Bldg.
300 Alameana Blvd.
9th Floor
Honolulu, HI 96850

Region X: ID, OR, WA
Emergency Response Team
Environmental Protection Agency
1200 Sixth Avenue
Seattle, WA 98101
(206) 442-1196

Thirteenth Coast Guard District
Federal Building
915 Second Avenue
Seattle, WA 98174
(206) 442-5233

Region X: AK
Alaska Operations Office
Environmental Protection Agency
Box 19
701 C Street
Anchorage, AK
(907) 271-5083

Seventeenth Coast Guard District
P. O. Box 3-5000
Juneau, AK 99802
(907) 586-7195

State Emergency Response Commissions

ALABAMA   Region IV
Alabama Emergency Response
  Commission
Alabama Department of
  Environmental Management
1751 Federal Drive
Montgomery, AL 36130
(205) 271-7700, (205) 271-7931

ALASKA   Region X
Alaska State Emergency Response
  Commission
P. O. Box O
Juneau, AK 99811
(907) 465-2600

AMERICAN SAMOA   Region IX
American Samoa Emergency

Response Commission
Territory Emergency Management
  Coordination Office
Government of American Samoa
Pago Pago, American Samoa
  96799
(684) 633-2331

ARIZONA   Region IX
Arizona Emergency Response
  Commission
Division of Emergency Service
5636 East McDowell Road
Phoenix, AZ 85008
(602) 244-0504, (602) 262-8012

ARKANSAS   Region VI
Arkansas Hazardous Materials
  Emergency Response
  Commission
P. O. Box 9583
Little Rock, AR 72205
(501) 661-2000, (501) 661-2136

CALIFORNIA   Region IX
Chemical Emergency Planning and
  Response Commission
Office of Emergency Services
Hazardous Material Division
2800 Meadowview Road
Sacramento, CA 95832
(916) 427-4201, (800) 852-7550

COLORADO   Region VIII
Colorado Emergency Planning and
  Community Right-to-Know
  Commission
Division of Disaster Emergency
  Services
Camp George West
1500 Golden Road
Golden, CO 80401
(303) 273-1614, (303) 377-6326

CONNECTICUT   Region I
Connecticut Emergency Response
  Commission
Department of Environmental
  Protection
State Office Building, Room 161
165 Capitol Avenue
Hartford, CT 06106
(203) 566-4017, (203) 566-3338

DELAWARE   Region III
Delaware Commission on
  Hazardous Materials
Department of Public Safety
P. O. Box 818
Dover, DE 19901
(302) 736-4321, (800) 662-8802

DISTRICT OF
COLUMBIA   Region III
Office of Emergency Preparedness
2000 14th Street, NW
8th Floor
Washington, DC 20009
(202) 727-6161

FLORIDA   Region IV
Florida Emergency Response
  Commission
Florida Department of Community
  Affairs
2740 Centerview Drive
Tallahassee, FL 32399
(904) 487-4915, (904) 488-1320

GEORGIA   Region IV
Georgia Emergency Response
  Commission
Georgia Department of Natural
  Resources
205 Butler Street, S.E.
Floyd Towers East
Atlanta, GA 30334
(404) 656-4713, (404) 656-4300

GUAM   Region IX
Guam Emergency Services Office
Territory of Guam
P. O. Box 2877
Agana, Guam 96910
FTS 550-7230

HAWAII   Region IX
Hawaii Emergency Response
    Commission
Hawaii Department of Health
Environmental Epidemiology
    Program
P. O. Box 3378
Honolulu, HI 96801
(808) 548-2076, (808) 548-5832

IDAHO   Region X
Idaho Emergency Response
    Commission
Department of Health & Welfare
Statehouse
Boise, ID 83720
(208) 334-5898, (800) 632-8000

ILLINOIS   Region V
Illinois Emergency Response
    Commission
Illinois Emergency Services and
    Disaster Agency
110 East Adams Street
Springfield, IL 62706
(217) 782-4694

INDIANA   Region V
Indiana Department of
    Environmental Management
Emergency Response Management
5500 West Bradbury Street
Indianapolis, IN 46241
(317) 243-5176

IOWA   Region VII
Iowa Emergency Response Com-

mission
301 East 7th Street
Des Moines, IA 50319
(515) 281-6175, (515) 281-8694

KANSAS   Region VII
State Emergency Response Com-
    mission
Kansas Department of Health and
    Environment
Forbes Field
Topeka, KS 66620
(913) 296-1500

KENTUCKY   Region IV
Kentucky Emergency Response
    Commission
Kentucky Disaster and Emergency
    Services
Boone National Guard Center
Frankfort, KY 40601
(502) 564-8682, (502) 564-7815

LOUISIANA   Region VI
Louisiana Emergency Response
    Commission
Department of Public Safety and
    Correction
Office of Public Safety
P. O. Box 66614
Baton Rouge, LA 70896
(504) 925-6117, (504) 925-6595

MAINE   Region I
Maine Emergency Management
    Agency
State Office Building, Station 72
Augusta, ME 04333
(207) 289-4082, (207) 289-4080

MARYLAND   Region III
Maryland Emergency Management
    and Civil Defense
2 Sudbrook Lane East

East Pikesville, MD 21208
(301) 486-4422

MASSACHUSETTS   Region I
Title III Emergency Response
  Commission
Department of Environmental
  Quality Engineering
One Winter Street
Boston, MA 02108
(617) 292-5993, (800) 525-5555

MICHIGAN   Region V
Michigan Department of Natural
  Resources
Environmental Response Division
Title III Notification
P. O. Box 30028
Lansing, MI 48909
(517) 373-9893, (517) 373-7660

MINNESOTA   Region V
Minnesota Emergency Response
  Commission
Division of Emergency Services
  Management
State Capitol
Room B-5
St. Paul, MN 55155
(612) 296-2233, (612) 778-0800

MISSISSIPPI   Region IV
Mississippi Emergency Response
  Commission
Mississippi Management Agency
P. O. Box 4501
Fondren Station
Jackson, MS 39216
(601) 352-9100, (800) 222-6362

MISSOURI   Region VII
Missouri Emergency Response
  Commission
Missouri Department of Natural

Resources
P. O. Box 3133
Jefferson City, MO 65102
(314) 751-7929, (314) 634-2436

MONTANA   Region VIII
Montana Emergency Response
  Commission
Environmental Sciences Division
Department of Health and
  Environmental Sciences
Cogswell Building A-107
Helena, MT 59620
(406) 444-3948

NEBRASKA   Region VII
Nebraska Emergency Response
  Commission
Nebraska Department of
  Environmental Control
P. O. Box 94877
State House Station
Lincoln, NB 68509
(402) 471-4230, (402) 471-4545

NEVADA   Region IX
Nevada Division of Emergency
  Management
Capitol Complex
2525 South Carson Street
Carson City, NV 89710
(702) 885-4240

NEW HAMPSHIRE   Region I
State Emergency Management
  Agency
State Office Park South
107 Pleasant Street
Concord, NH 03301
(603) 271-2231, (800) 852-3792

NEW JERSEY   Region II
New Jersey Emergency Response
  Commission

SARA Title III Project
Department of Environmental
  Protection
Division of Environmental Quality,
  CN-402
Trenton, NJ 08625
(609) 292-2885

NEW MEXICO   Region VI
New Mexico Emergency Response
  Commission
New Mexico State Police
Hazardous Materials Bureau
P. O. Box 1628
Santa Fe, NM 87504
(505) 827-9226

NEW YORK   Region II
New York Emergency Response
  Commission
New York State Department of
  Environmental Conservation
Bureau of Spill Prevention and
  Response
50 Wolf Road, Room 326
Albany, NY 12233
(518) 457-4107, (800) 457-7362

NORTH CAROLINA   Region IV
North Carolina Emergency
  Response Commission
Division of Emergency
  Management
116 West Jones Street
Raleigh, NC 27603
(919) 733-2126, (800) 662-7956

NORTH DAKOTA   Region VIII
North Dakota State Department of
  Health
P. O. Box 5520
Bismarck, ND 58502
(701) 224-2370

NORTHERN MARIANA
  ISLANDS   Region IX
Director, Civil Defense
Commonwealth of Northern
  Mariana-Saipan
Office of the Governor
Commonwealth of Northern
  Marianas
Saipan, CNMI 96950
(670) 322-9529

OHIO   Region V
Ohio Emergency Response
  Commission
Ohio Environmental Protection
  Agency
Office of Emergency Response
P. O. Box 1049
Columbus, OH 43266
(614) 481-4300, (800) 282-9378

OKLAHOMA   Region VI
Emergency Response Commission
Office of Civil Defense
P. O. Box 53365
Oklahoma City, OK 73152
(405) 521-2481

OREGON   Region X
Oregon Emergency Response
  Commission
c/o State Fire Marshall
3000 Market Street Plaza, Suite
  534
Salem, OR 97310
(503) 378-2885

PENNSYLVANIA   Region III
Pennsylvania Emergency Response
  Commission
SARA Title III Officer
FEMA Response and Recovery
P. O. Box 3321

Harrisburg, PA 17105
(717) 783-8150, (717) 783-8193

PUERTO RICO   Region II
Puerto Rico Emergency Response
   Commission
Environmental Quality Board
P. O. Box 11488
Santurce, Puerto Rico 00910
(809) 722-1175, (809) 722-2173

RHODE ISLAND   Region I
Rhode Island Emergency Response
   Commission
Rhode Island Emergency
   Management Agency
State House M.27
Providence, RI 02903
(401) 421-7333

SOUTH CAROLINA   Region IV
South Carolina Emergency
   Response Commission
Division of Public Safety Programs
Office of the Governor
1205 Pendleton Street
Columbia, SC 29201
(803) 734-0428

SOUTH DAKOTA   Region VIII
South Dakota Response
   Commission
Department of Water and Natural
   Resources
Joe Foss Building
523 East Capitol
Pierre, SD 57501
(605) 773-3151

TENNESSEE   Region IV
Tennessee Emergency Response
   Commission
Tennessee Emergency
   Management Agency

3041 Sidco Drive
Nashville, TN 37204
(615) 252-3300, (800) 322-8362

TEXAS   Region VI
Texas Emergency Response
   Commission
Division of Emergency
   Management
P. O. Box 4087
Austin, TX 78773
(512) 465-2138

UTAH   Region VIII
Utah Hazardous Chemical
   Emergency Response
   Commission
Department of Health
P. O. Box 16690
Salt Lake City, UT 84116
(801) 538-6101

Comprehensive Emergency
   Management
P. O. Box 8100
Salt Lake City, UT 84108
(801) 533-5271

VERMONT   Region I
Vermont Emergency Response
   Commission
Department of Labor and Industry
120 State Street
Montpelier, VT 05602
(802) 828-2286

VIRGINIA   Region III
Virginia Emergency Response
   Council
c/o Virginia Department of Waste
   Management
1205 East Main Street
Richmond, VA 23219
(804) 786-3017

VIRGIN ISLANDS   Region II
U.S. Virgin Islands Emergency
  Response Commission
Title III
179 Altona and Welgunst
Charlotte Amalie
St. Thomas, VI 00802
(809) 774-3320, ext. 169/170

WASHINGTON   Region X
Washington Emergency Response
  Commission
Division of Emergency
  Management
5220 East Martin Way Mail Stop
  PT-11
Olympia, WA 98504
(206) 753-5255

WEST VIRGINIA   Region III
West Virginia Emergency
  Response Commission
Department of Natural Resources

Capitol Building, Room 669
1800 Washington Street, East
Charleston, WV 25305
(304) 348-2754, (304) 348-5380

WISCONSIN   Region V
Department of Administration
Division of Emergency
  Government
P. O. Box 7865
Madison, WI 53707
(608) 266-0119, (608) 266-3232

WYOMING   Region VIII
Wyoming Emergency Response
  Management
Comprehensive Emergency
  Management
P. O. Box 1709
Cheyenne, WY 82003
(307) 777-7566

APPENDIX B.4

Sources of Packaged Training Programs

The following is a list of vendors who have advertised the availability of products designed to meet the training requirements of the Hazard Communication Standard. Some of these firms reportedly have, or soon will have, products designed to assist in meeting the requirements of SARA Title III as well. Types of training materials include film, slide, and video tape. Some suppliers also provide instructors' manuals, trainees' workbooks, information pamphlets, posters, etc.

This list has been compiled from several sources, but it probably is not complete. Some of the firms market their own products, and some market products produced by others. Therefore, it is likely that some products will be available from several sources.

The list is for information only. No endorsement of the products of any of these vendors is expressed or implied by the authors. Before purchasing such products, it is suggested that several vendors be contacted to obtain more detailed information. In some cases it may be possible to review the actual materials prior to making a purchase decision.

Some large corporations are known to have developed generic training programs for their own use and are willing to sell copies to others. If there are large manufacturing corporations located near you, it is suggested that you contact their personnel or safety departments to inquire about the availability of any training material they have. Even if they do not make their own materials available, they may be willing to tell you about their experience with materials offered by others.

Janice Adkins
431 Megan Court
Frederick, MD 21701
(301) 663-3640

American Scientific Products
American Hospital Supply Corp.
8855 McGaw Road
Columbia, MD 21045
(800) 638-4460

J. T. Baker
Office of Training Services
222 Red School Lane
Phillipsburg, NJ 08865
(201) 454-2500

Behr Consulting
366 Veterans Highway
Commack, NY 11725
(516) 864-1458

BNA Communications, Inc.
9439 Key West Avenue
Rockville, MD 20850
(800) 233-6067, (301) 948-0540

Channing L. Bete Co. Inc.
200 State Road South
Deerfield, MA 01373
(413) 665-7611

Bureau of Law & Business, Inc.
64 Wall Street
Madison, CT 06443
(800) 553-4569, (203) 245-7448

Business & Legal Reports
64 Wall Street
Madison, CT 06443-1513
(203) 245-7448

Carnow, Conibear Associates, Ltd.
333 West Wacker Drive
Chicago, IL 60606
(312) 782-4486

Carnow, Conibear Associates, Ltd.
1901 Pennsylvania Avenue, N.W.
Washington, DC 20006
(202) 861-0368

Center for Occupational Hazards
5 Beekman Street
New York, NY 10038
(212) 227-6220

Challenge Education Associates
16 Sylvia Lane
Lincoln, RI 02865
(401) 728-8888

Communications Concepts Inc.
5710 Harvey Way
Lakewood, CA 90713
(213) 420-9195

Comprehensive Loss Management
6601 Shingle Creek Parkway
Suite 800
Minneapolis, MN 55430
(612) 566-7270

Comprehensive Safety
    Compliance, Inc.
4943 Route 8
Gibsonia, PA 15044
(412) 443-2626

Coronet MTI Films & Video
108 Wilmot Road
Deerfield, IL 60015
(800) 255-0208, (312) 940-1260

Creative Media Development Inc.
710 Southwest 9th Avenue
Portland, OR 97205
(503) 223-6794

DuPont Company
F&FP, Training Materials
Barley Mill, P19-1210
Wilmington, DE 19898
(800) 532-7233

Educational Resources, Inc.
5534 Bush River Road
Columbia, SC 29210
(800) 845-8822

ELB/Monitor
605 Eastowne Drive
Chapel Hill, NC 27514
(800) 334-5478

Film Communicators
11136 Weddington Street
North Hollywood, CA 91601
(800) 423-2400, (818) 766-3747

Fireman's Fund Risk Management
    Services, Inc.
P.O. Box 3890
San Rafael, CA 94941
(800) 227-0765

Genium Publishing Corporation
1145 Catalyn Street
Schenectady, NY 12303
(518) 377-8854

Hazard Communication Program
10331 Morning Mist
Fort Wayne, IN 46804

HAZ COMP
17W703 Butterfield Road, Suite G
Oakbrook Terrace, IL 60181
(312) 620-6451

The HAZMAT Resource Inc.
2100 Raybrook, S.E., Suite 109
Grand Rapids, MI 49506
(800) 952-0335

Health Edutech, Inc.
7801 East Bush Lake Road
Suite 350
Minneapolis, MN 55435
(612) 831-0445

Howell Training Company
5227 Langfield Road
Houston, TX 77040-6618
(800) 527-1851

HOWSAFE!
5380 Naiman Parkway
Solon, OH 44139
(216) 349-3600

Human Resource Development
    Press
22 Amherst Road
Amherst, MA 01002
(800) 822-2801

Industrial Training, Inc.
P.O. Box 7186
Grand Rapids, MI 49510
(800) 253-4623

Industrial Training Systems Corp.
20 West Stow Road
Marlton, NJ 08053
(800) 922-0782, (609) 983-7300

Interactive Medical
  Communications
100 Fifth Avenue
Waltham, MA 02154
(617) 890-7707

International Film Bureau Inc.
332 South Michigan Avenue
Chicago, IL 60604
(312) 427-4545

J. J. Keller & Associates
P.O. Box 368
Neenah, WI 54956-0368
(800) 558-5011

Krames Communications
312 90th Street
Daly City, CA 94015-1898
(415) 994-8800

Lab Safety Supply Co.
P.O. Box 1368
Janesville, WI 53547-1368
(800) 356-0783, (608) 754-2345

Labelmaster
5724 North Pulaski Road
Chicago, IL 60646
(800) 621-5808

3M OH&SP Division
Bldg 220-7W
3M Center
St. Paul, MN 55144
(800) 328-1667, (612) 733-6486

MAR-COM Group Ltd
P.O. Box 9557
Wilmington, DE 19808-9557
(800) 654-2448, (302) 764-3400

Mentor Learning Systems, Inc.
1825 De La Cruz Boulevard
Santa Clara, CA 95050
(800) 554-1636

National Audiovisual Center
8700 Edgeworth Drive
Capitol Heights, MD 20743-3701
(301) 763-1896

National Safety Council
444 North Michigan Avenue
Chicago, IL 60611
(312) 527-4800

NATLSCO
K-3
Long Grove, IL 60049
(312) 540-3204

NUS Corporation
Park West Two
Cliff Mine Road
Pittsburgh, PA 15275
(800) 245-2730, (412) 788-1080

Occupational Health Training
  Foundation
120 Tremont Street Suite 321
Boston, MA 02108
(617) 638-8277

PACE Laboratories, Inc.
1710 Douglas Drive North
Minneapolis, MN 55422
(612) 544-5543

Pathfinder Associates, Inc.
P.O. Box 5240
North Muskegon, MI 49445
(616) 744-8462

ProAm Safety Inc.
Software Div.
4943 Route 8
Gibsonia, PA 15044
(800) 351-2477, (412) 443-4010

Professional Association of
  Regulatory Scientists
Suite A5, Hamilton Office Campus
1700 Whitehorse-Hamilton Square
  Road
Hamilton Square, NJ 08690
(609) 890-7277

Program in Occupational Health
Montefiore Medical Center
111 East 210th Street
Bronx, NY 10467
(212) 920-6204

Safety Training Systems
710 Southwest 9th Avenue
Portland, OR 97205
(503) 223-6794

F. I. Scott & Associates
Box 86
Check, VA 24072
(703) 651-3153

Stevens and Associates
11289 Hadley
Overland Park, KS 66210
(913) 451-2842

TEL-A-TRAIN Inc.
Box 4752
Chattanooga, TN 37405
(800) 251-6018, (615) 266-0113

TPC Training Systems
P.O. Box 1030
Barrington, IL 60010
(312) 381-1840

Training Communications Corp.
212 Heather Glen Road
Sterling, VA 22170
(703) 450-0110

VisuCom Productions, Inc.
Box 5472
Redwood City, CA 94063
(415) 364-5566

Roy F. Weston Inc.
Weston Way
West Chester, PA 19380
(215) 692-3030

APPENDIX B.5

Sources of Computer Software for Data Management

The following is a list of vendors who have advertised the availability of computer software designed to manage data needed under either the HCS or SARA Title III. Some of this software contains MSDS data bases, some includes data management characteristics, and some includes both. This list has been compiled from several sources, but it probably is not complete.

The list is for information only. No endorsement of the products of any of these vendors is expressed or implied by the authors. Before purchasing such products, it is suggested that several vendors be contacted to obtain more detailed information. In some cases, it may be possible to review the actual materials prior to making a purchase decision.

Advanced Systems Laboratory Inc.
7137 West Main Street
Lima, New York 14485
(716) 624-3276

Aries Software Corporation
Hurstbourn Place
9300 Shelbyville Road
Suite 603
Louisville, KY 40222
(502) 426-0848

Azimuth Technologies, Inc.
P.O. Box 5787
Pasadena, CA 91107
(818) 405-0300

Benchmark Software Systems, Inc.
P.O. Box 3533
Oakwood, IL 60521
(312) 345-3265

Carlton Industries, Inc.
P.O. Box 280
LaGrange, TX 78945
(409) 242-5055, (800) 231-5988

Clough Management Services
4 Montgomery Street
Rouse's Point, NY 12979-0625
(518) 298-4350

CSS
Post Office Box 3008
Jackson, MI 49203

Datalogix
200 Central Park Avenue
Hartsdale, NY 10530
(914) 997-1627

Envirogenics, Inc.
10 Hillsborough Road
Flemington, NJ 08822
(800) 527-7213

Environmental Health Associates,
   Inc.
520 Third Street
Suite 208
Oakland, CA 94607
(800) 922-4636

ERM Computer Services, Inc.
999 West Chester Pike
West Chester, PA 19382
(215) 431-3800

Flow General, Inc.
7655 Old Springhouse Road
McLean, VA 22102
(703) 893-5915

Fluor-Daniel
Daniel Building
Greenville, SC 29601-2170
(803) 281-4470

Genium Publishing Company
1145 Catalyn Street
Schenectady, NY 12303
(518) 377-8854

HazMat Control Systems, Inc.
5595 East 7th Street
Suite 654
Long Beach, CA 90804
(213) 498-8187

Hazox Corporation
1001 Wilmington Trust Center
11th & Market Street
Wilmington, DE 19801
(800) 558-6942, (215) 388-2030

Health and Hazard Systems
West Foothill Boulevard
Claremont, CA 91711

HRD Press
22 Amherst Road
Amherst, MA 01002
(800) 822-2801

Industrial Health & Safety
   Consultants
65 Walnut Tree Hill Road
Huntington, CT 06484
(203) 929-1131

Instant Reference Sources, Inc.
7605 Rockpoint Drive
Austin, TX 78731

J. J. Keller & Associates, Inc.
P.O. Box 368
Neenah, WI 54946-0368
(800) 558-5001

Labelmaster, Inc.
5724 North Pulaski Road
Chicago, IL 60646
(800) 621-5808, (312) 478-0900

Man Guard Systems, Inc.
25972 Novi Road
Suite 203
Novi, MI 48050
(313) 349-3830

MAR-COM Group Ltd
P.O. Box 9557
Wilmington, DE 19808-9557
(800) 654-2448, (302) 764-3400

MSDS Inc.
2674 East Main Stret
Suite C-107
Ventura, CA 93003-2899
(805) 648-6800

National Safety Council
P.O. Box 11933
Chicago, IL 60611
(312) 527-4800, (800) 621-7619

North Star Data Systems
P.O. Box 7646
St. Paul, MN 55119
(612) 529-7477

NUS Corporation
Park West Two
Cliff Mine Road
Pittsburgh, PA 15275
(800) 245-2730, (412) 788-1080

O.K. REMS Corporation
10 Hillsborough Road
Flemington, NJ 08822
(201) 782-3398

OSHA-Soft Corporation
31 Industrial Park Drive
Concord, NH 03301
(603) 228-3610

Pacific Micro Systems Engineering
6511 Salt Lake Avenue
Bell, CA 90201
(213) 434-0011

Parallax Computer Corporation
3490 Route 1, Building 19-1
Princeton, NJ 08540
(609) 452-8471

PICO Information Services
4217 Main Street
Suite A
Springfield, OR 97478
(503) 726-7823

PRO-AM Safety Inc. Software
   Div.
4943 Route 8
Gibsonia, PA 15044
(800) 351-2477, (412) 443-0410

PROFITMASTER Computer
   Systems, Inc.
11006 Metric Boulevard
Austin, TX 78758
(512) 835-7085

Radian Corporation
Box 9948
Austin, TX 78766
(512) 454-4797

Research Alternatives, Inc.
966 Hungerford Drive
Suite 3
Rockville, MD 20850
(301) 424-2803

Resource Consultants
P. O. Box 1848
Brentwood, TN 37027
(615) 373-5040

Safety Sciences, Inc.
7586 Trade Street
San Diego, CA 92121
(619) 578-8400

Science Information Systems
1815 Ranstead Street
Philadelphia, PA 19103
(215) 972-8150

Techna Corporation
5909 West Michigan Avenue
Ypsilanti, MI 48197
(313) 572-1390

Versar, Inc.
ESM Operations
9200 Rumsey Road
Columbia, MD 21045-1934
(301) 964-9200

VNR Information Services
115 Fifth Avenue
New York, NY 10003
(212) 254-3232

XL Datacorp
P. O. Box 28, Station 16
Denver, CO 80228

APPENDIX C.1

Text of the Hazard Communication Standard
29 CFR 1910.1200

GUIDE TO THE CONTENTS OF THE STANDARD

(a) Purpose
  (1) Ensure evaluation and communication of hazards
  (2) State and federal jurisdiction
(b) Scope and application
  (1) Manufacturers, importers, distributors, manufacturing employers
  (2) General applicability to chemicals
  (3) Applicability to laboratories
  (4) Partial exemptions when containers remain sealed
  (5) Exemptions from labeling requirements
  (6) Excluded materials
(c) Definitions [Article, Assistant Secretary, Chemical, Chemical manufacturer, Chemical name, Combustible liquid, Common name, Compressed gas, Container, Designated representative, Director, Distributor, Employee, Employer, Explosive, Exposure, Flammable, Flashpoint, Foreseeable emergency, Hazard warning, Hazardous chemical, Health hazard, Identity, Immediate use, Importer, Label, Material safety data sheet, Mixture, Organic peroxide, Oxidizer, Physical hazard, Produce, Pyrophoric, Responsible party, Specific chemical identity, Trade secret, Unstable (reactive), Use, Water-reactive, Work area, Workplace]
(d) Hazard determination
  (1) Responsibility

Text of the Standard

§1910.1200   Hazard Communication

(a) *Purpose.* (1) The purpose of this section is to ensure that the hazards of all chemicals produced or imported are evaluated, and that information concerning their hazards is transmitted to employers and employees. This transmittal of information is to be accomplished by means of comprehensive hazard communication programs, which are to include container labeling and other forms of warning, material safety data sheets and employee training.

(2) This occupational safety and health standard is intended to address comprehensively the issue of evaluating the potential hazards of chemicals, and communicating information concerning hazards and appropriate protective measures to employees, and to preempt any legal requirements of a state, or political subdivision of a state, pertaining to the subject. Evaluating the potential hazards of chemicals, and communicating information concerning hazards and appropriate protective measures to employees, may include, for example, but is not limited to, provisions for: developing and maintaining a written hazard communication program for the workplace, including lists of hazardous chemicals present; labeling of containers of chemicals in the workplace, as well as of containers of chemicals being shipped to other workplaces; preparation and distribution of material safety data sheets to employees and downstream employers; and development and implementation of employee training programs regarding hazards of chemicals and protective measures. Under section 18 of the Act, no state or political subdivision of a state may adopt or enforce, through any court or agency, any requirement relating to the issue addressed by this Federal standard, except pursuant to a Federally-approved state plan.

(b) *Scope and application.* (1) This section requires chemical manufacturers or importers to assess the hazards of chemicals which they produce or import, and all employers to provide information to their employees about the hazardous chemicals to which they are exposed, by means of a hazard communication program, labels and other forms of warning, material safety data sheets, and information and training. In addition, this section requires distributors to transmit the required information to employers.

(2) This section applies to any chemical which is known to be present in the workplace in such a manner that employees may be exposed under normal conditions of use or in a foreseeable emergency.

(3) This section applies to laboratories only as follows:

(i) Employers shall ensure that labels on incoming containers of hazardous chemicals are not removed or defaced;

(ii) Employers shall maintain any material safety data sheets that are received with incoming shipments of hazardous chemicals, and ensure that they are readily accessible to laboratory employees; and,

(iii) Employers shall ensure that laboratory employees are apprised of the hazards of the chemicals in their workplaces in accordance with paragraph (h) of this section.

(4) In work operations where employees only handle chemicals in sealed containers which are not opened under normal conditions of use (such as are found in marine cargo handling, warehousing, or retail sales), this section applies to these operations only as follows:

(i) Employers shall ensure that labels on incoming containers of hazardous chemicals are not removed or defaced;

(ii) Employers shall maintain copies of any material safety data sheets that are received with incoming shipments of the sealed containers of hazardous chemicals, shall obtain a material safety data sheet for sealed containers of hazardous chemicals received without a material safety data sheet if an employee requests the material safety data sheet, and shall ensure that the material safety data sheets are readily accessible during each work shift to employees when they are in their work area(s); and,

(iii) Employers shall ensure that employees are provided with information and training in accordance with paragraph (h) of this section (except for the location and availability of the written hazard communication program under paragraph (h)(1)(iii)), to the extent necessary to protect them in the event of a spill or leak of a hazardous chemical from a sealed container.

(5) This section does not require labeling of the following chemicals:

(i) Any pesticide as such term is defined in the Federal Insecticide, Fungicide, and Rodenticide Act (7 U.S.C. 136 et seq.), when subject to the labeling requirements of that Act and labeling regulations issued under that Act by the Environmental Protection Agency;

(ii) Any food, food additive, color additive, drug, cosmetic, or medical or veterinary device, including materials intended for use as ingredients in such products (e.g. flavors and fragrances), as such terms are defined in the Federal Food, Drug, and Cosmetic Act (21 U.S.C. 301 et seq.) and regulations issued under that Act, when they are subject to the labeling requirements under that Act by the Food and Drug Administration;

(iii) Any distilled spirits (beverage alcohols), wine, or malt beverage intended for nonindustrial use, as such terms are defined in the Federal Alcohol Administration Act (27 U.S.C. 201 et seq.) and regulations issued under that Act, when subject to the labeling requirements of that Act and labeling regulations issued under that Act by the Bureau of Alcohol Tobacco, and Firearms; and,

(iv) Any consumer product or hazardous substance as those terms are defined in the Consumer Product Safety Act (15 U.S.C. 2051 et seq.) and Federal Hazardous Substances Act (15 U.S.C. 1261 et seq.) respectively, when subject to a consumer product safety standard or labeling requirement of those Acts, or regulations issued under those Acts by

the Consumer Product Safety Commission.

(6) This section does not apply to:

(i) Any hazardous waste as such term is defined by the Solid Waste Disposal Act, as amended by the Resource Conservation and Recovery Act of 1976, as amended (42 U.S.C. 6901 et seq.), when subject to regulations issued under that Act by the Environmental Protection Agency;

(ii) Tobacco or tobacco products;

(iii) Wood or wood products;

(iv) Articles;

(v) Food, drugs, cosmetics, or alcoholic beverages in a retail establishment which are packaged for sale to consumers;

(vi) Foods, drugs, or cosmetics intended for personal consumption by employees while in the workplace;

(vii) Any consumer product or hazardous substance, as those terms are defined in the Consumer Product Safety Act (15 U.S.C. 2051 et seq.) and Federal Hazardous Substances Act (15 U.S.C. 1261 et seq.) respectively, where the employer can demonstrate it is used in the workplace in the same manner as normal consumer use, and which use results in a duration and frequency of exposure which is not greater than exposures experienced by consumers; and,

(viii) Any drug, as that term is defined in the Federal Food, Drug, and Cosmetic Act (21 U.S.C. 301 et seq.) when it is in solid, final form for direct administration to the patient (i.e. tablets or pills).

(c) *Definitions.*

"Article" means a manufactured item: (i) Which is formed to a specific shape or design during manufacture; (ii) which has end use function(s) dependent in whole or in part upon its shape or design during end use; and (iii) which does not release, or otherwise result in exposure to, a hazardous chemical, under normal conditions of use.

"Assistant Secretary" means the Assistant Secretary of Labor for Occupational Safety and Health, U.S. Department of Labor, or designee.

"Chemical" means any element,

chemical compound or mixture of elements and/or compounds.

"Chemical manufacturer" means an employer with a workplace where chemical(s) are produced for use or distribution.

"Chemical name" means the scientific designation of a chemical in accordance with the nomenclature system developed by the International Union of Pure and Applied Chemistry (IUPAC) or the Chemical Abstracts Service (CAS) rules of nomenclature, or a name which will clearly identify the chemical for the purpose of conducting a hazard evaluation.

"Combustible liquid" means any liquid having a flashpoint at or above 100 °F (37.8 °C), but below 200 ° F (93.3 ° C), except any mixture having components with flashpoints of 200 °F (93.3 °C), or higher, the total volume of which make up 99 percent or more of the total volume of the mixture.

"Common name" means any designation or identification such as code name, code number, trade name, brand name or generic name used to identify a chemical other than by its chemical name.

"Compressed gas" means:

(i) A gas or mixture of gases having, in a container, an absolute pressure exceeding 40 psi at 70 °F (21.1 °C); or

(ii) a gas or mixture of gases having, in a container, an absolute pressure exceeding 104 psi at 130 °F (54.4 °C) regardless of the pressure at 70 °F (21.1 °C); or

(iii) A liquid having a vapor pressure exceeding 40 psi at 100 °F (37.8 °C) as determined by ASTM D–323–72.

"Container" means any bag, barrel, bottle, box, can, cylinder, drum, reaction vessel, storage tank, or the like that contains a hazardous chemical. For purposes of this section, pipes or piping systems, and engines, fuel tanks, or other operating systems in a vehicle, are not considered to be containers.

"Designated representative" means any individual or organization to whom an employee gives written authorization to exercise such employee's rights under this section. A recognized or certified collective bargaining agent shall be treated automatically as a designated representative without regard to written employee authorization.

"Director" means the Director, National Institute for Occupational Safety and Health, U.S. Department of Health and Human Services, or designee.

"Distributor" means a business, other than a chemical manufacturer or importer, which supplies hazardous chemicals to other distributors or to employers.

"Employee" means a worker who may be exposed to hazardous chemicals under normal operating conditions or in foreseeable emergencies. Workers such as office workers or bank tellers who encounter hazardous chemicals only in non-routine, isolated instances are not covered.

"Employer" means a person engaged in a business where chemicals are either used, distributed, or are produced for use or distribution, including a contractor or subcontractor.

"Explosive" means a chemical that causes a sudden, almost instantaneous release of pressure, gas, and heat when subjected to sudden shock, pressure, or high temperature.

"Exposure" or "exposed" means that an employee is subjected to a hazardous chemical in the course of employment through any route of entry (inhalation, ingestion, skin contact or absorption, etc.), and includes potential (e.g. accidental or possible) exposure.

"Flammable" means a chemical that falls into one of the following categories:

(i) "Aerosol, flammable" means an aerosol that, when tested by the method described in 16 CFR 1500.45, yields a flame projection exceeding 18 inches at full valve opening, or a flashback (a flame extending back to the valve) at any degree of valve opening;

(ii) "Gas, flammable" means:

(A) A gas that, at ambient temperature and pressure, forms a flammable mixture with air at a concentration of thirteen (13) percent by volume or less; or

(B) A gas that, at ambient temperature and pressure, forms a range of flammable mixtures with air wider than twelve (12) percent by volume, regardless of the lower limit;

(iii) "Liquid, flammable" means any liquid having a flashpoint below 100 °F (37.8 °C), except any mixture having components with flashpoints of 100 °F (37.8 °C) or higher, the total of which make up 99 percent or more of the total volume of the mixture;

(iv) "Solid, flammable" means a solid, other than a blasting agent or explosive as defined in § 190.109(a), that is liable to cause fire through friction, absorption of moisture, spontaneous chemical change, or retained heat from manufacturing or processing, or which can be ignited readily and when ignited burns so vigorously and persistently as to create a serious hazard. A chemical shall be considered to be a flammable solid if, when tested by the method described in 16 CFR 1500.44, it ignites and burns with a self-sustained flame at a rate greater than one-tenth of an inch per second along its major axis.

"Flashpoint" means the minimum temperature at which a liquid gives off a vapor in sufficient concentration to ignite when tested as follows:

(i) Tagliabue Closed Tester (See American National Standard Method of Test for Flash Point by Tag Closed Tester, Z11.24–1979 (ASTM D 56–79)) for liquids with a viscosity of less than 45 Saybolt University Seconds (SUS) at 100 °F (37.8 °C), that do not contain suspended solids and do not have a tendency to form a surface film under test; or

(ii) Pensky-Martens Closed Tester (See American National Standard Method of Test for Flash Point by Pensky-Martens Closed Tester, Z11.7–1979 (ASTM D 93–79)) for liquids with a viscosity equal to or greater than 45 SUS at 100 °F (37.8 °C), or that contain suspended solids, or that have a tendency to form a surface film under test; or

(iii) Setaflash Closed Tester (see American National Standard Method of Test for Flash Point by Setaflash Closed Tester (ASTM D 3278–78))

Organic peroxides, which undergo autoaccelerating thermal decomposition, are excluded from any of the flashpoint determination methods specified above.

"Foreseeable emergency" means any potential occurrence such as, but not limited to, equipment failure, rupture of containers, or failure of control equipment which could result in an uncontrolled release of a hazardous chemical into the workplace.

"Hazardous chemical" means any chemical which is a physical hazard or a health hazard.

"Hazard warning" means any words, pictures, symbols, or combination thereof appearing on a label or other appropriate form of warning which convey the hazard(s) of the chemical(s) in the container(s).

"Health hazard" means a chemical for which there is statistically significant evidence based on at least one study conducted in accordance with established scientific principles that acute or chronic health effects may occur in exposed employees. The term "health hazard" includes chemicals which are carcinogens, toxic or highly toxic agents, reproductive toxins, irritants, corrosives, sensitizers, hepatotoxins, nephrotoxins, neurotoxins, agents which act on the hematopoietic system, and agents which damage the lungs, skin, eyes, or mucous membranes. Appendix A provides further definitions and explanations of the scope of health hazards covered by this section, and Appendix B describes the criteria to be used to determine whether or not a chemical is to be considered hazardous for purposes of this standard.

"Identity" means any chemical or common name which is indicated on the material safety data sheet (MSDS) for the chemical. The identity used shall permit cross-references to be made among the required list of hazardous chemicals, the label and the MSDS.

"Immediate use" means that the hazardous chemical will be under the control of and used only by the person who transfers it from a labeled

232 HANDBOOK OF RIGHT-TO-KNOW AND EMERGENCY PLANNING

container and only within the work shift in which it is transferred.

"Importer" means the first business with employees within the Customs Territory of the United States which receives hazardous chemicals produced in other countries for the purpose of supplying them to distributors or employers within the United States.

"Label" means any written, printed, or graphic material, displayed on or affixed to containers of hazardous chemicals.

"Material safety data sheet (MSDS)" means written or printed material concerning a hazardous chemical which is prepared in accordance with paragraph (g) of this section.

"Mixture" means any combination of two or more chemicals if the combination is not, in whole or in part, the result of a chemical reaction.

"Organic peroxide" means an organic compound that contains the bivalent -O-O-structure and which may be considered to be a structural derivative of hydrogen peroxide where one or both of the hydrogen atoms has been replaced by an organic radical.

"Oxidizer" means a chemical other than a blasting agent or explosive as defined in § 1910.109(a), that initiates or promotes combustion in other materials, thereby causing fire either of itself or through the release of oxygen or other gases.

"Physical hazard" means a chemical for which there is scientifically valid evidence that it is a combustible liquid, a compressed gas, explosive, flammable, an organic peroxide, an oxidizer, pyrophoric, unstable (reactive) or water-reactive.

"Produce" means to manufacture. process, formulate, or repackage.

"Pyrophoric" means a chemical that will ignite spontaneously in air at a temperature of 130 °F (54.4 °C) or below.

"Responsible party" means someone who can provide additional information on the hazardous chemical and appropriate emergency procedures, if necessary.

"Specific chemical identity" means the chemical name, Chemical Abstracts Service (CAS) Registry Number, or any

other information that reveals the precise chemical designation of the substance.

"Trade secret" means any confidential formula, pattern, process, device, information or compilation of information that is used in an employer's business, and that gives the employer an opportunity to obtain an advantage over competitors who do not know or use it. Appendix D sets out the criteria to be used in evaluating trade secrets.

"Unstable (reactive)" means a chemical which in the pure state, or as produced or transported, will vigorously polymerize, decompose, condense, or will become self-reactive under conditions of shocks, pressure or temperature.

"Use" means to package, handle. react, or transfer.

"Water-reactive" means a chemical that reacts with water to release a gas that is either flammable or presents a health hazard.

"Work area" means a room or defined space in a workplace where hazardous chemicals are produced or used, and where employees are present.

"Workplace" means an establishment, job site, or project, at one geographical location containing one or more work areas.

(d) *Hazard determination.* (1) Chemical manufacturers and importers shall evaluate chemicals produced in their workplaces or imported by them to determine if they are hazardous. Employers are not required to evaluate chemicals unless they choose not to rely on the evaluation performed by the chemical manufacturer or importer for the chemical to satisfy this requirement.

(2) Chemical manufacturers. importers or employers evaluating chemicals shall identify and consider the available scientific evidence concerning such hazards. For health hazards, evidence which is statistically significant and which is based on at least one positive study conducted in accordance with established scientific principles is considered to be sufficient to establish a

hazardous effect if the results of the study meet the definitions of health hazards in this section. Appendix A shall be consulted for the scope of health hazards covered, and Appendix B shall be consulted for the criteria to be followed with respect to the completeness of the evaluation, and the data to be reported.

(3) The chemical manufacturer, importer or employer evaluating chemicals shall treat the following sources as establishing that the chemicals listed in them are hazardous:

(i) 29 CFR Part 1910, Subpart Z, Toxic and Hazardous Substances, Occupational Safety and Health Administration (OSHA); or,

(ii) *Threshold Limit Values for Chemical Substances and Physical Agents in the Work Environment,* American Conference of Governmental Industrial Hygienists (ACGIH) (latest edition).

The chemical manufacturer, importer, or employer is still responsible for evaluating the hazards associated with the chemicals in these source lists in accordance with the requirements of this standard.

(4) Chemical manufacturers, importers and employers evaluating chemicals shall treat the following sources as establishing that a chemical is a carcinogen or potential carcinogen for hazard communication purposes:

(i) National Toxicology Program (NTP), *Annual Report on Carcinogens* (latest edition);

(ii) International Agency for Research on Cancer (IARC) *Monographs* (latest editions); or

(iii) 29 CFR Part 1910, Subpart Z, Toxic and Hazardous Substances, Occupational Safety and Health Administration.

Note.—The *Registry of Toxic Effects of Chemical Substances* published by the National Institute for Occupational Safety and Health indicates whether a chemical has been found by NTP or IARC to be a potential carcinogen.

(5) The chemical manufacturer, importer or employer shall determine

the hazards of mixtures of chemicals as follows:

(i) If a mixture has been tested as a whole to determine its hazards, the results of such testing shall be used to determine whether the mixture is hazardous;

(ii) If a mixture has not been tested as a whole to determine whether the mixture is a health hazard, the mixture shall be assumed to present the same health hazards as do the components which comprise one percent (by weight or volume) or greater of the mixture, except that the mixture shall be assumed to present a carcinogenic hazard if it contains a component in concentrations of 0.1 percent or greater which is considered to be a carcinogen under paragraph (d)(4) of this section;

(iii) If a mixture has not been tested as a whole to determine whether the mixture is a physical hazard, the chemical manufacturer, importer, or employer may use whatever scientifically valid data is available to evaluate the physical hazard potential of the mixture; and,

(iv) If the chemical manufacturer, importer, or employer has evidence to indicate that a component present in the mixture in concentrations of less than one percent (or in the case of carcinogens, less than 0.1 percent) could be released in concentrations which would exceed an established OSHA permissible exposure limit or ACGIH Threshold Limit Value, or could present a health hazard to employees in those concentrations, the mixture shall be assumed to present the same hazard.

(6) Chemical manufacturers, importers, or employers evaluating chemicals shall describe in writing the procedures they use to determine the hazards of the chemical they evaluate. The written procedures are to be made available, upon request, to employees, their designated representatives, the Assistant Secretary and the Director. The written description may be incorporated into the written hazard communication program required under paragraph (e) of this section.

(e) *Written hazard communication*

program. (1) Employers shall develop, implement, and maintain at the workplace, a written hazard communication program for their workplaces which at least describes how the criteria specified in paragraphs (f), (g), and (h) of this section for labels and other forms of warning, material safety data sheets, and employee information and training will be met, and which also includes the following:

(i) A list of the hazardous chemicals known to be present using an identity that is referenced on the appropriate material safety data sheet (the list may be compiled for the workplace as a whole or for individual work areas); and,

(ii) The methods the employer will use to inform employees of the hazards of non-routine tasks (for example, the cleaning of reactor vessels), and the hazards associated with chemicals contained in unlabeled pipes in their work areas.

(2) *Multi-employer workplaces.* Employers who produce, use, or store hazardous chemicals at a workplace in such a way that the employees of other employer(s) may be exposed (for example, employees of a construction contractor working on-site) shall additionally ensure that the hazard communication programs developed and implemented under this paragraph (e) include the following:

(i) The methods the employer will use to provide the other employer(s) with a copy of the material safety data sheet, or to make it available at a central location in the workplace, for each hazardous chemical the other employer(s)' employees may be exposed to while working;

(ii) The methods the employer will use to inform the other employer(s) of any precautionary measures that need to be taken to protect employees during the workplace's normal operating conditions and in foreseeable emergencies; and,

(iii) The methods the employer will use to inform the other employer(s) of the labeling system used in the workplace.

(3) The employer may rely on an existing hazard communication program to comply with these requirements, provided that it meets the criteria established in this paragraph (e).

(4) The employer shall make the written hazard communication program available, upon request, to employees, their designated representatives, the Assistant Secretary and the Director, in accordance with the requirements of 29 CFR 1910.20(e).

(f) *Labels and other forms of warning.* (1) The chemical manufacturer, importer, or distributor shall ensure that each container of hazardous chemicals leaving the workplace is labeled, tagged or marked with the following information:

(i) Identity of the hazardous chemical(s);

(ii) Appropriate hazard warnings; and

(iii) Name and address of the chemical manufacturer, importer, or other responsible party.

(2) For solid metal (such as a steel beam or a metal casting) that is not exempted as an article due to its downstream use, the required label may be transmitted to the customer at the time of the intial shipment, and need not be included with subsequent shipments to the same employer unless the information on the label changes. The label may be transmitted with the initial shipment itself, or with the material safety data sheet that is to be provided prior to or at the time of the first shipment. This exception to requiring labels on every container of hazardous chemicals is only for the solid metal itself and does not apply to hazardous chemicals used in conjunction with, or known to be present with, the metal and to which employees handling the metal may be exposed (for example, cutting fluids or lubricants).

(3) Chemical manufacturers, importers, or distributors shall ensure that each container of hazardous chemicals leaving the workplace is labeled, tagged, or marked in accordance with this section in a manner which does not conflict with the requirements of the Hazardous

Materials Transportation Act (49 U.S.C. 1801 *et seq.*) and regulations issued under that Act by the Department of Transportation.

(4) If the hazardous chemical is regulated by OSHA in a substance-specific health standard, the chemical manufacturer, importer, distributor or employer shall ensure that the labels or other forms of warning used are in accordance with the requirements of that standard.

(5) Except as provided in paragraphs (f)(6) and (f)(7) the employer shall ensure that each container of hazardous chemicals in the workplace is labeled, tagged or marked with the following information:

(i) Identity of the hazardous chemical(s) contained therein; and

(ii) Appropriate hazard warnings.

(6) The employer may use signs, placards, process sheets, batch tickets, operating procedures, or other such written materials in lieu of affixing labels to individual stationary process containers, as long as the alternative method identifies the containers to which it is applicable and conveys the information required by paragraph (f)(5) of this section to be on a label. The written materials shall be readily accessible to the employees in their work area throughout each work shift.

(7) The employer is not required to label portable containers into which hazardous chemicals are transferred from labeled containers, and which are intended only for the immediate use of the employee who performs the transfer.

(8) The employer shall not remove or deface existing labels on incoming containers of hazardous chemicals, unless the container is immediately marked with the required information.

(9) The employer shall ensure that labels or other forms of warning are legible, in English, and prominently displayed on the container, or readily available in the work area throughout each work shift. Employers having employees who speak other languages may add the information in their language to the material presented, as long as the information is presented in English as well.

(10) The chemical manufacturer, importer, distributor or employer need not affix new labels to comply with this section if existing labels already convey the required information.

(g) *Material safety data sheets.* (1) Chemical manufacturers and importers shall obtain or develop a material safety data sheet for each hazardous chemical they produce or import. Employers shall have a material safety data sheet for each hazardous chemical which they use.

(2) Each material safety data sheet shall be in English and shall contain at least the following information:

(i) The identity used on the label, and, except as provided for in paragraph (i) of this section on trade secrets:

(A) If the hazardous chemical is a single substance, its chemical and common name(s);

(B) If the hazardous chemical is a mixture which has been tested as a whole to determine its hazards, the chemical and common name(s) of the ingredients which contribute to these known hazards, and the common name(s) of the mixture itself; or,

(C) If the hazardous chemical is a mixture which has not been tested as a whole:

(*1*) The chemical and common name(s) of all ingredients which have been determined to be health hazards, and which comprise 1% or greater of the composition, except that chemicals identified as carcinogens under paragraph (d)(4) of this section shall be listed if the concentrations are 0.1% or greater; and,

(*2*) The chemical and common name(s) of all ingredients which have been determined to be health hazards, and which comprise less than 1% (0.1% for carcinogens) of the mixture, if there is evidence that the ingredient(s) could be released from the mixture in concentrations which would exceed an established OSHA permissible exposure limit or ACGIH Threshold Limit Value, or could present a health hazard to employees; and,

(*3*) The chemical and common name(s)

of all ingredients which have been determined to present a physical hazard when present in the mixture;

(ii) Physical and chemical characteristics of the hazardous chemical (such as vapor pressure, flash point);

(iii) The physical hazards of the hazardous chemical, including the potential for fire, explosion, and reactivity;

(iv) The health hazards of the hazardous chemical, including signs and symptoms of exposure, and any medical conditions which are generally recognized as being aggravated by exposure to the chemical;

(v) The primary route(s) of entry;

(vi) The OSHA permissible exposure limit, ACGIH Threshold Limit Value, and any other exposure limit used or recommended by the chemical manufacturer, importer, or employer preparing the material safety data sheet, where available;

(vii) Whether the hazardous chemical is listed in the National Toxicology Program (NTP) *Annual Report on Carcinogens* (latest edition) or has been found to be a potential carcinogen in the International Agency for Research on Cancer (IARC) *Monographs* (latest editions), or by OSHA;

(viii) Any generally applicable precautions for safe handling and use which are known to the chemical manufacturer, importer or employer preparing the material safety data sheet, including appropriate hygienic practices, protective measures during repair and maintenance of contaminated equipment, and procedures for clean-up of spills and leaks;

(ix) Any generally applicable control measures which are known to the chemical manufacturer, importer or employer preparing the material safety data sheet, such as appropriate engineering controls, work practices, or personal protective equipment;

(x) Emergency and first aid procedures;

(xi) The date of preparation of the material safety data sheet or the last change to it; and,

(xii) The name, address and telephone number of the chemical manufacturer, importer, employer or other responsible party preparing or distributing the material safety data sheet, who can provide additional information on the hazardous chemical and appropriate emergency procedures, if necessary.

(3) If not relevant information is found for any given category on the material safety data sheet, the chemical manufacturer, importer or employer preparing the material safety data sheet shall mark it to indicate that no applicable information was found.

(4) Where complex mixtures have similar hazards and contents (i.e. the chemical ingredients are essentially the same, but the specific composition varies from mixture to mixture), the chemical manufacturer, importer or employer may prepare one material safety data sheet to apply to all of these similar mixtures.

(5) The chemical manufacturer, importer or employer preparing the material safety data sheet shall ensure that the information recorded accurately reflects the scientific evidence used in making the hazard determination. If the chemical manufacturer, importer or employer preparing the material safety data sheet becomes newly aware of any significant information regarding the hazards of a chemical, or ways to protect against the hazards, this new information shall be added to the material safety data sheet within three months. If the chemical is not currently being produced or imported the chemical manufacturer or importer shall add the information to the material safety data sheet before the chemical is introduced into the workplace again.

(6) Chemical manufacturers or importers shall ensure that distributors and employers are provided an appropriate material safety data sheet with their intitial shipment, and with the first shipment after a material safety data sheet is updated. The chemical manufacturer or importer shall either provide material safety data sheets with the shipped containers or send them to the employer prior to or at the time of

the shipment. If the material safety data sheet is not provided with a shipment that has been labeled as a hazardous chemical, the employer shall obtain one from the chemical manufacturer, importer, or distributor as soon as possible.

(7) Distributors shall ensure that material safety data sheets, and updated information, are provided to other distributors and employers. Retail distributors which sell hazardous chemicals to commercial customers shall provide a material safety data sheet to such employers upon request, and shall post a sign or otherwise inform them that a material safety data sheet is available. Chemical manufacturers, importers, and distributors need not provide material safety data sheets to retail distributors which have informed them that the retail distributor does not sell the product to commercial customers or open the sealed container to use it in their own workplaces.

(8) The employer shall maintain copies of the required material safety data sheets for each hazardous chemical in the workplace, and shall ensure that they are readily accessible during each work shift to employees when they are in their work area(s).

(9) Where employees must travel between workplaces during a workshift, *i.e.*, their work is carried out at more than one geographical location, the material safety data sheets may be kept at a central location at the primary workplace facility. In this situation, the employer shall ensure that employees can immediately obtain the required information in an emergency.

(10) Material safety data sheets may be kept in any form, including operating procedures, and may be designed to cover groups of hazardous chemicals in a work area where it may be more appropriate to address the hazards of a process rather than individual hazardous chemicals. However, the employer shall ensure that in all cases the required information is provided for each hazardous chemical, and is readily accessible during each work shift to employees when they are in in their work areas(s).

(11) Material safety data sheets shall also be made readily available, upon request, to designated representatives and to the Assistant Secretary, in accordance with the requirements of 29 CFR 1910.20 (e). The Director shall also be given access to material safety data sheets in the same manner.

(h) *Employee information and training.* Employers shall provide employees with information and training on hazardous chemicals in their work area at the time of their initial assignment, and whenever a new hazard is introduced into their work area.

(1) *Information.* Employees shall be informed of:

(i) The requirements of this section;

(ii) Any operations in their work area where hazardous chemicals are present; and,

(iii) The location and availability of the written hazard communication program, including the required list(s) of hazardous chemicals, and material safety data sheets required by this section.

(2) *Training.* Employee training shall include at least:

(i) Methods and observations that may be used to detect the presence or release of a hazardous chemical in the work area (such as monitoring conducted by the employer, continuous monitoring devices, visual appearance or odor of hazardous chemicals when being released, etc.);

(ii) The physical and health hazards of the chemicals in the work area;

(iii) The measures employees can take to protect themselves from these hazards, including specific procedures the employer has implemented to protect employees from exposure to hazardous chemicals, such as appropriate work practices, emergency procedures, and personal protective equipment to be used; and,

(iv) The details of the hazard communication program developed by the employer, including an explanation of the labeling system and the material safety data sheet, and how employees can obtain and use the appropriate hazard information.

(i) *Trade secrets.* (1) The chemical manufacturer, importer, or employer may withhold the specific chemical identity, including the chemical name and other specific identification of a hazardous chemical, from the material safety data sheet, provided that:

(i) The claim that the information withheld is a trade secret can be supported;

(ii) Information contained in the material safety data sheet concerning the properties and effects of the hazardous chemical is disclosed;

(iii) The material safety data sheet indicates that the specific chemical identity is being withheld as a trade secret; and,

(iv) The specific chemical identity is made available to health professionals, employees, and designated representatives in accordance with the applicable provisions of this paragraph.

(2) Where a treating physician or nurse determines that a medical emergency exists and the specific chemical identity of a hazardous chemical is necessary for emergency or first-aid treatment, the chemical manufacturer, importer, or employer shall immediately disclose the specific chemical identity of a trade secret chemical to that treating physician or nurse, regardless of the existence of a written statement of need of a confidentiality agreement. The chemical manufacturer, importer, or employer may require a written statement of need and confidentiality agreement, in accordance with the provisions of paragraphs (i)(3) and (4) of this section, as soon as circumstances permit.

(3) In non-emergency situations, a chemical manufacturer, importer, or employer shall, upon request, disclose a specific chemical identity, otherwise permitted to be withheld under paragraph (i)(1) of this section, to a health professional (i.e. physician, industrial hygienist, toxicologist, epidemiologist, or occupational health nurse) providing medical or other occupational health services to exposed employee(s), and to employees or designated representatives, if:

(i) The request is in writing;

(ii) The request describes with reasonable detail one or more of the following occupational health needs for the information:

(A) To assess the hazards of the chemicals to which employees will be exposed;

(B) To conduct or assess sampling of the workplace atmosphere to determine employee exposure levels;

(C) To conduct pre-assignment or periodic medical surveillance of exposed employees;

(D) To provide medical treatment to exposed employees;

(E) To select or assess appropriate personal protective equipment for exposed employees;

(F) To design or assess engineering controls or other protective measures for exposed employees; and,

(G) To conduct studies to determine the health effects of exposure.

(iii) The request explains in detail why the disclosure of the specific chemical identity is essential and that, in lieu thereof, the disclosure of the following information to the health professional, employee, or designated representative, would not satisfy the purposes described in paragraph (i)(3)(ii) of this section:

(A) The properties and effects of the chemical;

(B) Measures for controlling workers' exposure to the chemical;

(C) Methods of monitoring and analyzing worker exposure to the chemical; and,

(D) Methods of diagnosing and treating harmful exposures to the chemical;

(iv) The request includes a description of the procedures to be used to maintain the confidentiality of the disclosed information; and,

(v) The health professional, and the employer or contractor of the services of the health professional (i.e. downstream employer, labor organization, or individual employee), employee, or designated representative, agree in a written confidentiality agreement that the health professional, employee, or designated representative, will not use the trade secret information for any purpose other than the health need(s) asserted and agree not to release the information under any circumstances other than to OSHA, as provided in paragraph (i)(6) of this section, except as authorized by the terms of the agreement or by the chemical manufacturer, importer, or employer.

(4) The confidentiality agreement authorized by paragraph (i)(3)(iv) of this section:

(i) May restrict the use of the information to the health purposes indicated in the written statement of need;

(ii) May provide for appropriate legal remedies in the event of a breach of the agreement, including stipulation of a reasonable pre-estimate of likely damages; and,

(iii) May not include requirements for the posting of a penalty bond.

(5) Nothing in this standard is meant to preclude the parties from pursuing non-contractual remedies to the extent permitted by law.

(6) If the health professional, employee, or designated representative receiving the trade secret information decides that there is a need to disclose it to OSHA, the chemical manufacturer, importer, or employer who provided the information shall be informed by the health professional, employee, or designated representative prior to, or at the same time as, such disclosure.

(7) If the chemical manufacturer, importer, or employer denies a written request for disclosure of a specific chemical identity, the denial must:

(i) Be provided to the health professional, employee, or designated representative, within thirty days of the request;

(ii) Be in writing;

(iii) Include evidence to support the claim that the specific chemical identity is a trade secret;

(iv) State the specific reasons why the request is being denied; and,

(v) Explain in detail how alternative information may satisfy the specific medical or occupational health need without revealing the specific chemical identity.

(8) The health professional, employee, or designated representative whose request for information is denied under paragraph (i)(3) of this section may refer the request and the written denial of the request to OSHA for consideration.

(9) When a health professional, employee, or designated representative refers the denial to OSHA under paragraph (i)(8) of this section, OSHA shall consider the evidence to determine if:

(i) The chemical manufacturer, importer, or employer has supported the claim that the specific chemical identity is a trade secret;

(ii) The health professional, employee, or designated representative has supported the claim that there is a medical or occupational health need for the information; and,

(iii) The health professional, employee, or designated representative has demonstrated adequate means to protect the confidentiality.

(10)(i) If OSHA determines that the specific chemical identity requested under paragraph (i)(3) of this section is not a *bona fide* trade secret, or that it is a trade secret, but the requesting health professional, employee, or designated representative has a legitimate medical or occupational health need for the information, has executed a written confidentiality agreement, and has shown adequate means to protect the confidentiality of the information, the chemical manufacturer, importer, or employer will be subject to citation by OSHA.

(ii) If a chemical manufacturer, importer, or employer demonstrates to OSHA that the execution of a confidentiality agreement would not provide sufficient protection against the potential harm from the unauthorized disclosure of a trade secret specific chemical identity, the Assistant Secretary may issue such orders or impose such additional limitations or conditions upon the disclosure of the requested chemical information as may be appropriate to assure that the occupational health services are provided without an undue risk of harm to the chemical manufacturer, importer, or employer.

(11) If a citation for a failure to release specific chemical identity information is contested by the chemical manufacturer, importer, or employer, the matter will be adjudicated before the Occupational Safety and Health Review Commission in accordance with the Act's enforcement scheme and the applicable Commission rules of procedure. In accordance with the Commission rules, when a chemical manufacturer, importer, or employer continues to withhold the information during the contest, the Administrative Law Judge may review the citation and supporting documentation *in camera* or issue appropriate orders to protect the confidentiality or such matters.

(12) Notwithstanding the existence of a trade secret claim, a chemical manufacturer, importer, or employer shall, upon request, disclose to the Assistant Secretary any information which this section requires the chemical manufacturer, importer, or employer to make available. Where there is a trade secret claim, such claim shall be made no later than at the time the information is provided to the Assistant Secretary so that suitable determinations of trade secret status can be made and the necessary protections can be implemented.

(13) Nothing in this paragraph shall be construed as requiring the disclosure under any circumstances of process or percentage of mixture information which is a trade secret.

(j) *Effective dates.* (1) Chemical manufacturers, importers, and distributors shall ensure that material safety data sheets are provided with the next shipment of hazardous chemicals to employers after September 23, 1987.

(2) Employers in the non-manufacturing sector shall be in compliance with all provisions of this section by May 23, 1988. (Note: Employers in the manufacturing sector (SIC Codes 20 through 39) are already required to be in compliance with this section.)

Appendix A to §1910.1200, Health Hazard Definitions (Mandatory)

Although safety hazards related to the physical characteristics of a chemical can be objectively defined in terms of testing requirements (e.g. flammability), health hazard definitions are less precise and more subjective. Health hazards may cause measurable changes in the body—such as decreased pulmonary function. These changes are generally indicated by the occurrence of signs and symptoms in the exposed employees—such as shortness of breath, a non-measurable, subjective feeling. Employees exposed to such hazards must be apprised of both the change in body function and the signs and symptoms that may occur to signal that change.

The determination of occupational health hazards is complicated by the fact that many of the effects or signs and symptoms occur commonly in non-occupationally exposed populations, so that effects of exposure are difficult to separate from normally occurring illnesses. Occasionally, a substance causes an effect that is rarely seen in the population at large, such as angiosarcomas caused by vinyl chloride exposure, thus making it easier to ascertain that the occupational exposure was the primary causative factor. More often, however, the effects are common, such as lung cancer. The situation is further complicated by the fact that most chemicals have not been adequately tested to determine their health hazard potential, and data do not exist to substantiate these effects.

There have been many attempts to categorize effects and to define them in various ways. Generally, the terms "acute" and "chronic" are used to delineate between effects on the basis of severity or duration. "Acute" effects usually occur rapidly as a result of short-term exposures, and are of short duration. "Chronic" effects generally occur as a result of long-term exposure, and are of long duration.

The acute effects referred to most frequently are those defined by the American National Standards Institute (ANSI) standard for Precautionary Labeling of Hazardous Industrial Chemicals (Z129.1–1982)—irritation, corrosivity, sensitization and lethal dose. Although these are important health effects, they do not adequately cover the considerable range of acute effects which may occur as a result of occupational exposure, such as, for example, narcosis.

Similarly, the term chronic effect is often used to cover only carcinogenicity, teratogenicity, and mutagenicity. These effects are obviously a concern in the workplace, but again, do not adequately cover the area of chronic effects, excluding, for example, blood dyscrasias (such as enemia), chronic bronchitis and liver atrophy.

The goal of defining precisely, in measurable terms, every possible health effect that may occur in the workplace as a result of chemical exposures cannot realistically be accomplished. This does not negate the need for employees to be informed of such effects and protected from them. Appendix B, which is also mandatory, outlines the principles and procedures of hazardous assessment.

For purposes of this section, any chemicals which meet any of the following definitions, as determined by the criteria set forth in Appendix B are health hazards:

1. *Carcinogen:* A chemical is considered to be a carcinogen if:

(a) It has been evaluated by the International Agency for Research on Cancer (IARC), and found to be a carcinogen or potential carcinogen; or

(b) It is listed as a carcinogen or potential carcinogen in the *Annual Report on Carcinogens* published by the National Toxicology Program (NTP) (latest edition); or,

(c) It is regulated by OSHA as a carcinogen.

2. *Corrosive:* A chemical that causes visible destruction of, or irreversible alterations in, living tissue by chemical action at the site of contact. For example, a chemical is considered to be corrosive if, when tested on the intact skin of albino rabbits by the method described by the U.S. Department of Transportation in Appendix A to 49 CFR Part 173, it destroys or changes irreversibly the structure of the tissue at the site of contact following an exposure period of four hours. This term shall not refer to action on inanimate surfaces.

3. *Highly toxic:* A chemical falling within any of the following categories:

(a) A chemical that has a median lethal dose (LD_{50}) of 50 milligrams or less per kilogram of body weight when administered orally to albino rats weighing between 200 and 300 grams each.

(b) A chemical that has a median lethal does (LD_{50}) of 200 milligrams or less per kilogram of body weight when administered by continuous contact for 24 hours (or less if death occurs within 24 hours) with the bare skin of albino rabbits weighing between two and three kilograms each.

(c) A chemical that has a median lethal concentration (LC_{50}) in air of 200 parts per million by volume or less of gas or vapor, or 2 milligrams per liter or less of mist, fume, or dust, when administered by continuous inhalation for one hour (or less if death occurs within one hour) to albino rats weighing between 200 and 300 grams each.

4. *Irritant:* A chemical, which is not corrosive, but which causes a reversible inflammatory effect on living tissue by chemical action at the site of contact. A chemical is a skin irritant if, when tested on the intact skin of albino rabbits by the methods of 16 CFR 1500.41 for four hours exposure or by other appropriate techniques, it results in an empirical score of five or more. A chemical is an eye irritant if so determined under the procedure listed in 16 CFR 1500.42 or other appropriate techniques.

5. *Sensitizer:* A chemical that causes a substantial proportion of exposed people or animals to develop an allergic reaction in normal tissue after repeated exposure to the chemical.

6. *Toxic.* A chemical falling within any of the following categories:

(a) A chemical that has a median lethal dose (LD_{50}) of more than 50 milligrams per kilogram but not more than 500 milligrams per kilogram of body weight when administered orally to albino rats weighing between 200 and 300 grams each.

(b) A chemical that has a median lethal dose (LD_{50}) of more than 200 milligrams per kilogram but not more than 1,000 milligrams per kilogram of body weight when administered by continuous contact for 24 hours (or less if death occurs within 24 hours) with the bare skin of albino rabbits weighing between two and three kilograms each.

(c) A chemical that has a median lethal concentration (LC_{50}) in air of more than 200 parts per million but not more than 2,000 parts per million by volume of gas or vapor, or more than two milligrams per liter but not more than 20 milligrams per liter of mist, fume, or dust, when administered by continuous inhalation for one hour (or less if death occurs within one hour) to albino rats weighing between 200 and 300 grams each.

7. *Target organ effects.* The following is a target organ categorization of effects which may occur, including examples of signs and symptoms and chemicals which have been found to cause such effects. These examples are presented to illustrate the range and diversity of effects and hazards found in the

workplace, and the broad scope employers must consider in this area, but are not intended to be all-inclusive.

a. Hepatotoxins: Chemicals which produce liver damage
    Signs & Symptoms: Jaundice; liver enlargement
    Chemicals: Carbon tetrachloride; nitrosamines
b. Nephrotoxins: Chemicals which produce kidney damage
    Signs & Symptoms: Edema; proteinuria
    Chemicals: Halogenated hydrocarbons; uranium
c. Neurotoxins: Chemicals which produce their primary toxic effects on the nervous system
    Signs & Symptoms: Narcosis; behavioral changes; decrease in motor functions
    Chemicals: Mercury; carbon disulfide
d. Agents which act on the blood or hematopoietic system: Decrease hemoglobin function; deprive the body tissues of oxygen
    Signs & Symptoms: Cyanosis; loss of consciousness
    Chemicals: Carbon monoxide; cyanides
e. Agents which damage the lung: Chemicals which irritate or damage the pulmonary tissue
    Signs & Symptoms: Cough; tightness in chest; shortness of breath
    Chemicals: Silica; asbestos
f. Reproductive toxins: Chemicals which affect the reproductive capabilities including chromosomal damage (mutations) and effects on fetuses (teratogenesis)
    Signs & Symptoms: Birth defects; sterility
    Chemicals: Lead; DBCP
g. Cutaneous hazards: Chemicals which affect the dermal layer of the body
    Signs & Symptoms: Defatting of the skin; rashes; irritation
    Chemicals: Ketones; chlorinated compounds
h. Eye hazards: Chemicals which affect the eye or visual capacity
    Signs & Symptoms: Conjunctivitis; corneal damage
    Chemicals: Organic solvents; acids

Appendix B to §1910.1200, Hazard Determination (Mandatory)

The quality of a hazard communication program is largely dependent upon the adequacy and accuracy of the hazard determination. The hazard determination requirement of this standard is performance-oriented. Chemical manufacturers, importers, and employers evaluating chemicals are not required to follow any specific methods for determining hazards, but they must be able to demonstrate that they have adequately ascertained the hazards of the chemicals produced or imported in accordance with the criteria set forth in this Appendix.

Hazard evaluation is a process which relies heavily on the professional judgment of the evaluator, particularly in the area of chronic hazards. The performance-orientation of the hazard determination does not diminish the duty of the chemical manufacturer, importer or employer to conduct a thorough evaluation, examining all relevant data and producing a scientifically defensible evaluation. For purposes of this standard, the following criteria shall be used in making hazard determinations that meet the requirements of this standard.

1. *Carcinogenicity:* As described in paragraph (d)(4) and Appendix A of this section, a determination by the National Toxicology Program, the International Agency for Research on Cancer, or OSHA that a chemical is a carcinogen or potential carcinogen will be considered conclusive evidence for purposes of this section.

2. *Human data:* Where available, epidemiological studies and case reports of adverse health effects shall be considered in the evaluation.

3. *Animal data:* Human evidence of health effects in exposed populations is generally not available for the majority of chemicals produced or used in the workplace. Therefore, the available results of toxicological testing in animal populations shall be used to predict the health effects that may be experienced by exposed workers. In particular, the definitions of certain acute hazards refer to specific animal testing results (see Appendix A).

4. *Adequacy and reporting of data.* The results of any studies which are designed and conducted according to established scientific principles, and which report statistically significant conclusions regarding the health effects of a chemical, shall be a sufficient basis for a hazard determination and reported on any material safety data sheet. The chemical manufacturer, importer, or employer may also report the results of other scientifically valid studies which tend to refute the findings of hazard.

Appendix C to §1910.1200 Information Sources (Advisory)

The following is a list of available data sources which the chemical manufacturer, importer, distributor, or employer may wish to consult to evaluate the hazards of chemicals they produce or import:

—Any information in their own company files, such as toxicity testing results or illness experience of company employees.

—Any information obtained from the supplier of the chemical, such as material safety data sheets or product safety bulletins.

—Any pertinent information obtained from the following source list (latest editions should be used):

Condensed Chemical Dictionary
  Van Nostrand Reinhold Co., 135 West 50th Street, New York, NY 10020.

The Merck Index: An Encyclopedia of Chemicals and Drugs
  Merck and Company, Inc., 126 E. Lincoln Ave., Rahway, NJ 07065.

IARC Monographs on the Evaluation of the Carcinogenic Risk of Chemicals to Man
  Geneva: World Health Organization, International Agency for Research on Cancer, 1972–Present. (Multivolume work). Summaries are available in supplement volumes. 49 Sheridan Street, Albany, NY 12210.

Industrial Hygiene and Toxicology, by F.A. Patty
  John Wiley & Sons, Inc., New York, NY (Multivolume work).

Clinical Toxicology of Commercial Products
  Gleason, Gosselin, and Hodge

Casarett and Doull's Toxicology; The Basic Science of Poisons
  Doull, Klaassen, and Amdur, Macmillan Publishing Co., Inc., New York, NY.

Industrial Toxicology, by Alice Hamilton and Harriet L. Hardy
  Publishing Sciences Group, Inc., Acton, MA.

Toxicology of the Eye, by W. Morton Grant
  Charles C. Thomas, 301–327 East Lawrence Avenue, Springfield, IL.

Recognition of Health Hazards in Industry
  William A. Burgess, John Wiley and Sons, 605 Third Avenue, New York, NY 10158.

Chemical Hazards of the Workplace
  Nick H. Proctor and James P. Hughes, J.P. Lipincott Company, 6 Winchester Terrace, New York, NY 10022.

Handbook of Chemistry and Physics
  Chemical Rubber Company, 18901 Cranwood Parkway, Cleveland, OH 44128.

Threshold Limit Values for Chemical Substances and Physical Agents in the Work Environment and Biological Exposure Indices with Intended Changes
  American Conference of Governmental Industrial Hygienists (ACGIH), 6500 Glenway Avenue, Bldg. D–5, Cincinnati, OH 45211.

Information on the physical hazards of chemicals may be found in publications of the National Fire Protection Association, Boston, MA.

Note.—The following documents may be purchased from the Superintendent of Documents, U.S. Government Printing Office, Washington, DC 20402.

Occupational Health Guidelines
  NIOSH/OSHA (NIOSH Pub. No. 81–123)
NIOSH Pocket Guide to Chemical Hazards
  NIOSH Pub. No. 85–114
Registry of Toxic Effects of Chemical Substances
  NIOSH Pub. No. 80–102
Miscellaneous Documents published by the National Institute for Occupational Safety and Health:
  Criteria documents.
  Special Hazard Reviews.
  Occupational Hazard Assessments.
  Current Intelligence Bulletins.
OSHA's General Industry Standards (29 CFR Part 1910)
NTP Annual Report on Carcinogens and *Summary of the Annual Report on Carcinogens.*
  National Technical Information Service (NTIS), 5285 Port Royal Road, Springfield, VA 22161; (703) 487–4650.

BIBLIOGRAPHIC DATA BASES

Service provider	File name
Bibliographic Retrieval Services (BRS), 1200 Route 7, Latham, NY 12110.	Biosis Previews CA Search Medlars NTIS Hazardline American Chemical Society Journal Excerpta Medica IRCS Medical Science Journal Pre-Med Intl Pharmaceutical Abstracts Paper Chem
Lockheed—DIALOG Information Service, Inc., 3460 Hillview Avenue, Palo Alto, CA 94304.	Biosis Prev. Files CA Search Files CAB Abstracts Chemical Exposure Chemname Chemsis Files Chemzero Embase Files Environmental Bibliographies Enviroline Federal Research in Progress IRL Life Science Collection NTIS Occupational Safety and Health (NIOSH) Paper Chem
SDC—Orbit, SDC Information Service, 2500 Colorado Avenue, Santa Monica, CA 90406.	CAS Files Chemdex, 2, 3 NTIS

National Library of Medicine, Department of Health and Human Services, Public Health Service, National Institutes of Health, Bethesda, MD 20209.	Hazardous Substances Data Bank (NSDB) Medline files Toxline Files Cancerlit RTECS Chemline Laboratory Hazard Bulletin	
Pergamon International Information Corp., 1340 Old Chain Bridge Rd., McLean, VA 22101.		
Questel, Inc., 1625 Eye Street, NW., Suite 818, Washington, DC 20006	CIS/ILO Cancernet	
Chemical Information System ICI (ICIS), Bureau of National Affairs, 1133 15th Street, NW., Suite 300, Washington, DC 20005	Structure and Nomenclature Search System (SANSS) Acute Toxicity (RTECS) Clinical Toxicology of Commercial Products Oil and Hazardous Materials Technical Assistance Data System CCRIS CESARS	
Occupational Health Services, 400 Plaza Drive, Secaucus, NJ 07094.	MSDS Hazardline	

Appendix D to §1910.1200 Definition of "Trade Secret" (Mandatory)

The following is a reprint of the *Restatement of Torts* section 757, comment *b* (1939):

b. Definition of trade secret. A trade secret may consist of any formula, pattern, device or compilation of information which is used in one's business, and which gives him an opportunity to obtain an advantage over competitors who do not know or use it. It may be a formula for a chemical compound, a process of manufacturing, treating or preserving materials, a pattern for a machine or other device, or a list of customers. It differs from other secret information in a business (see § 759 of the *Restatement of Torts* which is not included in this Appendix) in that it is not simply information as to single or ephemeral events in the conduct of the business, as, for example, the amount or other terms of a secret bid for a contract or the salary of certain employees, or the security investments made or contemplated, or the date fixed for the announcement of a new policy or for bringing out a new model or the like. A trade secret is a process or device for continuous use in the operations of the business. Generally it relates to the production of goods, as, for example, a machine or formula for the production of an article. It may, however, relate to the sale of goods or to other operations in the business, such as a code for determining discounts, rebates or other concessions in a price list or catalogue, or a list of specialized customers, or a method of bookkeeping or other office

management.

Secrecy. The subject matter of a trade secret must be secret. Matters of public knowledge or of general knowledge in an industry cannot be appropriated by one as his secret. Matters which are completely disclosed by the goods which one markets cannot be his secret. Substantially, a trade secret is known only in the particular business in which it is used. It is not requisite that only the proprietor of the business know it. He may, without losing his protection, communicate it to employees involved in its use. He may likewise communicate it to others pledged to secrecy. Others may also know of it independently, as, for example, when they have discovered the process or formula by independent invention and are keeping it secret. Nevertheless, a substantial element of secrecy must exist, so that, except by the use of improper means, there would be difficulty in acquiring the information. An exact definition of a trade secret is not possible. Some factors to be considered in determining whether given information is one's trade secret are: (1) The extent to which the information is known outside of his business; (2) the extent to which it is known by employees and others involved in his business; (3) the extent of measures taken by him to guard the secrecy of the information; (4) the value of the information to him and his competitors; (5) the amount of effort or money expended by him in developing the information; (6) the ease or difficulty with which the information could be properly acquired or duplicated by others.

Novelty and prior art. A trade secret may be a device or process which is patentable; but it need not be that. It may be a device or process which is clearly anticipated in the prior art or one which is merely a mechanical improvement that a good mechanic can make. Novelty and invention are not requisite for a trade secret as they are for patentability. These requirements are essential to patentability because a patent protects against unlicensed use of the patented device or process even by one who discovers it properly through independent research. The patent monopoly is a reward to the inventor. But such is not the case with a trade secret. Its protection is not based on a policy of rewarding or otherwise encouraging the development of secret processes or devices. The protection is merely against breach of faith and reprehensible means of learning another's secret. For this limited protection it is not appropriate to require also the kind of novelty and invention which is a requisite of patentability. The nature of the secret is, however, an important factor in determining the kind of relief that is appropriate against one who is subject to liability under the rule stated in this section. Thus, if the secret consists of a device or process which is a novel invention, one who acquires the secret wrongfully is ordinarily enjoined from further use of it and is required to account for the profits derived from his past use. If, on the other hand, the secret consists of mechanical improvements that a good mechanic can make without resort to the secret, the wrongdoer's liability may be limited to damages, and an injunction against future use of the improvements made with the aid of the secret may be inappropriate.

(Approved by the Office of Management and Budget under Control No. 1218-0072, except for: (1) The requirement that material safety data sheets be provided on multi-employer worksites; (2) coverage of any consumer product excluded from the definition of "hazardous chemical" under Section 311(e)(3) of the Superfund Amendments and Reauthorization Act of 1986; and (3) coverage of any drugs regulated by the Food and Drug Administration in the non-manufacturing sector.)

[53 FR 15035, April 27, 1988]

APPENDIX C.2

Text of SARA Title III

TITLE III — EMERGENCY PLANNING AND COMMUNITY RIGHT-TO-KNOW

Sec. 300. Short Title; Table of Contents.
(a) SHORT TITLE.— This title may be cited as the "Emergency Planning and Community Right-To-Know Act of 1986"
(b) TABLE OF CONTENTS.— The table of contents of this title is as follows:
Sec. 300. Short title; table of contents.

Subtitle A — Emergency Planning and Notification

Sec. 301. Establishment of State Commissions, Planning Districts, and Local Committees.
(a) ESTABLISHMENT OF STATE EMERGENCY RESPONSE COMMISSIONS.— Not later than six months after the date of the enactment of this title, the Governor of each State shall appoint a State emergency response commission. The Governor may designate as the State emergency response commission one or more existing emergency response organizations that are State-sponsored or appointed. The Governor shall, to the extent practicable, appoint persons to the State emergency response commission who have technical expertise in the emergency response field. The State emergency response commission shall appoint local emergency planning committees under subsection (c) and shall supervise and coordinate the activities of such committees. The State emergency response commission shall establish procedures for receiving and processing requests from the public for information under section 324, including tier II information under section 312. Such procedures shall include the designation of an official to serve as coordinator for information. If the Governor of any State does not designate a State emergency response commission within such period, the Governor shall operate as the State emergency response commission until the

247

Governor makes such designation.

(b) ESTABLISHMENT OF EMER-GENCY PLANNING DISTRICTS.— Not later than nine months after the date of the enactment of this title, the State emergency response commission shall designate emergency planning districts in order to facilitate preparation and implementation of emergency plans. Where appropriate, the State emergency response commission may designate existing political subdivisions or multijurisdictional planning organizations as such districts. In emergency planning areas that involve more than one State, the State emergency response commissions of all potentially affected States may designate emergency planning districts and local emergency planning committees by agreement. In making such designation, the State emergency response commission shall indicate which facilities subject to the requirements of this subtitle are within such emergency planning district.

(c) ESTABLISHMENT OF LOCAL EMERGENCY PLANNING COMMITTEES.— Not later than 30 days after designation of emergency planning districts or 10 months after the date of the enactment of this title, whichever is earlier, the State emergency response commission shall appoint members of a local emergency planning committee for each emergency planning district. Each committee shall include, at a minimum, representatives from each of the following groups or organizations: elected State and local officials; law enforcement, civil defense, firefighting, first aid, health, local environmental, hospital, and transportation personnel; broadcast and print media; community groups; and owners and operators of facilities subject to the requirements of this subtitle. Such committee shall appoint a chairperson and shall establish rules by which the committee shall function. Such rules shall include provisions for public notification of committee activities, public meetings to discuss the emergency plan, public comments, response to such comments by the committee, and distribution of the emergency plan. The local emergency planning com-

mittee shall establish procedures for receiving and processing requests from the public for information under section 324, including tier II information under section 312. Such procedures shall include the designation of an official to serve as coordinator for information.

(d) REVISIONS.— A State emergency response commission may revise its designations and appointments under subsections (b) and (c) as it deems appropriate. Interested persons may petition the State emergency response commission to modify the membership of a local emergency planning committee.

Sec. 302. Substances and Facilities Covered and Notification.

(a) Substances Covered.—

(1) In General.— A substance is subject to the requirements of this subtitle if the substance is on the list published under paragraph (2).

(2) List Of Extremely Hazardous Substances.— Within 30 days after the date of the enactment of this title, the Administrator shall publish a list of extremely hazardous substances. The list shall be the same as the list of substances published in November 1985 by the Administrator in Appendix A of the "Chemical Emergency Preparedness Program Interim Guidance".

(3) Thresholds. — (A) At the time the list referred to in paragraph (2) is published the Administrator shall—

(i) publish an interim final regulation establishing a threshold planning quantity for each substance on the list, taking into account the criteria described in paragraph (4), and

(ii) initiate a rulemaking in order to publish final regulations establishing a threshold planning quantity for each substance on the list.

(B) The threshold planning quantities may, at the Administrator's discretion, be based on classes of chemicals or categories of facilities.

(C) If the Administrator fails to publish an interim final regulation establishing a threshold planning quantity for a substance within 30 days after the date of the enactment of this title, the threshold planning quantity for the substance shall be 2 pounds until such time as the Administrator publishes regulations establishing a threshold for the substance.

(4) Revisions. — The Administrator may revise the list and thresholds under paragraphs (2) and (3) from time to time. Any revisions to the list shall take into account the toxicity, reactivity, volatility, dispersability, combustability, or flammability of a substance. For purposes of the preceding sentence, the term "toxicity" shall include any short- or long-term health effect which may result from a short-term exposure to the substance.

(b) Facilities Covered. — (1) Except as provided in section 304, a facility is subject to the requirements of this subtitle if a substance on the list referred to in subsection (a) is present at the facility in an amount in excess of the threshold planning quantity established for such substance.

(2) For purposes of emergency planning, a Governor or a State emergency response commission may designate additional facilities which shall be subject to the requirements of this subtitle, if such designation is made after public notice and opportunity for comment. The Governor or State emergency response commission shall notify the facility concerned of any facility designation under this paragraph.

(c) Emergency Planning Notification. — Not later than seven months after the date of the enactment of this title, the owner or operator of each facility subject to the requirements of this subtitle by reason of subsection (b)(1) shall notify the State emergency response commission for the State in which such facility is located that such facility is subject to the requirements of this subtitle. Thereafter, if a substance on the list of extremely hazardous substances referred to in subsection (a) first becomes present at such facility in excess of the threshold planning quantity established for such substance, or if there is a revision of such list and the facility has present a substance on the revised list in excess of the threshold planning quantity established for such substance, the owner or operator of the facil-

ity shall notify the State emergency response commission and the local emergency planning committee within 60 days after such acquisition or revision that such facility is subject to the requirements of this subtitle.

(d) Notification of Administrator. — The State emergency response commission shall notify the Administrator of facilities subject to the requirements of this subtitle by notifying the Administrator of—

(1) each notification received from a facility under subsection (c), and

(2) each facility designated by the Governor or State emergency response commission under subsection (b)(2).

Sec. 303. Comprehensive Emergency Response Plans.

(a) Plan Required. — Each local emergency planning committee shall complete preparation of an emergency plan in accordance with this section not later than two years after the date of the enactment of this title. The committee shall review such plan once a year, or more frequently as changed circumstances in the community or at any facility may require.

(b) Resources. — Each local emergency planning committee shall evaluate the need for resources necessary to develop, implement, and exercise the emergency plan, and shall make recommendations with respect to additional resources that may be required and the means for providing such additional resources.

(c) Plan Provisions. — Each emergency plan shall include (but is not limited to) each of the following:

(1) Identification of facilities subject to the requirements of this subtitle that are within the emergency planning district, identification of routes likely to be used for the transportation of substances on the list of extremely hazardous substances referred to in section 302(a), and identification of additional facilities contributing or subjected to additional risk due to their proximity to facilities subject to the requirements of this subtitle, such as hospitals or natural gas facilities.

(2) Methods and procedures to be followed by facility owners and operators and local emergency and medical personnel to

respond to any release of such substances.

(3) Designation of a community emergency coordinator and facility emergency coordinators, who shall make determinations necessary to implement the plan.

(4) Procedures providing reliable, effective, and timely notification by the facility emergency coordinators and the community emergency coordinator to persons designated in the emergency plan, and to the public, that a release has occurred (consistent with the emergency notification requirements of section 304).

(5) Methods for determining the occurrence of a release, and the area or population likely to be affected by such release.

(6) A description of emergency equipment and facilities in the community and at each facility in the community subject to the requirements of this subtitle, and an identification of the persons responsible for such equipment and facilities.

(7) Evacuation plans, including provisions for a precautionary evacuation and alternative traffic routes.

(8) Training programs, including schedules for training of local emergency response and medical personnel.

(9) Methods and schedules for exercising the emergency plan.

(d) PROVIDING OF INFORMATION. — For each facility subject to the requirements of this subtitle:

(1) Within 30 days after establishment of a local emergency planning committee for the emergency planning district in which such facility is located, or within 11 months after the date of the enactment of this title, whichever is earlier, the owner or operator of the facility shall notify the emergency planning committee (or the Governor if there is no committee) of a facility representative who will participate in the emergency planning process as a facility emergency coordinator.

(2) The owner or operator of the facility shall promptly inform the emergency planning committee of any relevant changes occurring at such facility as such changes occur or are expected to occur.

(3) Upon request from the emergency planning committee, the owner or operator of the facility shall promptly provide information to such committee necessary for developing and implementing the emergency plan.

(e) REVIEW BY THE STATE EMERGENCY RESPONSE COMMISSION. — After completion of an emergency plan under subsection (a) for an emergency planning district, the local emergency planning committee shall submit a copy of the plan to the State emergency response commission of each State in which such district is located. The commission shall review the plan and make recommendations to the committee on revisions of the plan that may be necessary to ensure coordination of such plan with emergency response plans of other emergency planning districts. To the maximum extent practicable, such review shall not delay implementation of such plan.

(f) GUIDANCE DOCUMENTS. — The national response team, as established pursuant to the National Contingency Plan as established under section 105 of the Comprehensive Environmental Response, Compensation, and Liability Act of 1980 (42 U.S.C. 9601 et seq.), shall publish guidance documents for preparation and implementation of emergency plans. Such documents shall be published not later than five months after the date of the enactment of this title.

(g) REVIEW OF PLANS BY REGIONAL RESPONSE TEAMS. — The regional response teams, as established pursuant to the National Contingency Plan as established under section 105 of the Comprehensive Environmental Response, Compensation, and Liability Act of 1980 (42 U.S.C. 9601 et seq.), may review and comment upon an emergency plan or other issues related to preparation, implementation, or exercise of such a plan upon request of a local emergency planning committee. Such review shall not delay implementation of the plan.

Sec. 304. Emergency Notification.

(a) TYPES OF RELEASES. —

(1) 302(a) SUBSTANCE WHICH REQUIRES CERCLA NOTICE. — If a release of an extremely hazardous sub-

stance referred to in section 302(a) occurs from a facility at which a hazardous chemical is produced, used, or stored, and such release requires a notification under section 103(a) of the Comprehensive Environmental Response, Compensation, and Liability Act of 1980 (hereafter in this section referred to as "CERCLA") (42 U.S.C. 9601 et seq.), the owner or operator of the facility shall immediately provide notice as described in subsection (b).

(2) OTHER 302(a) SUBSTANCE. — If a release of an extremely hazardous substance referred to in section 302(a) occurs from a facility at which a hazardous chemical is produced, used, or stored, and such release is not subject to the notification requirements under section 103(a) of CERCLA, the owner or operator of the facility shall immediately provide notice as described in subsection (b), but only if the release—

(A) is not a federally permitted release as defined in section 101(10) of CERCLA,

(B) is in an amount in excess of a quantity which the Administrator has determined (by regulation) requires notice, and

(C) occurs in a manner which would require notification under section 103(a) of CERCLA.

Unless and until superseded by regulations establishing a quantity for an extremely hazardous substance described in this paragraph, a quantity of 1 pound shall be deemed that quantity the release of which requires notice as described in subsection (b).

3. NON-302(a) SUBSTANCE WHICH REQUIRES CERCLA NOTICE. — If a release of a substance which is not on the list referred to in section 302(a) occurs at a facility at which a hazardous chemical is produced, used, or stored, and such release requires notification under section 103(a) of CERCLA, the owner or operator shall provide notice as follows:

(A) If the substance is one for which a reportable quantity has been established under section 102(a) of CERCLA, the owner or operator shall provide notice as described in subsection (b).

(B) If the substance is one for which a reportable quantity has not been established under section 102(a) of CERCLA—

(i) Until April 30, 1988, the owner or operator shall provide, for releases of one pound or more of the substance, the same notice to the community emergency coordinator for the local emergency planning committee, at the same time and in the same form, as notice is provided to the National Response Center under section 103(a) of CERCLA.

(ii) On and after April 30, 1988, the owner or operator shall provide, for releases of one pound or more of the substance, the notice as described in subsection (b).

4. EXEMPTED RELEASES. — This section does not apply to any release which results in exposure to persons solely within the site or sites on which a facility is located.

(b) NOTIFICATION. —

(1) RECIPIENTS OF NOTICE. — Notice required under subsection (a) shall be given immediately after the release by the owner or operator of a facility (by such means as telephone, radio, or in person) to the community emergency coordinator for the local emergency planning committees, if established pursuant to section 301(c), for any area likely to be affected by the release and to the State emergency planning commission of any State likely to be affected by the release. With respect to transportation of a substance subject to the requirements of this section, or storage incident to such transportation, the notice requirements of this section with respect to a release shall be satisfied by dialing 911 or, in the absence of a 911 emergency telephone number, calling the operator.

(2) CONTENTS. — Notice required under subsection (a) shall include each of the following (to the extent known at the time of the notice and so long as no delay in responding to the emergency results):

(A) The chemical name or identity of

any substance involved in the release.

(B) An indication of whether the substance is on the list referred to in section 302(a).

(C) An estimate of the quantity of any such substance that was released into the environment.

(D) The time and duration of the release.

(E) The medium or media into which the release occurred.

(F) Any known or anticipated acute or chronic health risks associated with the emergency and, where appropriate, advice regarding medical attention necessary for exposed individuals.

(G) Proper precautions to take as a result of the release, including evacuation (unless such information is readily available to the community emergency coordinator pursuant to the emergency plan).

(H) The name and telephone number of the person or persons to be contacted for further information.

(c) FOLLOWUP EMERGENCY NOTICE. — As soon as practicable after a release which requires notice under subsection (a), such owner or operator shall provide a written followup emergency notice (or notices, as more information becomes available) setting forth and updating the information required under subsection (b), and including additional information with respect to—

(1) actions taken to respond to and contain the release,

(2) any known or anticipated acute or chronic health risks associated with the release, and

(3) where appropriate, advice regarding medical attention necessary for exposed individuals.

(d) TRANSPORTATION EXEMPTION NOT APPLICABLE. — The exemption provided in section 327 (relating to transportation) does not apply to this section.

Sec. 305. Emergency Training and Review of Emergency Systems.

(a) EMERGENCY TRAINING.—

(1) PROGRAMS. — Officials of the United States Government carrying out existing Federal programs for emergency training are authorized to specifically provide training and education programs for Federal, State, and local personnel in hazard mitigation, emergency preparedness, fire prevention and control, disaster response, long-term disaster recovery, national security, technological and natural hazards, and emergency processes. Such programs shall provide special emphasis for such training and education with respect to hazardous chemicals.

(2) STATE AND LOCAL PROGRAM SUPPORT. — There is authorized to be appropriated to the Federal Emergency Management Agency for each of the fiscal years 1987, 1988, 1989, and 1990, $5,000,000 for making grants to support programs of State and local governments, and to support university-sponsored programs, which are designed to improve emergency planning, preparedness, mitigation, response, and recovery capabilities. Such programs shall provide special emphasis with respect to emergencies associated with hazardous chemicals. Such grants may not exceed 80 percent of the cost of any such program. The remaining 20 percent of such costs shall be funded from non-Federal sources.

(3) OTHER PROGRAMS. — Nothing in this section shall affect the availability of appropriations to the Federal Emergency Management Agency for any programs carried out by such agency other than the programs referred to in paragraph (2).

(b) REVIEW OF EMERGENCY SYSTEMS.—

(1) REVIEW. — The Administrator shall initiate, not later than 30 days after the date of the enactment of this title, a review of emergency systems for monitoring, detecting, and preventing releases of extremely hazardous substances at representative domestic facilities that produce, use, or store extremely hazardous substances. The Administrator may select representative extremely hazardous substances from the substances on the list referred to in section 302(a) for the purposes of this review. The Administrator shall report interim findings to the Congress not later than seven months after such date of enactment, and issue a final report of findings and recommendations to

the Congress not later than 18 months after such date of enactment. Such report shall be prepared in consultation with the States and appropriate Federal agencies.

(2) REPORT. — The report required by this subsection shall include the Administrator's findings regarding each of the following:

(A) The status of current technological capabilities to (i) monitor, detect, and prevent, in a timely manner, significant releases of extremely hazardous substances, (ii) determine the magnitude and direction of the hazard posed by each release, (iii) identify specific substances, (iv) provide data on the specific chemical composition of such releases, and (v) determine the relative concentrations of the constituent substances.

(B) The status of public emergency alert devices or systems for providing timely and effective public warning of an accidental release of extremely hazardous substances into the environment, including releases into the atmosphere, surface water, or groundwater from facilities that produce, store, or use significant quantities of such extremely hazardous substances.

(C) The technical and economic feasibility of establishing, maintaining, and operating perimeter alert systems for detecting releases of such extremely hazardous substances into the atmosphere, surface water, or groundwater, at facilities that manufacture, use, or store significant quantities of such substances.

(3) RECOMMENDATIONS. — The report required by this subsection shall also include the Administrator's recommendations for—

(A) initiatives to support the development of new or improved technologies or systems that would facilitate the timely monitoring, detection, and prevention of releases of extremely hazardous substances, and

(B) improving devices or systems for effectively alerting the public in a timely manner, in the event of an accidental release of such extremely hazardous substances.

Subtitle B — Reporting Requirements

Sec. 311. Material Safety Data Sheets.

(a) Basic Requirement.—

(1) Submission Of MSDS Or List. — The owner or operator of any facility which is required to prepare or have available a material safety data sheet for a hazardous chemical under the Occupational Safety and Health Act of 1970 and regulations promulgated under that Act (15 U.S.C. 651 et seq.) shall submit a material safety data sheet for each such chemical, or a list of such chemicals as described in paragraph (2), to each of the following:

(A) The appropriate local emergency planning committee.

(B) The State emergency response commission.

(C) The fire department with jurisdiction over the facility.

(2) Contents Of List. — (A) The list of chemicals referred to in paragraph (1) shall include each of the following:

(i) A list of the hazardous chemicals for which a material safety data sheet is required under the Occupational Safety and Health Act of 1970 and regulations promulgated under that Act, grouped in categories of health and physical hazards as set forth under such Act and regulations promulgated under such Act, or in such other categories as the Administrator may prescribe under subparagraph (B).

(ii) The chemical name or the common name of each such chemical as provided on the material safety data sheet.

(iii) Any hazardous component of each such chemical as provided on the material safety data sheet.

(B) For purposes of the list under this paragraph, the Administrator may modify the categories of health and physical hazards as set forth under the Occupational Safety and Health Act of 1970 and regulations promulgated under that Act by requiring information to be reported in terms of groups of hazardous chemicals which present similar hazards in an emergency.

(3) Treatment of mixtures. — An owner or operator may meet the requirements of this section with respect to a hazardous chemical which is a mixture by doing one of the following:

(A) Submitting a material safety data sheet for, or identifying on a list, each element or compound in the mixture which is a hazardous chemical. If more than one mixture has the same element or compound, only one material safety data sheet, or one listing, of the element or compound is necessary.

(B) Submitting a material safety data sheet for, or identifying on a list, the mixture itself.

(b) Thresholds. — The Administrator may establish threshold quantities for hazardous chemicals below which no facility shall be subject to the provisions of this section. The threshold quantities may, in the Administrator's discretion, be based on classes of chemicals or categories of facilities.

(c) Availability of MSDS on Request.—

(1) To local emergency planning committee. — If an owner or operator of a facility submits a list of chemicals under subsection (a)(1), the owner or operator, upon request by the local emergency planning committee, shall submit the material safety data sheet for any chemical on the list to such committee.

(2) To public. — A local emergency planning committee, upon request by any person, shall make available a material safety data sheet to the person in accordance with section 324. If the local emergency planning committee does not have the requested material safety data sheet, the committee shall request the sheet from the facility owner or operator and then make the sheet available to the person in accordance with section 324.

(d) Initial Submission and Updating. — (1) The initial material safety data sheet or list required under this section with respect to a hazardous chemical shall be provided before the later of—

(A) 12 months after the date of the enactment of this title, or

(B) 3 months after the owner or operator of a facility is required to prepare or have available a material safety data sheet for the chemical under the Occupational Safety and Health Act of 1970 and regulations promulgated under that Act.

(2) Within 3 months following discovery by an owner or operator of significant new information concerning an aspect of a hazardous chemical for which a material safety data sheet was previously submitted to the local emergency planning committee under subsection (a), a revised sheet shall be provided to such person.

(e) Hazardous Chemical Defined. — For purposes of this section, the term "hazardous chemical" has the meaning given such term by section 1910.1200(c) of title 29 of the Code of Federal Regulations, except that such term does not include the following:

(1) Any food, food additive, color additive, drug, or cosmetic regulated by the Food and Drug Administration.

(2) Any substance present as a solid in any manufactured item to the extent exposure to the substance does not occur under normal conditions of use.

(3) Any substance to the extent it is used for personal, family, or household purposes, or is present in the same form and concentration as a product packaged for distribution and use by the general public.

(4) Any substance to the extent it is used in a research laboratory or a hospital or other medical facility under the direct supervision of a technically qualified individual.

(5) Any substance to the extent it is used in routine agricultural operations or is a fertilizer held for sale by a retailer to the ultimate customer.

Sec. 312. Emergency and Hazardous Chemical Inventory Forms.

(a) Basic Requirements. — (1) The owner or operator of any facility which is required to prepare or have available a material safety data sheet for a hazardous chemical under the Occupational Safety and Health Act of 1970 and regulations promulgated under that Act shall prepare and submit an emergency and hazardous chemical inventory form (hereafter in this title referred to as an "inventory form") to each of the following:

(A) The appropriate local emergency planning committee.

(B) The State emergency response commission.

(C) The fire department with jurisdiction over the facility.

(2) The inventory form containing tier I information (as described in subsection (d)(1)) shall be submitted on or before March 1, 1988, and annually thereafter on March 1, and shall contain data with respect to the preceding calendar year. The preceding sentence does not apply if an owner or operator provides, by the same deadline and with respect to the same calendar year, tier II information (as described in subsection (d)(2) to the recipients described in paragraph (1).

(3) An owner or operator may meet the requirements of this section with respect to a hazardous chemical which is a mixture by doing one of the following:

(A) Providing information on the inventory form on each element or compound in the mixture which is a hazardous chemical. If more than one mixture has the same element or compound, only one listing on the inventory form for the element or compound at the facility is necessary.

(B) Providing information on the inventory form on the mixture itself.

(b) Thresholds. — The Administrator may establish threshold quantities for hazardous chemicals covered by this section below which no facility shall be subject to the provisions of this section. The threshold quantities may, in the Administrator's discretion, be based on classes of chemicals or categories of facilities.

(c) Hazardous Chemicals Covered. — A hazardous chemical subject to the requirements of this section is any hazardous chemical for which a material safety data sheet or a listing is required under section 311.

(d) CONTENTS OF FORM.—

(1) TIER I INFORMATION.—

(A) Aggregate information by category. — An inventory form shall provide the information described in subparagraph (B) in aggregate terms for hazardous chemicals in categories of health and physical hazards as set forth under the Occupational Safety and Health Act of 1970 and regulations promulgated under that Act.

(B) Required information. — The information referred to in subparagraph (A) is the following:

(i) An estimate (in ranges) of the maximum amount of hazardous chemicals in each category present at the facility at any time during the preceding calendar year.

(ii) An estimate (in ranges) of the average daily amount of hazardous chemicals in each category present at the facility during the preceding calendar year.

(iii) The general location of hazardous chemicals in each category

(C) Modifications.— For purposes of reporting information under this paragraph, the Administrator may—

(i) modify the categories of health and physical hazards as set forth under the Occupational Safety and Health Act of 1970 and regulations promulgated under that Act by requiring information to be reported in terms of groups of hazardous chemicals which present similar hazards in an emergency, or

(ii) require reporting on individual hazardous chemicals of special concern to emergency response personnel.

(2) TIER II INFORMATION.— An inventory form shall provide the following additional information for each hazardous chemical present at the facility, but only upon request and in accordance with subsection (e):

(A) The chemical name or the common name of the chemical as provided on the material safety data sheet.

(B) An estimate (in ranges) of the maximum amount of the hazardous chemical present at the facility at any time during the preceding calendar year.

(C) An estimate (in ranges) of the average daily amount of the hazardous chemical present at the facility during the preceding calendar year.

(D) A brief description of the manner of storage of the hazardous chemical.

(E) The location at the facility of the hazardous chemical.

(F) An indication of whether the owner elects to withhold location information of a specific hazardous chemical from disclosure to the public under section 324.

(e) AVAILABILITY OF TIER II INFORMATION.—

(1) AVAILABILITY TO STATE COMMISSIONS, LOCAL COMMITTEES, AND FIRE DEPARTMENTS.— Upon request by a State emergency planning commission, a local emergency planning committee, or a fire department with jurisdiction over the facility, the owner or operator of a facility shall provide tier II information, as described in subsection (d), to the person making the request. Any such request shall be with respect to a specific facility.

(2) AVAILABILITY TO OTHER STATE AND LOCAL OFFICIALS.— A State or local official acting in his or her official capacity may have access to tier II information by submitting a request to the State emergency response commission or the local emergency planning committee. Upon receipt of a request for tier II information, the State commission or local committee shall, pursuant to paragraph (1), request the facility owner or operator for the tier II information and make available such information to the official.

(3) AVAILABILITY TO PUBLIC.—

(A) IN GENERAL.— Any person may request a State emergency response commission or local emergency planning committee for tier II information relating to the preceding calendar year with respect to a facility. Any such request shall be in writing and shall be with respect to a specific facility.

(B) AUTOMATIC PROVISION OF INFORMATION TO PUBLIC.— Any tier II information which a State emergency response commission or local emergency planning committee has in its possession shall be made available to a person making a request under this paragraph in accordance with section 324. If the state emergency response commission or local emergency planning committee does not

have the tier II information in its possession, upon a request for tier II information the State emergency response commission or local emergency planning committee shall, pursuant to paragraph (1), request the facility owner or operator for tier II information with respect to a hazardous chemical which a facility has stored in an amount in excess of 10,000 pounds present at the facility at any time during the preceding calendar year and make such information available in accordance with section 324 to the person making the request.

(C) DISCRETIONARY PROVISION OF INFORMATION TO PUBLIC.— In the case of tier II information which is not in the possession of a State emergency response commission or local emergency planning committee and which is with respect to a hazardous chemical which a facility has stored in an amount less than 10,000 pounds present at the facility at any time during the preceding calendar year, a request from a person must include the general need for the information. The State emergency response commission or local emergency planning committee may, pursuant to paragraph (1), request the facility owner or operator for the tier II information on behalf of the person making the request. Upon receipt of any information requested on behalf of such person, the State emergency response commission or local emergency planning committee shall make the information available in accordance with section 324 to the person.

(D) RESPONSE IN 45 DAYS.— A State emergency response commission or local emergency planning committee shall respond to a request for tier II information under this paragraph no later than 45 days after the date of receipt of the request.

(f) FIRE DEPARTMENT ACCESS.— Upon request to an owner or operator of a facility which files an inventory form under this section by the fire department with jurisdiction over the facility, the owner or operator of the facility shall allow the fire department to conduct an on-site inspection of the facility and shall provide to the fire department specific location information on hazardous chemicals at the facility.

(g) FORMAT OF FORMS.— The Administrator shall publish a uniform format for inventory forms within three months after the date of the enactment of this title. If the Administrator does not publish such forms, owners and operators of facilities subject to the requirements of this section shall provide the information required under this section by letter.

Sec. 313. Toxic Chemical Release Forms.

(a) BASIC REQUIREMENT. — The owner or operator of a facility subject to the requirements of this section shall complete a toxic chemical release form as published under subsection (g) for each toxic chemical listed under subsection (c) that was manufactured, processed, or otherwise used in quantities exceeding the toxic chemical threshold quantity established by subsection (f) during the preceding calendar year at such facility. Such form shall be submitted to the Administrator and to an official or officials of the State designated by the Governor on or before July 1, 1988, and annually thereafter on July 1 and shall contain data reflecting releases during the preceding calendar year.

(b) COVERED OWNERS AND OPERATORS OF FACILITIES. —

(1) IN GENERAL. — (A) The requirements of this section shall apply to owners and operators of facilities that have 10 or more full-time employees and that are in Standard Industrial Classification Codes 20 through 39 (as in effect on July 1, 1985) and that manufactured, processed, or otherwise used a toxic chemical listed under subsection (c) in excess of the quantity of that toxic chemical established under subsection (f) during the calendar year for which a release form is required under this section.

(B) The Administrator may add or delete Standard Industrial Classification Codes for purposes of subparagraph (A), but only to the extent necessary to provide that each Standard Industrial Code to which this section applies is relevant to the purposes of this section.

(C) For purposes of this section —

(i) The term "manufacture" means to produce, prepare, import, or compound a toxic chemical.

(ii) The term "process" means the preparation of a toxic chemical, after its manufacture, for distribution in commerce —

(I) in the same form or physical state as, or in a different form or physical state from, that in which it was received by the person so preparing such chemical, or

(II) as part of an article containing the toxic chemical.

(2) DISCRETIONARY APPLICATION TO ADDITIONAL FACILITIES. —

The Administrator, on his own motion or at the request of a Governor of a State (with regard to facilities located in that State), may apply the requirements of this section to the owners and operators of any particular facility that manufactures, processes, or otherwise uses a toxic chemical listed under subsection (c) if the Administrator determines that such action is warranted on the basis of toxicity of the toxic chemical, proximity to other facilities that release the toxic chemical or to population centers, the history of releases of such chemical at such facility, or such other factors as the Administrator deems appropriate.

(c) TOXIC CHEMICALS COVERED. — The toxic chemicals subject to the requirements of this section are those chemicals on the list in Committee Print Number 99–169 of the Senate Committee on Environment and Public Works, titled "Toxic Chemicals Subject to Section 313 of the Emergency Planning and Community Right-To-Know Act of 1986" (including any revised version of the list as may be made pursuant to subsection (d) or (e)).

(d) REVISIONS BY ADMINISTRATOR. —

(1) IN GENERAL. — The Administrator may by rule add or delete a chemical from the list described in subsection (c) at any time.

(2) ADDITIONS. — A chemical may be added if the Administrator determines, in his judgment, that there is sufficient evidence to establish any one of the follow-

ing:

(A) The chemical is known to cause or can reasonably be anticipated to cause significant adverse acute human health effects at concentration levels that are reasonably likely to exist beyond facility site boundaries as a result of continuous, or frequently recurring, releases.

(B) The chemical is known to cause or can reasonably be anticipated to cause in humans —

(i) cancer or teratogenic effects, or

(ii) serious or irreversible —

(I) reproductive dysfunctions,

(II) neurological disorders,

(III) heritable genetic mutations, or

(IV) other chronic health effects.

(C) The chemical is known to cause or can reasonably be anticipated to cause, because of —

(i) its toxicity,

(ii) its toxicity and persistence in the environment, or

(iii) its toxicity and tendency to bioaccumulate in the environment,

a significant adverse effect on the environment of sufficient seriousness, in the judgment of the Administrator, to warrant reporting under this section. The number of chemicals included on the list described in subsection (c) on the basis of the preceding sentence may constitute in the aggregate no more than 25 percent of the total number of chemicals on the list. A determination under this paragraph shall be based on generally accepted scientific principles or laboratory tests, or appropriately designed and conducted epidemiological or other population studies, available to the Administrator.

(3) DELETIONS. — A chemical may be deleted if the Administrator determines there is not sufficient evidence to establish any of the criteria described in paragraph (2).

(4) EFFECTIVE DATE. — Any revision made on or after January 1 and before December 1 of any calendar year shall take effect beginning with the next calendar year. Any revision made on or after December 1 of any calendar year and before January 1 of the next calendar year shall take effect beginning with the

calendar year following such next calendar year.

(e) PETITIONS. —

(1) In General. — Any person may petition the Administrator to add or delete a chemical from the list described in subsection (c) on the basis of the criteria in subparagraph (A) or (B) of subsection (d)(2). Within 180 days after receipt of a petition, the Administrator shall take one of the following actions:

(A) Initiate a rulemaking to add or delete the chemical to the list, in accordance with subsection (d)(2) or (d)(3).

(B) Publish an explanation of why the petition is denied.

(2) Governor Petitions. — A State Governor may petition the Administrator to add or delete a chemical from the list described in subsection (c) on the basis of the criteria in subparagraph (A), (B), or (C) of subsection (d)(2). In the case of such a petition from a State Governor to delete a chemical, the petition shall be treated in the same manner as a petition received under paragraph (1) to delete a chemical. In the case of such a petition from a State Governor to add a chemical, the chemical will be added to the list within 180 days after receipt of the petition, unless the Administrator —

(A) initiates a rulemaking to add the chemical to the list, in accordance with subsection (d)(2), or

(B) publishes an explanation of why the Administrator believes the petition does not meet the requirements of subsection (d)(2) for adding a chemical to the list.

(f) Threshold for Reporting. —

(1) Toxic Chemical Threshold Amount. — The threshold amounts for purposes of reporting toxic chemicals under this section are as follows:

(A) With respect to a toxic chemical used at a facility, 10,000 pounds of the toxic chemical per year.

(B) With respect to a toxic chemical manufactured or processed at a facility —

(i) For the toxic chemical release form required to be submitted under this section on or before July 1, 1988, 75,000 pounds of the toxic chemical per year.

(ii) For the form required to be submitted on or before July 1, 1989, 50,000 pounds of the toxic chemical per year.

(iii) For the form required to be submitted on or before July 1, 1990, and for each form thereafter, 25,000 pounds of the toxic chemical per year.

(2) Revisions. — The Administrator may establish a threshold amount for a toxic chemical different from the amount established by paragraph (1). Such revised threshold shall obtain reporting on a substantial majority of total releases of the chemical at all facilities subject to the requirements of this section. The amounts established under this paragraph may, at the Administrator's discretion, be based on classes of chemicals or categories of facilities.

(g) Form. —

(1) Information Required. — Not later than June 1, 1987, the Administrator shall publish a uniform toxic chemical release form for facilities covered by this section. If the Administrator does not publish such a form, owners and operators of facilities subject to the requirements of this section shall provide the information required under this subsection by letter postmarked on or before the date on which the form is due. Such form shall —

(A) provide for the name and location of, and principal business activities at, the facility;

(B) include an appropriate certification, signed by a senior official with management responsibility for the person or persons completing the report, regarding the accuracy and completeness of the report; and

(C) provide for submission of each of the following items of information for each listed toxic chemical known to be present at the facility:

(i) Whether the toxic chemical at the facility is manufactured, processed, or otherwise used, and the general category or categories of use of the chemical.

(ii) An estimate of the maximum amounts (in ranges) of the toxic chemical present at the facility at any time during the preceding calendar year.

(iii) For each wastestream, the waste treatment or disposal methods employed, and an estimate of the treatment efficiency typically achieved by such methods for that wastestream.

(iv) The annual quantity of the toxic chemical entering each environmental medium.

(2) Use of Available Data. — In order to provide the information required under this section, the owner or operator of a facility may use readily available data (including monitoring data) collected pursuant to other provisions of law, or, where such data are not readily available, reasonable estimates of the amounts involved. Nothing in this section requires the monitoring or measurement of the quantities, concentration, or frequency of any toxic chemical released into the environment beyond that monitoring and measurement required under other provisions of law or regulation. In order to assure consistency, the Administrator shall require that data be expressed in common units.

(h) Use of Release Form. — The release forms required under this section are intended to provide information to the Federal, State, and local governments and the public, including citizens of communities surrounding covered facilities. The release form shall be available, consistent with section 324(a), to inform persons about releases of toxic chemicals to the environment; to assist governmental agencies, researchers, and other persons in the conduct of research and data gathering; to aid in the development of appropriate regulations, guidelines, and standards; and for other similar purposes.

(i) Modifications in Reporting Frequency. —

(1) In General. The Administrator may modify the frequency of submitting a report under this section, but the Administrator may not modify the frequency to be any more often than annually. A modification may apply, either nationally or in a specific geographic area, to the following:

(A) All toxic chemical release forms required under this section.

(B) A class of toxic chemicals or a

category of facilities.

(C) A specific toxic chemical.

(D) A specific facility.

(2) Requirements. — A modification may be made under paragraph (1) only if the Administrator —

(A) makes a finding that the modification is consistent with the provisions of subsection (h), based on—

(i) experience from previously submitted toxic chemical release forms, and

(ii) determinations made under paragraph (3), and

(B) the finding is made by a rulemaking in accordance with section 553 of title 5, United States Code.

(3) Determinations. — The Administrator shall make the following determinations with respect to a proposed modification before making a modification under paragraph (1):

(A) The extent to which information relating to the proposed modification provided on the toxic chemical release forms has been used by the Administrator or other agencies of the Federal Government, States, local governments, health professionals, and the public.

(B) The extent to which the information is (i) readily available to potential users from other sources, such as State reporting programs, and (ii) provided to the Administrator under another Federal law or through a State program.

(C) The extent to which the modification would impose additional and unreasonable burdens on facilities subject to the reporting requirements under this section.

(4) 5-year Review. — Any modification made under this subsection shall be reviewed at least once every 5 years. Such review shall examine the modification and ensure that the requirements of paragraphs (2) and (3) still justify continuation of the modification. Any change to a modification reviewed under this paragraph shall be made in accordance with this subsection.

(5) Notification to Congress. — The Administrator shall notify Congress of an intention to initiate a rulemaking for a modification under this subsection. After such notification, the Administrator shall

delay initiation of the rulemaking for at least 12 months, but no more than 24 months, after the date of such notification.

(6) Judicial Review. — In any judicial review of a rulemaking which establishes a modification under this subsection, a court may hold unlawful and set aside agency action, findings, and conclusions found to be unsupported by substantial evidence.

(7) Applicability. — A modification under this subsection may apply to a calendar year or other reporting period beginning no earlier than January 1, 1993.

(8) Effective Date. — Any modification made on or after January 1 and before December 1 of any calendar year shall take effect beginning with the next calendar year. Any modification made on or after December 1 of any calendar year and before January 1 of the next calendar year shall take effect beginning with the calendar year following such next calendar year.

(j) EPA Management of Data. — The Administrator shall establish and maintain in a computer data base a national toxic chemical inventory based on data submitted to the Administrator under this section. The Administrator shall make these data accessible by computer telecommunication and other means to any person on a cost reimbursable basis.

(k) Report. — Not later than June 30, 1991, the Comptroller General, in consultation with the Administrator and appropriate officials in the States, shall submit to the Congress a report including each of the following:

(1) A description of the steps taken by the Administrator and the States to implement the requirements of this section, including steps taken to make information collected under this section available to and accessible by the public.

(2) A description of the extent to which the information collected under this section has been used by the Environmental Protection Agency, other Federal agencies, the States, and the public, and the purposes for which the information has been used.

(3) An identification and evaluation of options for modifications to the require-

ments of this section for the purpose of making information collected under this section more useful.

(l) Mass Balance Study. —

(1) In General. — The Administrator shall arrange for a mass balance study to be carried out by the National Academy of Sciences using mass balance information collected by the Administrator under paragraph (3). The Administrator shall submit to Congress a report on such study no later than 5 years after the date of the enactment of this title.

(2) Purposes. — The purposes of the study are as follows:

(A) To assess the value of mass balance analysis in determining the accuracy of information on toxic chemical releases.

(B) To assess the value of obtaining mass balance information, or portions thereof, to determine the waste reduction efficiency of different facilities, or categories of facilities, including the effectiveness of toxic chemical regulations promulgated under laws other than this title.

(C) To assess the utility of such information for evaluating toxic chemical management practices at facilities, or categories of facilities, covered by this section.

(D) To determine the implications of mass balance information collection on a national scale similar to the mass balance information collection carried out by the Administrator under paragraph (3), including implications of the use of such collection as part of a national annual quantity toxic chemical release program.

(3) Information Collection. — (A) The Administrator shall acquire available mass balance information from States which currently conduct (or during the 5 years after the date of enactment of this title initiate) a mass balance-oriented annual quantity toxic chemical release program. If information from such States provides an inadequate representation of industry classes and categories to carry out the purposes of the study, the Administrator also may acquire mass balance information necessary for the study from a representative number of facilities in other States.

(B) Any information acquired under this section shall be available to the public, except that upon a showing satisfactory to the Administrator by any person that the information (or a particular part threreof) to which the Administrator or any officer, employee, or representative has access under this section if made public would divulge information entitled to protection under section 1905 of title 18, United States Code, such information or part shall be considered confidential in accordance with the purposes of that section, except that such information or part may be disclosed to other officers, employees, or authorized representatives of the United States concerned with carrying out this section.

(C) The Administrator may promulgate regulations prescribing procedures for collecting mass balance information under this paragraph.

(D) For purposes of collecting mass balance information under subparagraph (A), the Administrator may require the submission of information by a State or facility.

(4) Mass Balance Definition. — For purposes of this subsection, the term "mass balance" means an accumulation of the annual quantities of chemicals transported to a facility, produced at a facility, consumed at a facility, used at a facility, accumulated at a facility, released from a facility, and transported from a facility as a waste or as a commercial product or byproduct or component of a commercial product or byproduct.

Subtitle C — General Provisions

Sec. 321. Relationship to Other Law.

(a) In General. — Nothing in this title shall—

(1) preempt any State or local law,

(2) except as provided in subsection (b), otherwise affect any State or local law or the authority of any State or local government to adopt or enforce any State or local law, or

(3) affect or modify in any way the obligations or liabilities of any person under other Federal law.

(b) Effect on MSDS Requirements. —

Any State or local law enacted after August 1, 1985, which requires the submission of a material safety data sheet from facility owners or operators shall require that the data sheet be identical in content and format to the data sheet required under subsection (a) of section 311. In addition, a State or locality may require the submission of information which is supplemental to the information required on the data sheet (including information on the location and quantity of hazardous chemicals present at the facility), through additional sheets attached to the data sheet or such other means as the State or locality considers appropriate.

Sec. 322. Trade Secrets.

(a) Authority To Withhold Information. —

(1) General Authority. — (A) With regard to a hazardous chemical, an extremely hazardous substance, or a toxic chemical, any person required under section 303(d)(2), 303(d)(3), 311, 312, or 313 to submit information to any other person may withhold from such submittal the specific chemical identity (including the chemical name and other specific identification), as defined in regulations prescribed by the Administrator under. subsection (c), if the person complies with paragraph (2).

(B) Any person withholding the specific chemical identity shall, in the place on the submittal where the chemical identity would normally be included, include the generic class or category of the hazardous chemical, extremely hazardous substance, or toxic chemical (as the case may be).

(2) Requirements. — (A) A person is entitled to withhold information under paragraph (1) if such person —

(i) claims that such information is a trade secret, on the basis of the factors enumerated in subsection (b).

(ii) includes in the submittal referred to in paragraph (1) an explanation of the reasons why such information is claimed to be a trade secret, based on the factors enumerated in subsection (b), including a specific description of why such factors apply, and

(iii) submits to the Administrator a copy of such submittal, and the information withheld from such submittal.

(B) In submitting to the Administrator the information required by subparagraph (A)(iii), a person withholding information under this subsection may —

(i) designate, in writing and in such manner as the Administrator may prescribe by regulation, the information which such person believes is entitled to be withheld under paragraph (1), and

(ii) submit such designated information separately from other information submitted under this subsection.

(3) Limitation. — The authority under this subsection to withhold information shall not apply to information which the Administrator has determined, in accordance with subsection (c), is not a trade secret.

(b) Trade Secret Factors. — No person required to provide information under this title may claim that the information is entitled to protection as a trade secret under subsection (a) unless such person shows each of the following:

(1) Such person has not disclosed the information to any other person, other than a member of a local emergency planning committee, an officer or employee of the United States or a State or local government, an employee of such person, or a person who is bound by a confidentiality agreement, and such person has taken reasonable measures to protect the confidentiality of such information and intends to continue to take such measures.

(2) The information is not required to be disclosed, or otherwise made available, to the public under any other Federal or State law.

(3) Disclosure of the information is likely to cause substantial harm to the competitive position of such person.

(4) The chemical identity is not readily discoverable through reverse engineering.

(c) Trade secret regulations. — As soon as practicable after the date of enactment of this title, the Administrator shall prescribe regulations to implement this section. With respect to subsection (b)(4), such regulations shall be equivalent to comparable provisions in the Occupational Safety and Health Administration Hazard Communication Standard (29 C.F.R. 1910.1200) and any revisions of such standard prescribed by the Secretary of Labor in accordance with the final ruling of the courts of the United States in United Steelworkers of America, AFL-CIO-CLC v. Thorne G. Auchter.

(d) Petition for Review. —

(1) In general. — Any person may petition the Administrator for the disclosure of the specific chemical identity of a hazardous chemical, an extremely hazardous substance, or a toxic chemical which is claimed as a trade secret under this section. The Administrator may, in the absence of a petition under this paragraph, initiate a determination, to be carried out in accordance with this subsection, as to whether information withheld constitutes a trade secret.

(2) Initial review. — Within 30 days after the date of receipt of a petition under paragraph (1) (or upon the Administrator's initiative), the Administrator shall review the explanation filed by a trade secret claimant under subsection (a)(2) and determine whether the explanation presents assertions which, if true, are sufficient to support a finding that the specific chemical identity is a trade secret.

(3) Finding of Sufficient Assertions. —

(A) If the Administrator determines pursuant to paragraph (2) that the explanation presents sufficient assertions to support a finding that the specific chemical identity is a trade secret, the Administrator shall notify the trade secret claimant that he has 30 days to supplement the explanation with detailed information to support the assertions.

(B) If the Administrator determines, after receipt of any supplemental supporting detailed information under subparagraph (A), that the assetions in the explanation are true and that the specific chemical identity is a trade secret, the Administrator shall so notify the petitioner and the petitioner may seek judicial review of the determination.

(C) If the Administrator determines, after receipt of any supplemental supporting detailed information under subparagraph (A), that the assertions in the explanation are not true and that the specific chemical identity is not a trade secret, the Administrator shall notify the trade secret claimant that the Administrator intends to release the specific chemical identity. The trade secret claimant has 30 days in which he may appeal the Administrator's determination under this subparagraph to the Administrator. If the Administrator does not reverse his determination under this subparagraph in such an appeal by the trade secret claimant, the trade secret claimant may seek judicial review of the determination.

(4) Finding of insufficient assertions. —

(A) If the Administrator determines pursuant to paragraph (2) that the explanation presents insufficient assertions to support a finding that the specific chemical identity is a trade secret, the Administrator shall notify the trade secret claimant that he has 30 days to appeal the determination to the Administrator, or, upon a showing of good cause, amend the original explanation by providing supplementary assertions to support the trade secret claim.

(B) If the Administrator does not reverse his determination under subparagraph (A) after an appeal or an examination of any supplementary assertions under subparagraph (A), the Administrator shall so notify the trade secret claimant the trade secret claimant may seek judicial review of the determination.

(C) If the Administrator reverses his determination under subparagraph (A) after an appeal or an examination of any supplementary assertions under subparagraph (A), the procedures under paragraph (3) of this subsection apply.

(e) Exception for Information Provided to Health Professionals. — Nothing in this section, or regulations adopted pursuant to this section, shall authorize any person to withhold information which is required to be provided to a health professional, a doctor, or a nurse in accordance with section 323.

(f) Providing Information to the Administrator; Availability to Public. — Any information submitted to the Administrator under subsection (a)(2) or subsection (d)(3) (except a specific chemical identity) shall be available to the public, except that upon a showing satisfactory to the Administrator by any person that the information (or a particular part thereof) to which the Administrator has access under this section if made public would divulge information entitled to protection under section 1905 of title 18, United States Code, such information or part shall be considered confidential in accordance with the purposes of that section, except that such information or part may be disclosed to other officers, employees, or authorized representatives of the United States concerned with carrying out this title.

(g) Information Provided to State. — Upon request by a State, acting through the Governor of the State, the Administrator shall provide to the State any information obtained under subsection (a)(2) and subsection (d)(3).

(h) Information on Adverse Effects. — (1) In any case in which the identity of a hazardous chemical or an extremely hazardous substance is claimed as a trade secret, the Governor or State emergency response commission established under section 301 shall identify the adverse health effects associated with the hazardous chemical or extremely hazardous substance and shall assure that such information is provided to any person requesting information about such hazardous chemical or extremely hazardous substance.

(2) In any case in which the identity of a toxic chemical is claimed as a trade secret, the Administrator shall identify the adverse health and environmental effects associated with the toxic chemical and shall assure that such information is included in the computer database required by section 313(j) and is provided to any person requesting information about such toxic chemical.

(i) Information Provided to Congress. — Notwithstanding any limitation contained in this section or any other provision of law, all information reported to or otherwise obtained by the Administrator (or any representative of the Administrator) under this title shall be made available to a duly authorized committee of the Congress upon written request by such a committee.

SEC. 323. PROVISION OF INFORMATION TO HEALTH PROFESSIONALS, DOCTORS, AND NURSES.

(a) Diagnosis or Treatment by Health Professional. — An owner or operator of a facility which is subject to the requirements of section 311, 312, or 313 shall provide the specific chemical identity, if known, of a hazardous chemical, extremely hazardous substance, or a toxic chemical to any health professional who requests such information in writing if the health professional provides a written statement of need under this subsection and a written confidentiality agreement under subsection (d). The written statement of need shall be a statement that the health professional has a reasonable basis to suspect that—

(1) the information is needed for purposes of diagnosis or treatment of an individual,

(2) the individual or individuals being diagnosed or treated have been exposed to the chemical concerned, and

(3) knowledge of the specific chemical identity of such chemical will assist in diagnosis or treatment.

Following such a written request, the owner or operator to whom such request is made shall promptly provide the requested information to the health professional. The authority to withhold the specific chemical identity of a chemical under section 322 when such information is a trade secret shall not apply to information required to be provided under this subsection, subject to the provisions of subsection (d).

(b) Medical Emergency. — An owner or operator of a facility which is subject to the requirements of section 311, 312, or 313 shall provide a copy of a material safety data sheet, an inventory form, or a toxic chemical release form, including the specific chemical identity, if known, of a hazardous chemical, extremely hazardous substance, or a toxic chemical, to any treating physician or nurse who requests such information if such physician or nurse determines that—

(1) a medical emergency exists,

(2) the specific chemical identity of the chemical concerned is necessary for or will assist in emergency or first-aid diagnosis or treatment, and

(3) the individual or individuals being diagnosed or treated have been exposed to the chemical concerned.

Immediately following such a request, the owner or operator to whom such request is made shall provide the requested information to the physician or nurse. The authority to withhold the specific chemical identity of a chemical from a material safety data sheet, an inventory form, or a toxic chemical release form under section 322 when such information is a trade secret shall not apply to information required to be provided to a treating physician or nurse under this subsection. No written confidentiality agreement or statement of need shall be required as a precondition of such disclosure, but the owner or operator disclosing such information may require a written confidentiality agreement in accordance with subsection (d) and a statement setting forth the items listed in paragraphs (1) through (3) as soon as circumstances permit.

(c) Preventive Measures by Local Health Professionals.—

(1) Provision of Information.— An owner or operator of a facility subject to the requirements of section 311, 312, or 313 shall provide the specific chemical identity, if known, of a hazardous chemical, an extremely hazardous substance, or a toxic chemical to any health professional (such as a physician, toxicologist, or epidemiologist)—

(A) who is a local government employee or a person under contract with the local government, and

(B) who requests such information in writing and provides a written statement of need under paragraph (2) and a written confidentiality agreement under subsection (d).

Following such a written request, the owner or operator to whom such request is made shall promptly provide the requested information to the local health professional. The authority to withhold the specific chemical identity of a chemical under section 322 when such information is a trade secret shall not apply to information required to be provided under this subsection, subject to the provisions of subsection (d).

(2) Written statement of need. — The written statement of need shall be a statement that describes with reasonable detail one or more of the following health needs for the information:

(A) To assess exposure of persons living in a local community to the hazards of the chemical concerned.

(B) To conduct or assess sampling to determine exposure levels of various population groups.

(C) To conduct periodic medical surveillance of exposed population groups.

(D) To provide medical treatment to exposed individuals or population groups.

(E) To conduct studies to determine the health effects of exposure.

(F) To conduct studies to aid in the identification of a chemical that may reasonably be anticipated to cause an observed health effect.

(d) Confidentiality Agreement. — Any person obtaining information under subsection (a) or (c) shall, in accordance with such subsection (a) or (c), be required to agree in a written confidentiality agreement that he will not use the information for any purpose other than the health needs asserted in the statement of need, except as may otherwise be authorized by the terms of the agreement or by the person providing such information. Nothing in this subsection shall preclude the parties to a confidentiality agreement from pursuing any remedies to the extent permitted by law.

(e) REGULATIONS.— As soon as practicable after the date of the enactment of this title, the Administrator shall promulgate regulations describing criteria and parameters for the statement of need under subsection (a) and (c) and the confidentiality agreement under subsection (d).

Sec. 324. Public Availability of Plans, Data Sheets, Forms, and Followup Notices.

(a) AVAILABILITY TO PUBLIC.— Each emergency response plan, material safety data sheet, list described in section 311(a)(2), inventory form, toxic chemical release form, and followup emergency notice shall be made available to the general public, consistent with section 322, during normal working hours at the location or locations designated by the Administrator, Governor, State emergency response commission, or local emergency planning committee, as appropriate. Upon request by an owner or operator of a facility subject to the requirements of section 312, the State emergency response commission and the appropriate local emergency planning committee shall withhold from disclosure under this section the location of any specific chemical required by section 312(d)(2) to be continued in an inventory form as tier II information.

(b) NOTICE OF PUBLIC AVAILABILITY.— Each local emergency planning committee shall annually publish a notice in local newspapers that the emergency response plan, material safety data sheets, and inventory forms have been submitted under this section. The notice shall state that followup emergency notices may subsequently be issued. Such notice shall announce that members of the public who wish to review any such plan, sheet, form, or followup notice may do so at the location designated under subsection (a).

Sec. 325. Enforcement.

(a) CIVIL PENALTIES FOR EMERGENCY PLANNING.— The Administrator may order a facility owner or operator (except an owner or operator of a facility designated under section 302(b)(2)) to comply with section 302(c) and section 303(d). The United States district court for the district in which the facility is located shall have jurisdiction to enforce the order, and any person who violates or fails to obey such an order shall be liable to the United States for a civil penalty of not more than $25,000 for each day in which such violation occurs or such

failure to comply continues.

(b) CIVIL, ADMINISTRATIVE, AND CRIMINAL PENALTIES FOR EMERGENCY NOTIFICATION.—

(1) CLASS I ADMINISTRATIVE PENALTY.— (A) A civil penalty of not more than $25,000 per violation may be assessed by the Administrator in the case of a violation of the requirements of section 304.

(B) No civil penalty may be assessed under this subsection unless the person accused of the violation is given notice and opportunity for a hearing with respect to the violation.

(C) In determining the amount of any penalty assessed pursuant to this subsection, the Administrator shall take into account the nature, circumstances, extent and gravity of the violation or violations and, with respect to the violator, ability to pay, any prior history of such violations, the degree of culpability, economic benefit or savings (if any) resulting from the violation, and such other matters as justice may require.

(2) CLASS II ADMINISTRATIVE PENALTY.— A civil penalty of not more than $25,000 per day for each day during which the violation continues may be assessed by the Administrator in the case of a violation of the requirements of section 304. In the case of a second or subsequent violation the amount of such penalty may be not more than $75,000 for each day during which the violation continues. Any civil penalty under this subsection shall be assessed and collected in the same manner, and subject to the same provisions, as in the case of civil penalties assessed and collected under section 16 of the Toxic Substances Control Act. In any proceeding for the assessment of a civil penalty under this subsection the Administrator may issue subpoenas for the attendance and testimony of witnesses and the production of relevant papers, books, and documents and may promulgate rules for discovery procedures.

(3) JUDICIAL ASSESSMENT.— The Administrator may bring an action in the United States District court for the appropriate district to assess and collect a

penalty of not more than $25,000 per day for each day during which the violation continues in the case of a violation of the requirements of section 304. In the case of a second or subsequent violation, the amount of such penalty may be not more than $75,000 for each day during which the violation continues.

(4) CRIMINAL PENALTIES.— Any person who knowingly and willfully fails to provide notice in accordance with section 304 shall, upon conviction, be fined not more than $25,000 or imprisoned for not more than two years, or both (or in the case of a second or subsequent conviction, shall be fined not more than $50,000 or imprisoned for not more than five years, or both).

(c) Civil and Administrative Penalties for Reporting Requirements. — (1) Any person (other than a governmental entity) who violates any requirement of section 312 or 313 shall be liable to the United States for a civil penalty in an amount not to exceed $25,000 for each such violation.

(2) Any person (other than a governmental entity) who violates any requirement of section 311 or 323(b), and any person who fails to furnish to the Administrator information required under section 322(a)(2) shall be liable to the United States for a civil penalty in an amount not to exceed $10,000 for each such violation.

(3) Each day a violation described in paragraph (1) or (2) continues shall, for purposes of this subsection, continues a separate violation.

(4) The Administrator may assess any civil penalty for which a person is liable under this subsection by administrative order or may bring an action to assess and collect the penalty in the United States district court for the district in which the person from whom the penalty is sought resides or in which such person's principal place of business is located.

(d) Civil, Administrative, and Criminal Penalties With Respect to Trade Secrets.—

(1) Civil and Administrative Penalty for Frivolous Claims. — If the Administrator determines —

(A)(i) under section 322(d)(4) that an explanation submitted by a trade secret claimant presents insufficient assertions to support a finding that a specific chemical identity is a trade secret, or (ii) after receiving supplemental supporting detailed information under section 322(d)(3)(A), that the specific chemical identity is not a trade secret; and

(B) that the trade secret claim is frivolous,

the trade secret claimant is liable for a penalty of $25,000 per claim. The Administrator may assess the penalty by administrative order or may bring an action in the appropriate district court of the United States to assess and collect the penalty.

(2) Criminal Penalty for Disclosure of Trade Secret Information. — Any person who knowingly and willfully divulges or discloses any information entitled to protection under section 322 shall, upon conviction, be subject to fine of not more than $20,000 or to imprisonment not to exceed one year, or both.

(e) Special Enforcement Provisions for Section 323. — Whenever any facility owner or operator required to provide information under section 323 to a health professional who has requested such information fails or refuses to provide such information in accordance with such section, such health professional may bring an action in the appropriate United States district court to require such facility owner or operator to provide the information. Such court shall have jurisdiction to issue such orders and take such other action as may be necessary to enforce the requirements of section 323.

(f) Procedures for Administrative Penalties. —

(1) Any person against whom a civil penalty is assessed under this section may obtain review thereof in the appropriate district court of the United States by filing a notice of appeal in such court within 30 days after the date of such order and by simultaneously sending a copy of such notice by certified mail to the Administrator. The Administrator shall promptly file in such court a certified copy of the record upon which such violation was found or such penalty imposed. If any person fails

to pay an assessment of a civil penalty after it has become a final and unappealable order or after the appropriate court has entered final judgment in favor of the United States, the Administrator may request the Attorney General of the United States to institute a civil action in an appropriate district court of the United States to institute a civil action in an appropriate district court of the United decide any such action. In hearing such action, the court shall have authority to review the violation and the assessment of the civil penalty on the record.

(2) The Administrator may issue subpoenas for the attendance and testimony of witnesses and the production of relevant papers, books, or documents in connection with hearings under this section. In case of contumacy or refusal to obey a subpoena issued pursuant to this paragraph and served upon any person, the district court of the United States for any district in which such person is found, resides, or transacts business, upon application by the United States and after notice to such person, shall have jurisdiction to issue an order requiring such person to appear and give testimony before the administrative law judge or to appear and produce documents before the administrative law judge, or both, and any failure to obey such order of the court may be punished by such court as a contempt thereof.

Sec. 326. Civil Actions.

(a) Authority To Bring Civil Actions.—

(1) Citizen suits. — Except as provided in sebsection (e), any person may commence a civil action on his own behalf against the following:

(A) An owner or operator of a facililty for failure to do any of the following:

(i) Submit a followup emergency notice under section 304(c).

(ii) Submit a material safety data sheet or a list under section 311(a).

(iii) Complete and submit an inventory form under section 312(a) containing tier I information as described in section 312(d)(1) unless such requirement does not apply by reason of the second sentence of section 312(a)(2).

(iv) Complete and submit a toxic chemi-

cal release form under section 313(a).

(B) The Administrator for failure to do any of the following:

(i) Publish inventory forms under section 312(g).

(ii) Respond to a petition to add or delete a chemical under section 313(e)(1) within 180 days after receipt of the petition.

(iii) Publish a toxic chemical release form under 313(g).

(iv) Establish a computer database in accordance with section 313(j).

(v) Promulgate trade secret regulations under section 322(c).

(vi) Render a decision in response to a petition under section 322(d) within 9 months after receipt of the petition.

(C) The Administrator, a State Governor, or a State emergency response commission, for failure to provide a mechanism for public availability of information in accordance with section 324(a).

(D) A State Governor or a State emergency response commission for failure to respond to a request for tier II information under section 312(e)(3) within 120 days after the date of receipt of the request.

(2) State or local suits.—

(A) Any State or local government may commence a civil action against an owner or operator of a facility for failure to do any of the following:

(i) Provide notification to the emergency response commission in the State under section 302(c).

(ii) Submit a material safety data sheet or a list under section 311(a).

(iii) Make available information requested under section 311(c).

(iv) Complete and submit an inventory form under section 312(a) containing tier I information unless such requirement does not apply by reason of the second sentence of section 312(a)(2).

(B) Any State emergency response commission or local emergency planning committee may commence a civil action against an owner or operator of a facility for failure to provide information under section 303(d) or for failure to submit tier II information under section 312(e)(1).

(C) Any State may commence a civil

action against the Administrator for failure to provide information to the State under section 322(g).

(b) Venue.—

(1) Any action under subsection (a) against an owner or operator of a facility shall be brought in the district court for the district in which the alleged violation occurred.

(2) Any action under subsection (a) against the Administrator may be brought in the United States District Court for the District of Columbia.

(c) Relief. — The district court shall have jurisdiction in actions brought under subsection (a) against an owner or operator of a facility to enforce the requirement concerned and to impose any civil penalty provided for violation of that requirement. The district court shall have jurisdiction in actions brought under subsection (a) against the Administrator to order the Administrator to perform the act or duty concerned.

(d) Notice.—

(1) No action may be commenced under subsection (a)(1)(A) prior to 60 days after the plaintiff has given notice of the alleged violation to the Administrator, the State in which the alleged violation occurs, and the alleged violator. Notice under this paragraph shall be given in such manner as the Administrator shall prescribe by regulation.

(2) No action may be commenced under subsection (a)(1)(B) or (a)(1)(C) prior to 60 days after the date on which the plaintiff gives notice to the Administrator, State Governor, or State emergency response commission (as the case may be) that the plaintiff will commence the action. Notice under this paragraph shall be given in such manner as the Administrator shall prescribe by regulation.

(e) Limitation. — No action may be commenced under subsection (a) against an owner or operator of a facility if the Administrator has commenced and is diligently pursuing an administrative order or civil action to enforce the requirement concerned or to impose a civil penalty under this Act with respect to the violation of the requirement.

(f) Costs. — The court, in issuing any final order in any action brought pursuant to this section, may award costs of litigation (including reasonable attorney and expert witness fees) to the prevailing or the substantially prevailing party whenever the court determines such an award is appropriate. The court may, if a temporary restraining order or preliminary injunction is sought, require the filing of a bond or equivalent security in accordance with the Federal Rules of Civil Procedure.

(g) Other Rights. — Nothing in this section shall restrict or expand any right which any person (or class of persons) may have under any Federal or State statute or common law to seek enforcement of any requirement or to seek any other relief (including relief against the Administrator or a State agency).

(h) Intervention.—

(1) By the United States. — In any action under this section the United states or the State, or both, if not a party, may intervene as a matter of right.

(2) By persons. — In any action under this section, any person may intervene as a matter of right when such person has a direct interest which is or may be adversely affected by the action and the disposition of the action may, as a practical matter, impair or impede the person's ability to protect that interest unless the Administrator or the State shows that the person's interest is adequately represented by existing parties in the action.

Sec. 327. Exemption.

Except as provided in section 304, this title does not apply to the transportation, including the storage incident to such transportation, of any substance or chemical subject to the requirements of this title, including the transportation and distribution of natural gas.

Sec. 328. Regulations.

The Administrator may prescribe such regulations as may be necessary to carry out this title.

Sec. 329. Definitions.

For purposes of this title—

(1) Administrator. — The term "Administrator" means the Administrator of

the Environmental Protection Agency.

(2) Environment. — The term "environment" includes water, air, and land and the interrelationship which exists among and between water, air, and land and all living things.

(3) Extremely hazardous substance. — The term "extremely hazardous substance" means a substance on the list described in section 302(a)(2).

(4) Facility.— The term "facility" means all buildings, equipment, structures, and other stationary items which are located on a single site or on contiguous or adjacent sites and which are owned or operated by the same person (or by any person which controls, is controlled by, or under common control with, such person). For purposes of section 304, the term includes motor vehicles, rolling stock, and aircraft.

(5) Hazardous Chemical.— The term "hazardous chemical" has the meaning given such term by section 311(e).

(6) Material Safety Data Sheet.— The term "material safety data sheet" means the sheet required to be developed under section 1910.1200(g) of title 29 of the Code of Federal Regulations, as that section may be amended from time to time.

(7) Person.— The term "person" means any individual, trust, firm, joint stock company, corporation (including a government corporation), partnership, association, State, municipality, commission, political subdivision of a State, or interstate body.

(8) Release.— The term "release" means any spilling, leaking, pumping, pouring, emitting, emptying, discharging, injecting, escaping, leaching, dumping, or disposing into the environment (including the abandonment or discarding of barrels, containers, and other closed receptacles) of any hazardous chemical, extremely hazardous substance, or toxic chemical.

(9) State.— The term "State" means any State of the United States, the District of Columbia, the Commonwealth of Puerto Rico, Guam, American Samoa, the United States Virgin Islands, the Northern Mariana Islands, and any other territory or possession over which the United States has jurisdiction.

(10) Toxic Chemical.— The term "toxic chemical" means a substance on the list described in section 313(c).

Sec. 330. Authorization of Appropriations.

There are authorized to be appropriated for fiscal years beginning after September 30, 1986, such sums as may be necessary to carry out this title.

[Editor's note: Section 126 of PL 99-499 provides the following concerning worker protection standards for hazardous waste operations:

"Sec. 126. Worker Protection Standards.

(a) Promulgation.— Within one year after the date of the enactment of this section, the Secretary of Labor shall, pursuant to section 6 of the Occupational Safety and Health Act of 1970, promulgate standards for the health and safety protection of employees engaged in hazardous waste operations.

(b) Proposed Standards.— The Secretary of Labor shall issue proposed regulations on such standards which shall include, but need not be limited to, the following worker protection provisions:

(1) Site Analysis.— Requirements for a formal hazard analysis of the site and development of a site specific plan for worker protection.

(2) Training.— Requirements for contractors to provide initial and routine training of workers before such workers are permitted to engage in hazardous waste operations which would expose them to toxic substances.

(3) Medical Surveillance.— A program of regular medical examination, monitoring, and surveillance of workers engaged in hazardous waste operations which would expose them to toxic substances.

(4) Protective Equipment.— Requirements for appropriate personal protective equipment, clothing, and respirators for work in hazardous waste operations.

(5) Engineering Controls.— Requirements for engineering controls concerning the use of equipment and exposure of workers engaged in hazardous waste operations.

(6) Maximum Exposure Limits.— Re-

quirements for maximum exposure limitations for workers engaged in hazardous waste operations, including necessary monitoring and assessment procedures.

(7) Informational Program. — A program to inform workers engaged in hazardous waste operations of the nature and degree of toxic exposure likely as a result of such hazardous waste operations.

(8) Handling. — Requirements for the handling, transporting, labeling, and disposing of hazardous wastes.

(9) New Technology Program. — A program for the introduction of new equipment or technologies that will maintain worker protections.

(10) Decontamination Procedures. — Procedures for decontamination.

(11) Emergency Response. — Requirements for emergency response and protection of workers engaged in hazardous waste operations.

(c) Final Regulations. — Final regulations under subsection (a) shall take effect one year after the date they are promulgated. In promulgating final regulations on standards under subsection (a), the Secretary of Labor shall include each of the provisions listed in paragraphs (1) through (11) of subsection (b) unless the Secretary determines that the evidence in the public record considered as a whole does not support inclusion of any such provision.

(d) Specific Training Standards.—

(1) Offsite Instruction; Field Experience. — Standards promulgated under subsection (a) shall include training standards requiring that general site workers (such as equipment operators, general laborers, and other supervised personnel) engaged in hazardous substance removal or other activities which expose or potentially expose such workers to hazardous substances receive a minimum of 40 hours of initial instruction off the site, and a minimum of three days of actual field experience under the direct supervision of a trained, experienced supervisor, at the time of assignment. The requirements of the preceding sentence shall not apply to any general site worker who has received the equivalent of such training. Workers who may be exposed to unique or special hazards shall be provided additional training.

(2) Training Of Supervisors. — Standards promulgated under subsection (a) shall include training standards requiring that onsite managers and supervisors directly responsible for the hazardous waste operations (such as foremen) receive the same training as general site workers set forth in paragraph (1) of this subsection and at least eight additional hours of specialized training on managing hazardous waste operations. The requirements of the preceding sentence shall not apply to any person who has received the equivalent of such training.

(3) CERTIFICATION; ENFORCEMENT. — Such training standards shall contain provisions for certifying that general site workers, onsite managers, and supervisors have received the specified training and shall prohibit any individual who has not received the specified training from engaging in hazardous waste operations covered by the standard.

(4) TRAINING OF EMERGENCY RESPONSE PERSONNEL. — Such training standards shall set forth requirements for the training of workers who are responsible for responding to hazardous emergency situations who may be exposed to toxic substances in carrying out their responsibilities.

(e) INTERIM REGULATIONS. — The Secretary of Labor shall issue interim final regulations under this section within 60 days after the enactment of this section which shall provide no less protection under this section for workers employed by contractors and emergency response workers than the protections contained in the Environmental Protection Agency Manual (1981) "Health and Safety Requirements for Employees Engaged in Field Activities" and existing standards under the Occupational Safety and Health Act of 1970 found in subpart C of part 1926 of title 29 of the Code of Federal Regulations. Such interim final regulations shall take effect upon issuance and shall apply until final regulations become effective under subsection (c).

(f) COVERAGE OF CERTAIN STATE AND LOCAL EMPLOYEES. — Not later than 90 days after the promulgation of final regulations under subsection (a), the Administrator shall promulgate standards identical to those promulgated by the Secretary of Labor under subsection (a). Standards promulgated under this subsection shall apply to employees of State and local governments in each State which does not have in effect an approved State plan under section 18 of the Occupational Safety and Health Act of 1970 providing for standards for the health and safety protection of employees engaged in hazardous waste operations.

(g) GRANT PROGRAM.—

(1) GRANT PURPOSES. — Grants for the training and education of workers who are or may be engaged in activities related to hazardous waste removal or containment or emergency response may be made under this subsection.

(2) ADMINISTRATION. — Grants under this subsection shall be administered by the National Institute of Environmental Health Sciences.

(3) GRANT RECIPIENTS. — Grants shall be awarded to nonprofit organizations which demonstrate experience in implementing and operating worker health and safety training and education programs and demonstrate the ability to reach and involve in training programs target populations of workers who are or will be engaged in hazardous waste removal or containment or emergency response operations."]

APPENDIX C.3

List of Other Relevant Statutes and Regulations

EPA REGULATIONS UNDER SARA TITLE III

Trade Secret Claims for Emergency Planning and Community Right-to-Know Information; and Trade Secret Disclosures to Health Professionals, 52 FR 38312 (Oct. 15, 1987); 40 CFR Part 350 (proposed rule)

Emergency Planning and Notification, 52 FR 13377 (April 22, 1987); 40 CFR Part 355

Hazardous Chemical Reporting; Community Right-to-Know, 52 FR 38344 (Oct. 15, 1987); 40 CFR Part 370

Toxic Chemical Release Reporting; Community Right-to-Know, 53 FR 4500 (Feb. 16, 1988); 40 CFR Part 372

OTHER OSHA REGULATIONS RELATED TO RIGHT-TO-KNOW

Hazardous Waste Operations and Emergency Response,
  29 CFR § 1910.120 (interim final rule)
  52 FR 29620 (Aug. 10, 1987) (proposed rule)
Inspections, Citations and Proposed Penalties
  29 CFR Part 1903
Occupational Exposures to Toxic Substances in Laboratories, 51 FR 26660 (July 24, 1986); 29 CFR § 1910.1450 (proposed rule)

OTHER FEDERAL STATUTES AND REGULATIONS

Comprehensive Environmental Response, Compensation and Liability
    Act (Superfund)
    42 USC § 9601 *et seq.*
    40 CFR Part 300 (National Contingency Plan)
    40 CFR Part 302 (RQs and notification)

Consumer Product Safety Act
    15 USC § 2051 *et seq.*
    16 CFR Subchapter B, Part 1101 *et seq.*

Energy Reorganization Act of 1974
    42 USC § 5801 *et seq.*
    10 CFR § 50.47 (emergency plans around nuclear power plants)
    10 CFR Part 50, App. E

Federal Emergency Management Agency
    5 USC App., Reorganization Plan No. 3 of 1978
    44 CFR Subchapter A, Parts 0–25 (general)
    44 CFR Subchapter D, Parts 200–99 (disaster assistance)
    44 CFR Subchapter E, Parts 300–99 (preparedness)

Federal Food, Drug, and Cosmetic Act
    21 USC § 301 *et seq.*
    21 CFR Parts 101, 201, 501, 701, 801 (labeling)

Federal Hazardous Substances Act
    15 USC § 1261 *et seq.*
    16 CFR Subchapter C, Part 1500 *et seq.*

Federal Insecticide, Fungicide, and Rodenticide Act
    7 USC § 136 *et seq.*
    40 CFR § 162.10 (labeling requirements)

Hazardous Materials Transportation Act
    49 USC App. § 1801 *et seq.*
    49 CFR Subchapter C, Part 171 *et seq.*

Resource Conservation and Recovery Act
    42 USC § 6901 *et seq.*
    40 CFR Subchapter I, Parts 240–80

Toxic Substances Control Act of 1976
    15 USC § 2601 *et seq.*
    40 CFR Subchapter R, Part 702 *et seq.*

APPENDIX C.4

List of Cases Interpreting the HCS

Associated Builders and Contractors Inc. v. Brock, No. 87–1582 (D.C. Cir.) (expanded scope; case pending)

Associated General Contractors of Virginia v. OSHA, No. 87–1185 (D.C. Cir.) (expanded scope; case pending)

Manufacturer's Association of Tri-County v. Knepper, 623 F.Supp. 1066 (M.D.Pa. 1985), *aff'd in part,* 801 F.2d 130 (3d Cir. 1986) (federal/ state preemption; property taking; commerce clause)

National Grain and Feed Association v. OSHA, No. 87–1603 (D.C. Cir.) (expanded scope; case pending)

New Jersey State Chamber of Commerce v. Hughey, 600 F.Supp. 606 (D.N.J. 1985), *aff'd in part,* 774 F.2d 587 (3d Cir. 1985) (federal/ state preemption; property taking; commerce clause)

Ohio Manufacturer's Association v. City of Akron, 628 F.Supp. 623 (N.D. Ohio 1986), *rev'd* 801 F.2d 824 (6th Cir. 1986) (federal/local preemption)

United Steelworkers of America v. Auchter, 763 F.2d 728 (3d Cir. 1985) (federal/state preemption; trade secrets; limited scope)

United Steelworkers of America v. Pendergrass, 819 F.2d 1263 (3d Cir. 1987) (limited scope)

United Technologies Corp. v. OSHA, No. 87–4143 (D.C. Cir.) (expanded scope; case pending)

APPENDIX D

Regulated Materials

The four tables in this appendix include information as to what materials are regulated, and how, under various aspects of Right-to-Know statutes and regulations. Great care has been taken to minimize errors in these tables. However, regulatory agencies are constantly in the process of revising rules that affect these matters, so a printed list may not be completely up to date, and regulated parties have a responsibility to inquire further as to the current status of such regulations.

In addition, some materials are listed in a generic fashion, so that many individual materials are also regulated under the same heading even though they may not be specifically indicated in these tables. In general, it is well to consult with a chemist, or with a person who has had chemistry courses at least through organic chemistry, in order to assure compliance in such cases.

Since many of the regulated substances are known by a variety of different common, trade, and chemical names, all are tabulated according to their Chemical Abstracts Service Registry Numbers (CAS Nos.). The CAS numbers, when known, provide a simplified, more compact means of tabulating data than could be provided if all the pertinent names were used in all the tables. The first three tables give regulatory information, and the fourth table provides an index to the CAS numbers.

It should be noted that for many of the materials listed here, there are dozens of different names in use. This is particularly true for dyestuffs, insecticides, herbicides, and other agricultural chemicals. Persons concerned with such materials would be well advised to obtain a copy of the "Glossary of Synonyms" developed by EPA for those chemicals subject to Section 313 annual reporting. A copy may be obtained from the EPA TSCA Assistance Office by phoning (202) 554–1411.

In Table D.1 following the CAS No. column are two columns headed TLV and PEL. These columns merely indicate whether ACGIH has assigned a Threshold Limit Value or OSHA has assigned a Permissible Exposure Limit for the material. In either case, it is regulated by reference under the HCS.

The next three columns indicate whether the material is considered carcinogenic by OSHA, in which case it is also regulated by reference. The first of these, IARC, indicates the carcinogenicity ratings for categories 1 and 2 by that agency; category 3 carcinogens are not among those considered carcinogenic by OSHA. The NTP column simply indicates those materials listed as carcinogenic by the National Toxicology Program, in which case they are regulated as carcinogenic by OSHA. In the column headed OSHA are X marks for all the materials regulated by OSHA as carcinogens with individual standards. References to the locations of the standards in the regulations are indicated in Table 5, Chapter 5.

The last column in Table D.1 is headed SKIN, which means that the material may be absorbed through the skin into the bloodstream in amounts that may cause undesirable effects.

In Table D.2 following the CAS No. column are two columns of TPQs, or Threshold Planning Quantities. The first of these applies to materials that are used only in the solid state with particle size greater than 100 μm and that do not meet criteria for a NFPA 704M rating of 2, 3, or 4 for reactivity. The second of these columns is the TPQ for all other forms of the material.

The first column under Reportable Quantities is the amount released that escapes, or is likely to escape, the facility into the surrounding neighborhood. The second of these columns is the RQ under Superfund for releases wherever they occur. The EPA has indicated that there will be further rulemaking to consolidate these two types of RQs in order to reduce the confusion that they currently engender.

The final column of this table indicates that an Annual Toxic Chemical Release Inventory Report is required by facilities that produce, use, or otherwise have on their premises large amounts of these materials.

Table D.3 gives RCRA Hazardous Waste Identification Numbers (P- and U-lists only) for materials that are regulated under either HCS or SARA. Note that many other regulated materials will be classified as hazardous wastes if they are spilled or otherwise to be disposed of. Thus, the fact that a material does not have a number given under the RCRA heading in this list does not mean that it is free from regulation under RCRA, just that it does not have a listed number assigned to it.

Also in this table are shipping identification numbers for many of the regulated materials. Some of the other materials may be regulated in shipping by a defined hazard even though they do not have a number assigned. In many cases, more than one shipping number is listed. This usually means that the material is shipped in different forms, some or all of which have distinct numbers.

In Table D.4 are the CAS numbers for regulated substances listed in alphabetical order. Note that alphabetical ordering ignores numerals and some short abbreviations at the beginning of the name. The same numerals and short abbreviations within the names are also ignored in alphabetical sorting. In general, the first capitalized letter in the name is the beginning of the portion that is alphabetized. For the most part, names are listed in normal order rather than in a manner analogous to the last-name-first order commonly used for names of persons.

A number of the materials regulated by either HCS or SARA do not have CAS numbers assigned, or at least none were found. For ease of location, "Quasi-CAS" numbers were assigned to such materials. They have the same form as the CAS numbers except that each begins with the letter X in the first position of the number, whereas CAS numbers use only numerals and hyphens.

This form of number allows sorting by an ASCII protocol in which the Quasi-CAS numbers always follow the true CAS numbers. If a different computer sorting protocol, such as EBCDIC, is used, then it may be necessary to use a different lead character to obtain the same results. Any other kind of indexing identification system might be used also, but the CAS system has the advantage that many computer data bases, MSDSs, and other sources of information include the CAS number, thereby making their use compatible with others.

APPENDIX D.1

Materials Regulated by Reference Under the HCS

| CAS No. | TLV | PEL | Carcinogenic | | | SKIN |
			IARC	NTP	OSHA	
50-00-0	X		2B	X	X	
50-18-0			1	X		
50-29-3	X	X				X
50-32-8	X		2A	X		
50-55-5				X		
50-76-0			2B			
50-78-2	X					
51-52-5			2B			
51-75-2			2A			
51-79-6				X		
52-24-4				X		
53-70-3				X		
53-96-3				X	X	
54-11-5	X	X				X
55-18-5				X		
55-38-9	X					X
55-63-0	X	X				X
56-23-5	X	X	2B	X		X
56-38-2	X	X				X
56-53-1			1	X		
56-55-3				X		
56-75-7			2B			
56-81-5	X					
57-14-7	X	X				X
57-24-9	X	X				
57-41-0			2B	X		
57-50-1	X					
57-57-8	X			X	X	
57-74-9	X	X				X
58-89-9	X	X		X		X
59-89-2				X		
60-11-7				X	X	

281

| CAS No. | TLV | PEL | Carcinogenic | | | SKIN |
			IARC	NTP	OSHA	
60-29-7	X	X				
60-34-4	X					X
60-57-1	X	X				X
61-82-5	X		2B	X		
62-44-2			2A	X		
62-53-3	X	X				X
62-55-5				X		
62-56-6				X		
62-73-7	X					X
62-74-8	X	X				
62-75-9	X			X	X	X
63-25-2	X	X				
64-17-5	X	X				
64-18-6	X	X				
64-19-7	X	X				
64-67-5			2A			
67-56-1	X		X			
67-63-0	X	X				
67-64-1	X	X				
67-66-3	X	X	2B	X		
67-72-1	X	X				X
68-11-1	X					X
68-12-2	X	X				
71-23-8	X	X				X
71-36-3	X	X				X
71-43-2	X		1	X	X	
71-55-6	X	X				
72-20-8	X					X
72-43-5	X	X				
74-82-8	X					
74-83-9	X	X				X
74-84-0	X					
74-85-1	X					
74-86-2	X					
74-87-3	X					
74-88-4	X	X				X
74-89-5	X	X				
74-90-8	X	X				X
74-93-1	X	X				
74-96-4	X	X				
74-97-5	X	X				
74-98-6	X	X				
74-99-7	X					
75-00-3	X	X				
75-01-4	X		1	X	X	
75-04-7	X	X				
75-05-8	X	X				X
75-07-0	X	X				
75-08-1	X	X				
75-09-2	X					
75-12-7	X					
75-15-0	X	X				X
75-21-8	X		2B		X	
75-25-2	X	X				X
75-31-0	X	X				X

CAS No.	TLV	PEL	Carcinogenic			SKIN
			IARC	NTP	OSHA	
75-34-3	X	X				
75-35-4	X					
75-43-4	X	X				
75-44-5	X	X				
75-45-6	X					
75-47-8	X					
75-50-3	X					
75-52-5	X	X				
75-55-8	X	X				X
75-56-9	X	X				
75-61-6	X	X				
75-63-8	X	X				
75-65-0	X	X				
75-69-4	X					
75-71-8	X	X				
75-74-1	X	X				X
75-99-0	X					
76-03-9	X					
76-06-2	X	X				
76-11-9	X	X				
76-12-0	X	X				
76-13-1	X	X				
76-14-2	X	X				
76-15-3	X					
76-22-2	X	X				
76-44-8	X	X				X
77-47-4	X					
77-73-6	X					
77-78-1	X	X	2A	X		X
78-00-2	X	X				X
78-10-4	X	X				
78-30-8	X	X				X
78-34-2	X					X
78-59-1	X	X				
78-83-1	X	X				
78-87-5	X	X				
78-92-2	X	X				
78-93-3	X	X				
79-00-5	X	X				X
79-01-6	X					
79-04-9	X					
79-06-1	X	X				X
79-09-4	X					
79-10-7	X					
79-20-9	X	X				
79-24-3	X	X				
79-27-6	X	X				
79-34-5	X	X				X
79-41-4	X					
79-44-7	X					
79-46-9	X	X				
80-62-6	X	X				
81-07-2				X		
81-81-2	X	X				
83-26-1	X					

CAS No.	TLV	PEL	Carcinogenic			SKIN
			IARC	NTP	OSHA	
83-79-4	X	X				
84-66-2	X					
84-74-2	X	X				
85-00-7	X					
85-44-9	X	X				
86-50-0	X	X				X
86-88-4	X	X				
87-68-3	X					X
87-86-5	X	X				X
88-06-2			2B	X		
88-89-1	X	X				X
89-72-5	X					X
90-04-0	X	X		X		X
90-94-8				X		
91-20-3	X	X				
91-59-8	X		1	X	X	
91-94-1	X		2B	X	X	X
92-52-4	X	X				
92-67-1	X		1	X	X	X
92-84-2	X					X
92-87-5	X		1	X	X	X
92-93-3	X				X	
93-76-5	X	X	2B			
94-36-0	X	X				
94-59-7				X		
94-75-7	X		2B			
94-78-0			2B	X		
95-06-7				X		
95-13-6	X					
95-49-8	X					
95-50-1	X					
95-53-4	X	X	2A	X		X
95-57-8			2B			
95-80-7				X		
96-12-8				X	X	
96-18-4	X	X				X
96-22-0	X					
96-33-3	X	X				X
96-45-7			2B	X		
96-69-5	X					
97-77-8	X					
98-00-0	X	X				X
98-01-1	X	X				X
98-07-7			2B			
98-51-1	X	X				
98-82-8	X	X				X
98-83-9	X					
98-95-3	X	X				X
99-08-1	X	X				X
99-59-2				X		
99-65-0	X	X				X
100-00-5	X	X				X
100-01-6	X	X				X
100-25-4	X	X				X
100-37-8	X	X				X

| CAS No. | TLV | PEL | Carcinogenic | | | SKIN |
			IARC	NTP	OSHA	
100-41-4	X	X				
100-42-5	X	X				
100-44-7	X	X				
100-61-8	X					X
100-63-0	X	X				X
100-74-3	X					X
100-75-4				X		
101-14-4	X			X		X
101-61-1				X		
101-68-8	X					
101-77-9	X					X
101-84-8	X	X				
102-54-5	X					
102-81-8	X					X
105-30-6	X	X				X
105-46-4	X	X				
105-60-2	X					
106-35-4	X	X				
106-46-7	X	X				
106-48-9			2B			
106-49-0	X					X
106-50-3	X	X				X
106-51-4	X	X				
106-87-6	X					X
106-89-8	X	X	2B			X
106-92-3	X	X				X
106-93-4	X		2B	X		X
106-97-8	X					
106-99-0	X	X				
107-02-8	X	X				
107-05-1	X	X				
107-06-2	X			X		
107-07-3	X	X				X
107-13-1	X		2A	X	X	X
107-15-3	X	X				
107-18-6	X	X				X
107-19-7	X					X
107-20-0	X	X				
107-21-1	X					
107-30-2	X		1	X	X	
107-31-3	X	X				
107-41-5	X					
107-49-3	X	X				X
107-66-4	X	X				
107-87-9	X	X				
107-98-2	X					
108-03-2	X	X				
108-05-4	X					
108-10-1	X	X				
108-11-2	X	X				X
108-18-9	X	X				X
108-20-3	X	X				
108-21-4	X	X				
108-24-7	X	X				
108-31-6	X	X				

			Carcinogenic			
CAS No.	TLV	PEL	IARC	NTP	OSHA	SKIN
108-38-3	X	X				
108-43-0			2B			
108-44-1	X					X
108-46-3	X					
108-57-6	X					
108-83-8	X	X				
108-84-9	X	X				
108-87-2	X	X				
108-88-3	X					
108-90-7	X	X				
108-91-8	X					
108-93-0	X	X				
108-94-1	X	X				
108-95-2	X	X				X
108-98-5	X					
109-59-1	X					
109-60-4	X	X				
109-66-0	X	X				
109-73-9	X	X				X
109-79-5	X	X				
109-86-4	X	X				X
109-87-5	X	X				
109-89-7	X	X				
109-94-4	X	X				
109-99-9	X	X				
110-12-3	X					
110-19-0	X	X				
110-43-0	X	X				
110-49-6	X	X				X
110-54-3	X	X				
110-62-3	X					
110-80-5	X	X				X
110-82-7	X	X				
110-83-8	X	X				
110-86-1	X	X				
110-91-8	X	X				X
111-15-9	X	X				X
111-30-8	X					
111-40-0	X					X
111-42-2	X					
111-44-4	X	X				X
111-65-9	X	X				
111-76-2	X	X				X
111-84-2	X					
112-62-9	X					
114-26-1	X					
115-07-1	X					
115-29-7	X					X
115-77-5	X					
115-86-6	X	X				
115-90-2	X					
117-81-7	X	X		X		
118-52-5	X	X				
118-96-7	X	X				X
119-90-4			2B	X		

| CAS No. | TLV | PEL | Carcinogenic | | | SKIN |
			IARC	NTP	OSHA	
119-93-7	X			X		X
120-71-8				X		
120-80-9	X					
120-82-1	X					
121-14-2	X	X				X
121-44-8	X	X				
121-45-9	X					
121-69-7	X	X				X
121-75-5	X	X				X
121-82-4	X					X
122-39-4	X					
122-60-1	X	X				
122-66-7				X		
123-19-3	X					
123-31-9	X	X				
123-42-2	X	X				
123-51-3	X	X				
123-86-4	X	X				
123-91-1	X	X	2B	X		X
123-92-2	X	X				
124-38-9	X	X				
124-40-3	X	X				
126-72-7				X		
126-73-8	X	X				
126-98-7	X					
126-99-8	X	X				X
127-18-4	X					
127-19-5	X	X				X
128-37-0	X					
131-11-3	X	X				
133-06-2	X					
134-29-2				X		
134-32-7					X	
135-20-6				X		
135-88-6	X					
136-40-3				X		
136-78-7	X					
137-05-3	X					
137-26-8	X	X				
138-22-7	X					
139-13-9				X		
140-57-8				X		
140-88-5	X	X				X
141-32-2	X					
141-43-5	X	X				
141-66-2	X					X
141-78-6	X	X				
141-79-7	X	X				
142-64-3	X					
142-82-5	X	X				
142-92-7	X	X				
143-33-9	X	X				X
143-50-0				X		
144-62-7	X	X				
148-01-6	X					

| CAS No. | TLV | PEL | Carcinogenic | | | SKIN |
			IARC	NTP	OSHA	
148-82-3			1	X		
150-76-5	X					
151-56-4	X	X			X	X
154-93-8			2B			
156-10-5				X		
156-62-7	X					
189-55-9				X		
189-64-0				X		
193-39-5				X		
194-59-2				X		
205-99-2				X		
218-01-9	X					
224-42-0				X		
226-36-8				X		
287-92-3	X					
298-00-0	X					X
298-02-2	X					X
298-04-4	X					
299-75-2			X			
299-84-3	X	X				
299-86-5	X					
300-76-5	X					X
301-04-2				X		
302-01-2	X	X	2B	X		X
305-03-3			1	X		
309-00-2	X	X				X
314-40-9	X					
319-84-6	X	X		X		X
330-54-1	X					
333-41-5	X					X
334-88-3	X	X				
353-50-4	X					
366-70-1				X		
409-21-2	X					
420-04-2	X					
434-07-1			2A	X		
443-48-1			2B			
446-86-6			1			
460-19-5	X					
463-51-4	X	X				
479-45-8	X	X				X
492-80-8			1			
494-03-1			1	X		
504-29-0	X	X				
505-60-2			1	X		
506-77-4	X					
509-14-8	X	X				
528-29-0	X	X				X
532-27-4	X	X				
534-52-1	X	X				X
540-59-0	X	X				
540-88-5	X	X				
541-85-5	X	X				
542-75-6	X					X
542-88-1	X		1	X	X	

| | | | Carcinogenic | | | |
CAS No.	TLV	PEL	IARC	NTP	OSHA	SKIN
542-92-7	X	X				
546-93-0	X					
552-30-7	X					
556-52-5	X	X				
557-05-1	X					
558-13-4	X					
563-12-2	X					X
563-80-4	X					
583-60-8	X	X				X
584-84-9	X	X				
591-78-6	X	X				
593-60-2	X					
594-42-3	X	X				
594-72-9	X	X				
600-25-9	X					
603-34-9	X					
621-64-7				X		
624-83-9	X	X				X
626-17-5	X					
626-38-0	X	X				
627-13-4	X	X				
628-63-7	X	X				
628-96-6	X	X				X
630-08-0	X	X				
632-99-5			2A			
636-21-5	X	X		X		X
638-21-1	X					
671-16-9			2A	X		
680-31-9	X					X
681-84-5	X					
684-16-2	X					X
684-93-5				X		
759-73-9				X		
768-52-5	X					X
924-16-3				X		
930-55-2				X		
944-22-9	X					X
999-61-1	X					X
1116-54-7				X		
1120-71-4	X					
1189-85-1	X	X				X
1300-73-8	X	X				X
1303-86-2	X	X				
1303-96-4	X					
1304-82-1	X					
1305-62-0	X					
1305-78-8	X	X				
1306-19-0	X		2B	X		
1309-37-1	X	X				
1309-48-4	X	X				
1309-64-4	X	X				
1310-58-3	X					
1310-73-2	X	X				
1314-13-2	X	X				
1314-20-1				X		

CAS No.	TLV	PEL	Carcinogenic			SKIN
			IARC	NTP	OSHA	
1314-62-1	X	X				
1314-80-3	X	X				
1317-35-7	X					
1317-65-3	X					
1317-95-9	X	X				
1319-77-3	X	X				X
1321-64-8	X	X				X
1321-65-9	X	X				X
1321-74-0	X					
1327-53-3	X					
1330-20-7	X	X				
1330-43-4	X					
1331-28-8	X					
1332-29-2	X					
1333-74-0	X					
1333-86-4	X	X				
1335-87-1	X	X				X
1335-88-2	X	X				X
1336-36-3	X		2B	X		X
1338-23-4	X					
1344-28-1	X					
1344-38-3	X					
1344-90-7	X					
1344-95-2	X					
1395-21-7	X					
1464-53-5				X		
1477-55-0	X					X
1563-66-2	X					
1746-01-6			2B	X		
1836-75-5				X		
1912-24-9	X					
1918-02-1	X					
1929-82-4	X					
1937-37-7			2B	X		
2039-87-4	X					
2104-64-5	X	X				X
2179-59-1	X	X				
2234-13-1	X	X				X
2238-07-5	X	X				
2385-85-5				X		
2425-06-1	X					X
2426-08-6	X					
2551-62-4	X	X				
2602-46-2			2B	X		
2698-41-1	X					X
2699-79-8	X	X				
2921-88-2	X					X
2971-90-6	X					
3333-52-6	X	X				X
3383-96-8	X					
3689-24-5	X					X
4016-14-2	X	X				
4098-71-9	X	X				
4170-30-3	X	X				
4685-14-7	X	X				X

| CAS No. | TLV | PEL | Carcinogenic | | | SKIN |
			IARC	NTP	OSHA	
5124-30-1	X					
5714-22-7	X	X				
6423-43-4	X					X
6923-22-4	X					
7429-90-5	X					
7429-90-5	X					
7439-92-1	X					
7439-96-5	X	X				
7439-97-6	X					
7439-98-7	X					
7440-01-9	X					
7440-02-0	X	X				
7440-06-4	X					
7440-16-6	X	X				
7440-21-3	X					
7440-22-4	X	X				
7440-25-7	X	X				
7440-31-5	X					
7440-36-0	X	X				
7440-37-1	X					
7440-38-2	X		1	X		
7440-41-7	X	X	2A	X		
7440-43-9	X	X	2B	X		
7440-47-3	X	X				
7440-48-4	X	X				
7440-50-8	X	X				
7440-58-6	X	X				
7440-59-7	X					
7440-65-5	X	X				
7440-74-6	X					
7446-09-5	X	X				
7446-27-7				X		
7446-34-6	X	X		X		
7553-56-2	X	X				
7572-29-4	X					
7580-67-8	X					
7616-94-6	X	X				
7631-86-9	X	X				
7631-90-5	X					
7637-07-2	X	X				
7646-85-7	X	X				
7647-01-0	X	X				
7664-38-2	X	X				
7664-39-3	X					
7664-41-7	X	X				
7664-93-9	X	X				
7681-57-4	X					
7697-37-2	X	X				
7719-09-7	X					
7719-12-2	X	X				
7722-84-1	X	X				
7722-88-5	X					
7723-14-0	X	X				
7726-95-6	X	X				
7727-43-7	X					

CAS No.	TLV	PEL	Carcinogenic			SKIN
			IARC	NTP	OSHA	
7758-97-6	X					
7773-06-0	X	X				
7778-18-9	X					
7782-41-4	X	X				
7782-42-5	X	X				
7782-50-5	X	X				
7782-65-2	X					
7783-06-4	X					
7783-07-5	X	X				
7783-41-7	X	X				
7783-54-2	X	X				
7783-60-0	X					
7783-79-1	X	X				
7783-80-4	X	X				
7784-42-1	X	X				
7786-34-7	X					X
7789-30-2	X					
7790-91-2	X	X				
7803-51-2	X	X				
7803-52-3	X	X				
7803-62-5	X					
8001-35-2	X	X				X
8002-74-2	X					
8003-34-7	X	X				
8006-61-9	X					
8006-64-2	X	X				
8007-45-2	X	X				
8012-95-1	X	X				
8022-00-2	X					X
8030-30-6	X					
8030-31-7		X				
8052-41-3	X	X				
8052-42-4	X					
8065-48-3	X	X				X
9004-34-6	X					
9004-66-4				X		
9005-25-8	X					
10025-67-9	X	X				
10025-87-3	X					
10026-13-8	X	X				
10028-15-6	X	X				
10034-93-2				X		
10035-10-6	X	X				
10049-04-4	X	X				
10101-41-4	X					
10102-43-9	X	X				
10102-44-0	X	X				
10102-48-4	X					
10210-68-1	X					
10294-33-4	X					
11097-69-1	X	X				X
12001-25-2	X	X				
12001-28-4	X		1	X	X	
12001-29-5	X		1	X	X	
12035-72-2	X					

CAS No.	TLV	PEL	Carcinogenic			SKIN
			IARC	NTP	OSHA	
12079-65-1	X					X
12108-13-3	X					X
12125-02-9	X					
12172-73-5	X		1	X	X	
12179-04-3	X					
12604-58-9	X	X				
13010-47-4			2B			
13121-70-5	X					
13256-13-8				X		
13256-22-9				X		
13463-39-3	X	X				
13463-40-6	X					
13463-67-7	X	X				
13494-80-9	X	X				
13530-65-9	X					
14464-46-1	X	X				
14484-64-1	X	X				
14807-96-6	X					
14808-60-7	X	X				
14901-08-7				X		
14977-61-8	X					
15468-32-3	X	X				
15663-27-1			2B			
16071-86-6			2B			
16219-75-3	X					
16543-55-8				X		
16752-77-5	X					
16842-03-8	X					
17702-41-9	X	X				X
17804-35-2	X					
18883-66-4				X		
19287-45-7	X	X				
19624-22-7	X	X				
20816-12-0	X	X				
21087-64-9	X					
21351-79-1	X					
22224-92-6	X					X
23214-92-8			2B			
25013-15-4	X	X				
25551-13-7	X					
25639-42-3	X	X				
26140-60-3	X	X				
26628-22-8	X					
26952-21-6	X					
29191-52-4	X	X		X		X
34590-94-8	X	X				X
35400-43-2	X					
36355-01-8				X		
39156-41-7				X		
53469-21-9	X	X				X
55720-99-5	X	X				
60676-86-0	X	X				
61028-24-8	X					
61788-32-7	X					
65996-93-2	X	X				

			Carcinogenic			
CAS No.	TLV	PEL	IARC	NTP	OSHA	SKIN
68476-85-7	X	X				
68855-54-9	X	X				
X00001-00-3	X	X				*
X00001-01-4	X					
X00001-02-5	X					
X00001-03-6	X	X				X
X00001-04-7	X					
X00001-05-8	X	X				
X00001-06-9	X					
X00001-07-0	X	X	1	X		
X00001-08-1	X	X				
X00001-09-2	X	X	2A	X		
X00001-10-5	X					
X00001-12-7	X	X	1	X		
X00001-13-8	X		2B	X		
X00001-15-0	X		1	X		
X00001-16-1	X					
X00001-17-2	X	X				
X00001-18-3	X	X				
X00001-19-4	X					
X00001-20-7	X					
X00001-21-8	X	X				
X00001-22-9			2A			
X00001-23-0	X	X				
X00001-24-1			1			
X00001-25-2			1			
X00001-26-3			2B			
X00001-27-4			2B			
X00001-32-1	X	X				
X00001-33-2	X					
X00001-34-3	X					
X00001-35-4	X					
X00001-36-5	X					
X00001-37-6				X	X	
X00001-38-7	X					
X00001-39-8	X					
X00001-40-1	X					
X00001-41-2	X					
X00001-42-3	X					
X00001-43-4	X					
X00001-44-5	X	X				
X00001-45-6	X	X				
X00001-46-7	X	X				
X00001-47-8	X	X				
X00001-48-9	X					
X00001-50-3	X					
X00001-52-5	X					
X00001-54-7	X					
X00001-55-8	X					
X00001-56-9	X	X				
X00001-57-0	X	X				
X00001-58-1	X					
X00001-59-2	X					
X00001-60-5	X	X				
X00001-61-6	X	X				

| CAS No. | TLV | PEL | Carcinogenic | | | SKIN |
			IARC	NTP	OSHA	
X00001-62-7	X					
X00001-64-9	X	X				
X00001-65-0	X	X				X
X00001-67-2	X	X				
X00001-68-3	X	X				
X00001-71-8	X					
X00001-72-9	X					
X00001-73-0	X	X				
X00001-74-1	X	X				
X00001-76-3	X					
X00001-77-4	X					
X00001-78-5	X					
X00001-79-6	X					
X00001-82-1	X	X				

APPENDIX D.2

Materials Regulated Under CERCLA and SARA Title III

| CAS No. | Thresh. Planning Qty | | Reportable Quantity | | Annual Report |
	Lg. Gran.	Other	SARA	CERCLA	
50-00-0		500	1,000	1,000	X
50-07-7	10,000	500	1	1	
50-14-6	10,000	1,000	1		
50-18-0				1	
50-29-3				1	
50-32-8				1	
50-55-5				5,000	
51-21-8	10,000	500	1		
51-28-5				10	X
51-43-4				1,000	
51-75-2		10	1		X
51-79-6				1	X
51-83-2	10,000	500	1		
52-68-6			100	100	X
52-85-7				1,000	
53-70-3				1	
53-96-3				1	X
54-11-5		100	100	100	
54-62-6	10,000	500	1		
55-18-5				1	X
55-21-0					X
55-63-0				10	X
55-91-4		100	100	100	
56-04-2				1	
56-23-5				5,000	X
56-25-7	10,000	100	1		
56-38-2		100	1	1	X
56-49-5				1	

All quantities expressed in pounds

| | Thresh. Planning Qty | | Reportable Quantity | | Annual |
CAS No.	Lg. Gran.	Other	SARA	CERCLA	Report
56-53-1				1	
56-55-3				1	
56-72-4	10,000	100	10	10	
57-12-5				10	X
57-14-7		1,000	1	1	X
57-24-9	10,000	100	10	10	
57-47-6	10,000	100	1		
57-57-8		500	1		X
57-64-7	10,000	100	1		
57-74-9		1,000	1	1	X
57-97-6					
58-36-6	10,000	500	1		
58-89-9	10,000	1,000	1	1	
58-90-2				10	
59-50-7				5,000	
59-88-1	10,000	1,000	1		
59-89-2					X
60-00-4				5,000	
60-09-3					X
60-11-7				1	X
60-29-7				100	
60-34-4		500	10	10	X
60-35-5					X
60-41-3	10,000	100	1		
60-51-5	10,000	500	10	10	
60-57-1				1	
61-82-5				1	
62-38-4	10,000	500	100	100	
62-44-2				1	
62-50-0				1	
62-53-3		1,000	5,000	5,000	X
62-55-5				1	X
62-56-6				1	X
62-73-7		1,000	10	10	X
62-74-8	10,000	10	10	10	
62-75-9		1,000	1	1	X
63-25-2				100	X
64-00-6	10,000	500	1		
64-18-6				5,000	
64-19-7				5,000	
64-67-5					X
64-86-8	10,000	10	1		
65-30-5	10,000	100	1		
65-85-0				5,000	
66-75-1				1	
66-81-9	10,000	100	1		
67-56-1				5,000	X
67-63-0					X
67-64-1				5,000	X
67-66-3		10,000	5,000	5,000	X
67-72-1				1	X
68-76-8					X
70-25-7				1	
70-30-4				100	
70-69-9	10,000	100	1		

All quantities expressed in pounds

| CAS No. | Thresh. Planning Qty | | Reportable Quantity | | Annual Report |
	Lg. Gran.	Other	SARA	CERCLA	
71-36-3				5,000	X
71-43-2				1,000	X
71-55-6				1,000	X
71-63-6	10,000	100	1		
72-20-8	10,000	500	1	1	
72-43-5				1	X
72-54-8				1	
72-55-9				1	
72-57-1				1	
74-83-9		1,000	1,000	1,000	X
74-85-1					X
74-87-3				1	X
74-88-4				1	X
74-89-5				100	
74-90-8		100	10	10	X
74-93-1		500	100	100	
74-95-3				1,000	X
75-00-3				100	X
75-01-4				1	X
75-04-7				100	
75-05-8				5,000	X
75-07-0				1,000	X
75-09-2				1,000	X
75-15-0		10,000	100	100	X
75-18-3		100	1		
75-20-7				10	
75-21-8		1,000	1	1	X
75-25-2				100	X
75-27-4				5,000	X
75-34-3				1,000	
75-35-4				5,000	X
75-36-5				5,000	
75-44-5		10	10	10	X
75-50-3				100	
75-55-8		10,000	1	1	X
75-56-9		10,000	100	100	X
75-60-5				1	
75-64-9				1,000	
75-65-0					X
75-69-4				5,000	
75-71-8				5,000	
75-74-1		100	1		
75-77-4		1,000	1		
75-78-5		500	1		
75-79-6		500	1		
75-86-5		1,000	10	10	
75-87-6				1	
75-99-0				5,000	
76-01-7			1	1	
76-02-8		500	1		
76-44-8				1	X
77-13-1					X
77-47-4		100	1	1	X
77-78-1		500	1	1	X
77-81-6		10	1		

All quantities expressed in pounds

| CAS No. | Thresh. Planning Qty | | Reportable Quantity | | Annual Report |
	Lg. Gran.	Other	SARA	CERCLA	
78-00-2		100	10	10	
78-34-2		500	1		
78-53-5		500	1		
78-59-1				5,000	
78-71-7		500	1		
78-79-5				100	
78-81-9				1,000	
78-82-0		1,000			
78-83-1				5,000	
78-84-2					X
78-87-5				1,000	X
78-88-6				100	
78-92-2					X
78-93-3				5,000	X
78-94-4		10	1		
78-97-7		1,000	1		
78-99-9				1,000	
79-00-5				1	X
79-01-6				1,000	X
79-06-1	10,000	1,000	5,000	5,000	X
79-09-4				5,000	
79-10-7				5,000	X
79-11-8	10,000	100	1		X
79-19-6	10,000	100	100	100	
79-21-0		500	1		X
79-22-1		500	1,000	1,000	
79-31-2				5,000	
79-34-5				1	X
79-44-7				1	X
79-46-9				1	X
80-05-7					X
80-15-9				10	X
80-62-6				1,000	X
80-63-7		500	1		
81-07-2				1	X
81-81-2	10,000	500	100	100	
81-88-9					X
82-28-0					X
82-66-6	10,000	10	1		
82-68-8				1	X
83-32-9				100	
84-66-2				1,000	X
84-74-2				10	X
85-00-7				1,000	
85-01-8				5,000	
85-44-9				5,000	X
85-68-7				100	X
86-30-6				100	X
86-50-0	10,000	10	1	1	
86-73-7				5,000	
86-88-4	10,000	500	100	100	
87-62-7					X
87-65-0				100	
87-68-3				1	X
87-86-5			10	10	X

All quantities expressed in pounds

| CAS No. | Thresh. Planning Qty | | Reportable Quantity | | Annual Report |
	Lg. Gran.	Other	SARA	CERCLA	
88-05-1		500	1		
88-06-2				10	X
88-72-2				1,000	
88-75-5				100	X
88-85-7	10,000	100	1,000	1,000	
88-89-1					X
90-04-0					X
90-43-7					X
90-94-8					X
91-08-7		100	100	100	X
91-20-3				100	X
91-22-5				5,000	X
91-58-7				5,000	
91-59-8				1	X
91-80-5				5,000	
91-94-1				1	X
92-52-4					X
92-67-1					X
92-87-5				1	X
92-93-3					X
93-72-1				100	
93-76-5				1,000	
93-79-8				1,000	
94-11-1				100	
94-36-0					X
94-58-6				1	
94-59-7				1	X
94-75-7				100	X
94-79-1				100	
94-80-4				100	
95-47-6				1,000	X
95-48-7	10,000	1,000	1,000	1,000	X
95-50-1				100	X
95-53-4				1	X
95-57-8				100	
95-63-6			1		X
95-80-7				1	X
95-94-3				5,000	
95-95-4				10	X
96-09-3					X
96-12-8				1	X
96-33-3					X
96-45-7				1	X
97-18-7	10,000	100	1		
97-56-3					X
97-63-2				1,000	
98-01-1				5,000	
98-05-5	10,000	10	1		
98-07-7		100	1	1	X
98-09-9			100	100	
98-13-5		500	1		
98-16-8		500	1		
98-82-8				5,000	X
98-86-2				5,000	
98-87-3		500	5,000	5,000	X

All quantities expressed in pounds

| CAS No. | Thresh. Planning Qty | | Reportable Quantity | | Annual Report |
	Lg. Gran.	Other	SARA	CERCLA	
98-88-4				1,000	X
98-95-3		10,000	1,000	1,000	X
99-08-1				1,000	
99-35-4				10	
99-55-8				1	
99-59-2					X
99-65-0				100	
99-98-9	10,000	10	1		
99-99-0				1,000	
100-01-6				5,000	
100-02-7				100	X
100-14-1	10,000	500	1		
100-21-0					X
100-25-4				100	
100-41-4				1,000	X
100-42-5				1,000	X
100-44-7		500	100	100	X
100-47-0				5,000	
100-75-4				1	X
101-14-4				1	X
101-55-3				100	
101-61-1					X
101-68-8					X
101-77-9					X
101-80-4					X
102-36-3	10,000	500	1		
103-23-1					X
103-85-5	10,000	100	100	100	
104-94-9					X
105-46-4				5,000	
105-67-9				100	X
106-42-3				1,000	X
106-44-5				1,000	X
106-46-7				100	X
106-47-8				1,000	
106-49-0				1	
106-50-3					X
106-51-4				10	X
106-88-7					X
106-89-8		1,000	1,000	1,000	X
106-93-4				1,000	X
106-96-7		10	1		
106-99-0			1		X
107-02-8		500	1	1	X
107-05-1				1,000	X
107-06-2				5,000	X
107-07-3		500	1		
107-10-8				5,000	
107-11-9		500	1		
107-12-0		500	10	10	
107-13-1		10,000	100	100	X
107-15-3		10,000	5,000	5,000	
107-16-4		1,000	1		
107-18-6		1,000	100	100	
107-19-7				1,000	

All quantities expressed in pounds

| CAS No. | Thresh. Planning Qty | | Reportable Quantity | | Annual Report |
	Lg. Gran.	Other	SARA	CERCLA	
107-20-0			1,000	1,000	
107-21-1					
107-30-2		100	1	1	
107-31-3		10	1		
107-49-3		100	10	10	
107-92-6				5,000	
108-05-4		1,000	5,000	5,000	X
108-10-1				5,000	X
108-23-6		1,000	1		
108-24-7				5,000	
108-31-6				5,000	X
108-38-3				1,000	X
108-39-4				1,000	X
108-46-3				5,000	
108-60-1				1,000	X
108-78-1					X
108-88-3				1,000	X
108-90-7				100	X
108-91-8		10,000	1		
108-94-1				5,000	
108-95-2	10,000	500	1,000	1,000	X
108-98-5		500	100	100	
109-06-8				5,000	
109-61-5		500	1		
109-73-9				1,000	
109-77-3	10,000	500	1,000	1,000	
109-86-4					X
109-89-7				100	
109-99-9				1,000	
110-00-9		500	100	100	
110-16-7				5,000	
110-17-8				5,000	
110-19-0				5,000	
110-57-6		500	1		
110-75-8				1,000	
110-80-5				1	X
110-82-7				1,000	X
110-86-1				1,000	X
110-89-4		1,000	1		
111-42-2					X
111-44-4		10,000	1	1	X
111-54-6				5,000	
111-69-3		1,000	1		
111-91-1				1,000	
114-26-1					X
115-02-6				1	
115-07-1					X
115-21-9		500	1		
115-26-4		500	1		
115-29-7	10,000	10	1	1	
115-32-2				10	X
115-90-2		500	1		
116-06-3	10,000	100	1	1	
117-79-3					X
117-80-6				1	

All quantities expressed in pounds

| CAS No. | Thresh. Planning Qty | | Reportable Quantity | | Annual Report |
	Lg. Gran.	Other	SARA	CERCLA	
117-81-7				1	X
117-84-0				5,000	X
118-74-1				1	X
119-38-0		500	1		
119-90-4				1	X
119-93-7				1	X
120-12-7				5,000	X
120-58-1				1	
120-71-8					X
120-80-9					X
120-82-1				100	X
120-83-2				100	X
121-14-2				1,000	X
121-21-1				1	
121-29-9				1	
121-44-8				5,000	
121-69-7					X
121-75-5				100	
122-09-8				5,000	
122-14-5		500	1		
122-66-7				1	X
123-31-9	10,000	500	1		X
123-33-1				5,000	
123-38-6					X
123-62-6				5,000	
123-63-7				1,000	
123-72-8					X
123-73-9		1,000	100	100	
123-86-4				5,000	
123-91-1				1	X
123-92-2				5,000	
124-04-9				5,000	
124-40-3				1,000	
124-41-4				1,000	
124-48-1				100	
124-65-2	10,000	100	1		
124-87-8	10,000	500	1		
126-72-7				1	X
126-98-7		500	1	1,000	
126-99-8					X
127-18-4				1	X
127-82-2				5,000	
128-66-5					X
129-00-0	10,000	1,000	5,000	5,000	
129-06-6	10,000	100	1		
130-15-4				5,000	
131-11-3				5,000	X
131-52-2	10,000	100	1		
131-74-8				10	
131-89-5				100	
132-64-9					X
133-06-2				10	X
133-90-4					X
134-29-2					X
134-32-7				1	X

All quantities expressed in pounds

| CAS No. | Thresh. Planning Qty | | Reportable Quantity | | Annual Report |
	Lg. Gran.	Other	SARA	CERCLA	
135-20-6					X
137-26-8			10		
139-13-9					X
139-65-1					X
140-29-4		500	1		
140-76-1		500	1		
140-88-5				1,000	X
141-32-2					X
141-66-2		100	1		
141-78-6				5,000	
142-28-9				1,000	
142-71-2				100	
142-84-7				5,000	
143-33-9		100	10	10	
143-50-0				1	
144-49-0	10,000	10	1		
145-73-3				1,000	
148-82-3				1	
149-74-6		1,000	1		
151-38-2	10,000	500	1		
151-50-8		100	10	10	
151-56-4		500	1	1	X
152-16-9		100	100	100	
156-10-5					X
156-60-5				1,000	
156-62-7					X
189-55-9				1	
191-24-2				5,000	
193-39-5				1	
205-99-2				1	
206-44-0				100	
207-08-9				1	
208-96-8				5,000	
218-01-9				1	
225-51-4				1	
287-92-3			1		
297-78-9	10,000	100	1		
297-97-2		500	100	100	
298-00-0	10,000	100	100	100	
298-02-2		10	10	10	
298-04-4		500	1	1	
300-62-9		1,000	1		
300-76-5				10	
301-04-2				5,000	
302-01-2		1,000	1	1	X
303-34-4				1	
305-03-3				1	
309-00-2	10,000	500	1	1	X
311-45-5				100	
315-18-4	10,000	500	1,000	1,000	
316-42-7	10,000	1	1		
319-84-6				1	
319-85-7				1	
319-86-8				1	
327-98-0		500	1		

All quantities expressed in pounds

| | Thresh. Planning Qty | | Reportable Quantity | | Annual |
CAS No.	Lg. Gran.	Other	SARA	CERCLA	Report
329-71-5				10	
330-54-1				100	
333-41-5				1	
334-88-3					X
353-42-4		1,000	1		
353-50-4				1,000	
357-57-3				100	
359-06-8		10	1		
371-62-0		10	1		
379-79-3	10,000	500	1		
460-19-5				100	
463-58-1					X
465-73-6	10,000	100	1	1	
470-90-6		500	1		
492-80-8				1	X
494-03-1				1	
496-72-0				1	
502-39-6	10,000	500	1		
504-24-5	10,000	500	1,000	1,000	
504-60-9				100	
505-60-2		500	1		X
506-61-6		500	1	1	
506-64-9				1	
506-68-3	10,000	500	1,000	1,000	
506-77-4				10	
506-78-5	10,000	1,000	1		
506-87-6				5,000	
506-97-7				5,000	
509-14-8		500	10	10	
510-15-6				1	X
513-49-5				1,000	
514-73-8	10,000	500	1		
528-29-0				100	
532-27-4					X
534-07-6	10,000	10	1		
534-52-1	10,000	10	10	10	X
535-89-7	10,000	100	1		
538-07-8		500	500		
540-59-0					X
540-73-8				1	
540-88-5				5,000	
541-09-3				100	
541-25-3		10	1		
541-41-3					X
541-53-7	10,000	100	100	100	
541-73-1				100	X
542-62-1				10	
542-75-6				100	X
542-76-7		1,000	1,000	1,000	
542-88-1		100	1	1	X
542-90-5		10,000		1	
543-90-8				100	
544-18-3				1,000	
544-92-3				10	
554-84-7				100	

All quantities expressed in pounds

| CAS No. | Thresh. Planning Qty | | Reportable Quantity | | Annual Report |
	Lg. Gran.	Other	SARA	CERCLA	
555-77-1		100	1		
556-61-1		500	1		
556-64-9		10,000	1		
557-19-7				1	
557-21-7				10	
557-34-6				1,000	
557-41-5				1,000	
558-25-8		1,000	1		
563-12-2		1,000	10	10	
563-41-7	10,000	1,000	1		
563-68-8				100	
569-64-2					X
573-56-8				10	
584-84-9		500	100	100	X
591-08-2				1,000	
592-01-8				10	
592-04-1				1	
592-85-8				10	
592-87-0				100	
593-60-2					X
594-42-3		500	100	100	
594-72-9		100	1		
598-31-2				1,000	
606-20-2				1,000	X
608-93-5				10	
609-19-8				10	
610-39-9				1,000	
614-78-8	10,000	500	1		
615-05-4					X
615-53-2				1	
621-64-7				1	X
624-83-9		500	1	1	X
624-92-0		100	1		
625-16-1				5,000	
625-55-8		500	1		
626-38-0				5,000	
627-11-2		1,000	1		
628-63-7				5,000	
628-86-4				10	
630-10-4				1,000	
630-20-6				1	
630-60-4	10,000	100	1		
631-61-8				5,000	
636-21-5				1	X
639-58-7	10,000	500	1		
640-19-7	10,000	100	100	100	
644-64-4	10,000	500	1		
675-14-9		100	1		
676-97-1		100	1		
680-31-9					X
684-93-5				1	X
692-42-2				1	
696-26-6		500	1	1	
732-11-6	10,000	10	1		
757-58-4				100	

All quantities expressed in pounds

| CAS No. | Thresh. Planning Qty | | Reportable Quantity | | Annual Report |
	Lg. Gran.	Other	SARA	CERCLA	
759-73-9				1	X
760-93-0		500	1		
764-41-0				1	
765-33-4				1	
786-19-6		500	1		
810-49-3		500	1		
815-82-7				100	
823-40-5				1	
824-11-3	10,000	100	1		
842-07-9					X
900-95-8	10,000	500	1		
919-86-8		500	1		
920-46-7		100	1		
924-16-3				1	X
930-55-2				1	
933-75-5				10	
933-78-8				10	
944-22-9		500	1		
947-02-4	10,000	100	1		
950-10-7		500	1		
950-37-8	10,000	500	1		
959-98-8				1	
961-11-5					X
989-38-8					X
991-42-4	10,000	100	1		
998-30-1		500	1		
999-81-5	10,000	100	1		
1024-57-3				1	
1031-07-8				1	
1031-47-6	10,000	500	1		
1066-30-4				1,000	
1066-33-7				5,000	
1066-45-1	10,000	500	1		
1072-35-1				5,000	
1111-78-0				5,000	
1116-54-7				1	
1120-71-4				1	X
1122-60-7		500	1		
1124-33-0	10,000	500	1		
1129-41-5	10,000	100	1		
1163-19-5					X
1185-57-5				1,000	
1194-65-7				100	
1300-71-6				1,000	
1303-28-2	10,000	100	5,000	5,000	
1303-32-8				5,000	
1303-33-9				5,000	
1306-19-0	10,000	100	1		
1309-64-4				1,000	
1310-58-3				1,000	
1310-73-2				1,000	X
1313-27-5					X
1314-20-1					X
1314-32-5			100	100	
1314-56-3		10	1		

All quantities expressed in pounds

| CAS No. | Thresh. Planning Qty | | Reportable Quantity | | Annual Report |
	Lg. Gran.	Other	SARA	CERCLA	
1314-62-1	10,000	100	1,000	1,000	
1314-80-3				100	
1314-84-7		500	100	100	
1314-87-0				5,000	
1314-96-1				100	
1319-72-8				5,000	
1319-77-3				1,000	X
1320-18-9				100	
1321-12-6				1,000	
1327-52-2				1	
1327-53-3	10,000	100	5,000	5,000	
1330-20-7				1,000	X
1332-07-6				1,000	
1332-21-4				1	X
1333-83-1				100	
1335-32-6				1	
1335-87-1			1		X
1336-21-6				1,000	
1336-36-3				10	X
1338-23-4				10	
1338-24-5				100	
1341-49-7				100	
1344-28-1					X
1397-94-0	10,000	1,000	1		
1420-07-1	10,000	500	1		
1464-53-5		500	1	1	X
1558-25-4		100	1		
1563-66-2	10,000	10	10	10	
1582-09-8					X
1600-27-7	10,000	500	1		
1615-80-1				1	
1622-32-8		500	1		
1634-04-4					X
1642-54-2	10,000	100	1		
1746-01-6				1	
1752-30-3	10,000	1,000	1		
1762-95-4				5,000	
1836-75-5					X
1863-63-4				5,000	
1888-71-7				1,000	
1897-45-6					X
1910-42-5	10,000	10	1		
1918-00-9				1,000	
1928-38-7				100	
1928-47-8				1,000	
1928-61-6				100	
1929-73-3				100	
1937-37-7					X
1982-47-4	10,000	500	1		
2001-95-8	10,000	1,000	1		
2008-46-0				5,000	
2032-65-7	10,000	500	10	10	
2074-50-2	10,000	10	1		
2097-19-0	10,000	100	1		
2104-64-5	10,000	100	1		

All quantities expressed in pounds

| CAS No. | Thresh. Planning Qty | | Reportable Quantity | | Annual Report |
	Lg. Gran.	Other	SARA	CERCLA	
2164-17-2					X
2223-93-0	10,000	1,000	1		
2231-57-4	10,000	1,000	1		
2234-13-1					X
2238-07-5		1,000	1		
2275-18-5	10,000	100	1		
2303-16-4				1	X
2312-35-8				10	
2497-07-6		500	1		
2524-03-0		500	1		
2540-82-1		100	1		
2545-59-7				1,000	
2570-26-5	10,000	100	1		
2587-90-8		500	1		
2602-46-2					X
2631-37-0	10,000	500	1		
2636-26-2		1,000	1		
2642-71-9	10,000	100	1		
2650-18-2					X
2665-30-7		500	1		
2703-13-1		500	1		
2757-18-8	10,000	100	1		
2763-96-4		10,000	1,000	1,000	
2764-72-9				1,000	
2778-04-3	10,000	500	1		
2832-40-8					X
2921-88-2				1	
2944-67-4				1,000	
2971-38-2				100	
3012-65-5				5,000	
3037-72-7		1,000	1		
3118-97-6					X
3164-29-2				5,000	
3165-93-3				1	
3251-23-8				100	
3254-63-5		500	1		
3288-58-2				5,000	
3386-35-9				1,000	
3569-57-1		500	1		
3615-21-2	10,000	500	1		
3689-24-5		500	100	100	
3691-35-8	10,000	100	1		
3734-97-2	10,000	100	1		
3735-23-7		500	1		
3761-53-3					X
3813-14-7				5,000	
3844-45-9					X
3878-19-1	10,000	100	1		
4044-65-9	10,000	500	1		
4098-71-9		100	1		
4104-14-7	10,000	100	1		
4170-30-3		1,000	100	100	
4301-50-2	10,000	100	1		
4418-66-0	10,000	100	1		
4549-40-0				1	X

All quantities expressed in pounds

| CAS No. | Thresh. Planning Qty | | Reportable Quantity | | Annual Report |
	Lg. Gran.	Other	SARA	CERCLA	
4680-78-8					X
4835-11-4		500	1		
5281-13-0	10,000	100	1		
5333-41-5				1	
5344-82-1	10,000	100	100	100	
5836-29-3	10,000	500	1		
5893-66-3				100	
5972-73-6				5,000	
6009-70-7				5,000	
6369-96-6				5,000	
6369-97-7				5,000	
6484-52-2					X
6533-73-9	10,000	100	100	100	
6923-22-4	10,000	10	1		
7005-72-3				5,000	
7421-93-4				1	
7428-48-0				5,000	
7429-90-5					X
7439-92-1				1	X
7439-96-5					X
7439-97-6				1	X
7440-02-0			1	1	X
7440-22-4				1,000	X
7440-23-5				10	
7440-28-0				1,000	X
7440-36-0				5,000	X
7440-38-2				1	X
7440-39-3					X
7440-41-7				1	X
7440-43-9				1	X
7440-47-3				1	X
7440-48-4			1		X
7440-50-8				5,000	X
7440-62-2					X
7440-66-6				1,000	X
7446-08-4				10	
7446-09-5		500	1		
7446-11-9		100	1		
7446-14-2				100	
7446-18-6	10,000	100	100	100	
7446-27-7				1	
7447-39-4				10	
7487-94-7	10,000	500	1		
7488-56-4				1	
7550-45-0		100	1		X
7558-79-4				5,000	
7580-67-8		100	1		
7601-54-9				5,000	
7631-89-2	10,000	1,000	1,000	1,000	
7631-90-5				5,000	
7632-00-0				100	
7637-07-2		500	1		
7645-25-2				5,000	
7646-85-7				1,000	
7647-01-0		500	1	5,000	X

All quantities expressed in pounds

| CAS No. | Thresh. Planning Qty | | Reportable Quantity | | Annual Report |
	Lg. Gran.	Other	SARA	CERCLA	
7647-18-9				1,000	
7664-38-2				5,000	X
7664-39-3		100	100	100	X
7664-41-7		500	100	100	X
7664-93-9		1,000	1,000	1,000	X
7681-49-4				1,000	
7681-52-9				100	
7697-37-2		1,000	1,000	1,000	X
7699-45-8				1,000	
7705-08-0				1,000	
7718-54-9				5,000	
7719-12-2		1,000	1,000	1,000	
7720-78-7				1,000	
7722-64-7				100	
7722-84-1		1,000	1		
7723-14-0		100	1	1	X
7726-95-6		500	1		
7733-02-0				1,000	
7738-94-5				1,000	
7757-82-6					X
7758-29-4				5,000	
7758-94-3				100	
7758-95-4				100	
7758-98-7				10	
7761-88-8				1	
7773-06-0				5,000	
7775-11-3				1,000	
7778-39-4				1	
7778-44-1	10,000	500	1,000	1,000	
7778-50-9				1,000	
7778-54-3				10	
7779-86-4				1,000	
7779-88-6				1,000	
7782-41-4		500	10	10	
7782-49-2				100	X
7782-50-5		100	10	10	X
7782-63-0				1,000	
7782-82-3				100	
7782-86-7				10	
7783-00-8	10,000	1,000	10	10	
7783-06-4		500	100	100	
7783-07-5		10	1		
7783-18-8				5,000	
7783-20-2					X
7783-35-9				10	
7783-46-2				100	
7783-49-5				1,000	
7783-50-8				100	
7783-56-4				1,000	
7783-60-0		100	1		
7783-70-2		500	1		
7783-80-4		100	1		
7784-34-1		500	5,000	5,000	
7784-40-9				5,000	
7784-41-0				1,000	

All quantities expressed in pounds

| CAS No. | Thresh. Planning Qty | | Reportable Quantity | | Annual Report |
	Lg. Gran.	Other	SARA	CERCLA	
7784-42-1		100	1		
7784-46-5	10,000	500	1,000	1,000	
7785-84-4				5,000	
7786-34-7		500	10	10	
7786-81-4				5,000	
7787-47-5				5,000	
7787-49-7				5,000	
7787-55-5				5,000	
7788-98-9				1,000	
7789-00-6				1,000	
7789-06-2				1,000	
7789-09-5				1,000	
7789-42-6				100	
7789-43-7				1,000	
7789-61-9				1,000	
7790-94-5				1,000	
7791-12-0	10,000	100	100	100	
7791-23-3		500	1		
7803-51-2		500	100	100	
7803-55-6				1,000	
8001-35-2	10,000	500	1	1	X
8001-58-9				1	
8003-19-8				100	
8003-34-7				1	
8014-95-7				1,000	
8065-48-3		500	1		
9004-66-4				5,000	
10022-70-5				100	
10025-73-7	10,000	1	1		
10025-87-3		500	,000	1,000	
10025-91-9				1,000	
10026-11-6				5,000	
10026-13-8		500	1		
10028-15-6		100	1		
10028-22-5				1,000	
10031-59-1	10,000	100	100	100	
10034-93-2					X
10039-32-4				5,000	
10043-01-3				5,000	
10045-89-3				1,000	
10045-94-0				10	
10049-04-4					X
10049-05-5				1,000	
10099-74-8				100	
10101-53-8				1,000	
10101-63-0				100	
10101-89-0				5,000	
10102-06-4				100	
10102-18-8	10,000	100	100	100	
10102-20-2	10,000	500	1		
10102-43-9		100	10	10	
10102-44-0		100	10	10	
10102-45-1				100	
10102-48-4				5,000	
10108-64-2				100	

All quantities expressed in pounds

| CAS No. | Thresh. Planning Qty | | Reportable Quantity | | Annual Report |
	Lg. Gran.	Other	SARA	CERCLA	
10124-50-2	10,000	500	1,000	1,000	
10124-56-8				5,000	
10140-65-5				5,000	
10140-87-1		1,000	1		
10192-30-0				5,000	
10196-04-0				5,000	
10210-68-1	10,000	10	1		
10265-92-6	10,000	100	1		
10294-34-5		500	1		
10311-84-9	10,000	100	1		
10361-89-4				5,000	
10380-29-7				100	
10415-75-5				10	
10421-48-4				1,000	
10476-95-6		1,000	1		
10544-72-6				10	
10588-01-9				1,000	
11096-82-5				10	
11097-69-1				10	
11104-28-2				10	
11115-74-5				1,000	
11141-16-5				10	
12002-03-8	10,000	500	100	100	
12039-52-0				1,000	
12054-48-7				1,000	
12108-13-3		100	1		
12122-67-7					X
12125-01-8				100	
12125-02-9				5,000	
12135-76-1				100	
12427-38-2					X
12672-29-6				10	
12674-11-2				10	
12771-08-3				1,000	
13071-79-9		100	1		
13171-21-6		100	1		
13194-48-4		1,000	1		
13410-01-0	10,000	100	1		
13450-90-3	10,000	500	1		
13463-39-3		1	1	1	
13463-40-6		100	1		
13494-80-9	10,000	500	1		
13560-99-1				1,000	
13597-99-4				5,000	
13746-89-9				5,000	
13765-19-0				1,000	
13814-96-5				100	
13826-83-0				5,000	
13952-84-6				1,000	
14017-41-5				1,000	
14167-18-1	10,000	500	1		
14216-75-2				5,000	
14258-49-2				5,000	
14307-35-8				1,000	
14307-43-8				5,000	

All quantities expressed in pounds

| CAS No. | Thresh. Planning Qty | | Reportable Quantity | | Annual Report |
	Lg. Gran.	Other	SARA	CERCLA	
14639-97-5				5,000	
14639-98-6				5,000	
14644-61-2				5,000	
15271-41-7	10,000	500	1		
15699-18-0				5,000	
15739-80-7				100	
15950-66-0				10	
16071-86-6					X
16543-55-8					X
16721-80-5				5,000	
16752-77-5	10,000	500	100	100	
16871-71-9				5,000	
16919-19-0				1,000	
16923-95-8				1,000	
17702-41-9	10,000	500	1		
17702-57-7	10,000	100	1		
18883-66-4				1	
19287-45-7		100	1		
19624-22-7		500	1		
20816-12-0			1,000	1,000	X
20830-75-5	10,000	1	1		
20830-81-3				1	
20859-73-8		500	100	100	
21548-32-3		500	1		
21609-90-5	10,000	500	1		
21908-53-2	10,000	500	1		
21923-23-9		500	1		
22224-92-6	10,000	10	1		
23135-22-0	10,000	100	1		
23422-53-9	10,000	500	1		
23505-41-1		1,000	1		
23950-58-5				5,000	
24017-47-8		500	1		
24934-91-6		500	1		
25154-54-5				100	
25154-55-6				100	
25155-30-0				1,000	
25167-82-2				10	
25168-15-4				1,000	
25168-26-7				100	
25321-14-6				1,000	
25321-22-6				100	X
25376-45-8				1	X
25550-58-7				10	
26264-06-2				1,000	
26419-73-8	10,000	100	1		
26471-62-5				100	
26628-22-8		500	1,000	1,000	
26638-19-7				1,000	
26952-23-8				100	
27137-85-5		500	1		
27176-87-0				1,000	
27323-41-7				1,000	
27774-13-6				1,000	
28300-74-5				100	

All quantities expressed in pounds

| CAS No. | Thresh. Planning Qty | | Reportable Quantity | | Annual Report |
	Lg. Gran.	Other	SARA	CERCLA	
28347-13-9	10,000	100	1		
28772-56-7	10,000	100	1		
30525-89-4				1,000	
30674-80-7		100	1		
32534-95-5				100	
33213-65-9				1	
36355-01-8					X
36478-76-9				100	
37211-05-5				5,000	
39156-41-7					X
39196-18-4	10,000	100	100	100	
42504-46-1				1,000	
50782-69-9		100	1		
52628-25-8				1,000	
52652-59-2				5,000	
52740-16-6				1,000	
53467-11-1				100	
53469-21-9				10	
53558-25-1	10,000	100	1		
55488-87-4				1,000	
56189-09-4				5,000	
58270-08-9	10,000	100	1		
61792-07-2				1,000	
62207-76-5	10,000	100	1		
X00001-05-8					X
X00001-09-2					X
X00001-11-6	10,000	10	1		
X00001-14-9					X
X00001-28-5					X
X00001-29-6					X
X00001-30-9					X
X00001-31-0					X
X00001-37-6				1	
X00001-40-1					X
X00001-49-0					X
X00001-51-4					X
X00001-53-6					X
X00001-60-5					X
X00001-63-8				1	
X00001-66-1					X
X00001-69-4					X
X00001-70-7					X
X00001-75-2					X
X00001-80-9					X
X00001-81-0					X

All quantities expressed in pounds

APPENDIX D.3

Materials Regulated Under Other Federal Statutes

CAS No.	RCRA	Shipping Identification Number
50-00-0	U122	UN1198,UN2209
50-07-7	U010	
50-18-0	U058	
50-29-3	U061	NA2761
50-32-8	U022	
50-55-5	U200	
51-28-5	P048	UN0076,UN1320,UN1321
51-43-4	P042	
51-79-6	U238	
52-68-6		NA2783
52-85-7	P097	
53-70-3	U063	
53-96-3	U005	
54-11-5	P075	UN1654,UN1656
55-18-5	U174	
55-63-0	P081	UN0143,UN0144,NA1204
55-91-4	P043	
56-04-2	U164	
56-23-5	U211	UN1846,UN2516
56-38-2	P089	NA2783
56-49-5	U157	
56-53-1	U089	
56-55-3	U018	
56-72-4		NA2783
57-12-5	P030	UN1935
57-14-7	U098	UN1163,UN2382
57-24-9	P108	UN1692
57-74-9	U036	NA2762
57-97-6	U094	
58-89-9	U129	NA2761

CAS No.	RCRA	Shipping Identification Number
58-90-2	U212	
59-50-7	U039	UN2669
60-00-4		NA9117
60-11-7	U093	
60-29-7	U117	UN1155
60-34-4	P068	UN1244
60-51-5	P044	
60-57-1	P037	NA2761
61-82-5	U011	
62-38-4	P092	
62-44-2	U187	
62-50-0	U119	
62-53-3	U012	UN1547
62-55-5	U218	
62-56-6	U219	UN2877
62-73-7		NA2783
62-74-8	P058	UN2629
62-75-9	P082	
63-25-2		NA2757
64-17-5		UN1170
64-18-6	U123	UN1779
64-19-7		UN2789,UN2790
64-67-5		UN1594
65-85-0		NA9094
66-75-1	U237	
67-56-1	U154	UN1230
67-63-0		UN1219
67-64-1	U002	UN1090
67-66-3	U044	UN1888
67-72-1	U131	NA9037
68-11-1		UN1940
68-12-2		UN2265
70-25-7	U163	NA1325
70-30-4	U132	UN2875
71-23-8		UN1274
71-36-3	U031	UN1120
71-43-2	U019	UN1114
71-55-6	U226	UN2831
72-20-8	P051	NA2761
72-43-5	U247	NA2761
72-54-8	U060	NA2761
72-57-1	U236	
74-82-8		UN1971,UN1972
74-83-9	U029	UN1062
74-84-0		UN1035,UN1961
74-85-1		UN1038,UN1962
74-86-2		UN1001
74-87-3	U045	UN1063
74-88-4	U138	UN2644
74-89-5		UN1061,UN1235
74-90-8	P063	UN1051,UN1613,UN1614
74-93-1	U153	UN1064
74-95-3	U068	UN2664
74-96-4		UN1891
74-97-5		UN1887
74-98-6		UN1978
75-00-3		UN1037

CAS No.	RCRA	Shipping Identification Number
75-01-4	U043	UN1086
75-04-7		UN1036,UN2270
75-05-8	U003	UN1648,NA1648
75-07-0	U001	UN1089
75-08-1		UN2363
75-09-2	U080	UN1593
75-15-0	P022	UN1131
75-18-3		UN1164
75-21-8	U115	UN1040
75-25-2	U225	UN2515
75-31-0		UN1221
75-34-3	U076	UN2362
75-35-4	U078	UN1150,UN1303
75-36-5	U006	UN1717
75-43-4		UN1029
75-44-5	P095	UN1076
75-45-6		UN1018
75-50-3		UN1083,UN1297
75-52-5		UN1261
75-55-8	P067	UN1921
75-56-9		UN1280
75-60-5	U136	UN1572
75-61-6		UN1941
75-63-8		UN1009
75-65-0		UN1120
75-69-4	U121	
75-71-8	U075	UN1028
75-77-4		UN1298
75-78-5		UN1162
75-79-6		UN1250
75-86-5	P069	UN1541
75-87-6	U034	UN2075
75-99-0		NA1760
76-01-7	U184	UN1669
76-02-8		UN2442
76-03-9		UN1839,UN2564
76-06-2		UN1580,UN1583
76-14-2		UN1958
76-22-2		UN2717
76-44-8	P059	NA2761
77-47-4	U130	UN2646
77-73-6		UN2048
77-78-1	U103	UN1595
78-00-2	P110	NA1649
78-10-4		UN1292
78-30-8		UN2574
78-79-5		UN1218
78-81-9		UN1214
78-82-0		UN2284
78-83-1	U140	UN1212
78-84-2		UN2045
78-87-5	U083	UN1279
78-92-2		UN1120
78-93-3	U159	UN1193
78-94-4		UN1251
79-00-5	U227	
79-01-6	U228	UN1710

CAS No.	RCRA	Shipping Identification Number
79-04-9		UN1752
79-06-1	U007	UN2075
79-09-4		UN1848
79-10-7	U008	UN2218
79-19-6	P116	
79-20-9		UN1231
79-22-1	U156	UN1238
79-24-3		UN2842
79-27-6		UN2504
79-31-2		UN2529
79-34-5	U209	UN1702
79-41-4		UN2531
79-44-7	U097	UN2262
79-46-9	U171	UN2608
80-15-9	U096	UN2116
80-62-6	U162	UN1247
81-07-2	U202	
81-81-2	P001	
82-68-8	U185	
83-26-1		UN2472
84-66-2	U088	
84-74-2	U069	
85-00-7		NA2781
85-44-9	U190	UN2214
86-50-0		NA2783
86-88-4	P072	UN1651
87-62-7		UN1711
87-65-0	U082	
87-68-3	U128	UN2279
87-86-5	U242	NA2020
88-06-2	U231	
88-72-2		UN1664
88-75-5		UN1663
88-85-7	P020	
88-89-1		UN0154,NA1344,UN1344
89-72-5		UN2228,UN2229
91-08-7		UN2078
91-20-3	U165	UN1334,UN2304
91-22-5		UN2656
91-58-7	U047	
91-59-8	U168	UN1650,UN2077
91-80-5	U155	
91-94-1	U073	
92-87-5	U021	UN1885
93-72-1	U233	NA2765
93-76-5	U232	NA2765
93-79-8		NA2765
94-11-1		NA2765
94-36-0		UN2085,UN2086,UN2087,UN2088,UN2089,UN2090
94-58-6	U090	
94-59-7	U203	
94-75-7	U240	NA2765
94-79-1		NA2765
94-80-4		NA2765
95-47-6		UN1307
95-48-7	U052	
95-49-8		UN2238

CAS No.	RCRA	Shipping Identification Number
95-50-1	U070	UN1591
95-53-4	U328	UN1708
95-57-8	U048	UN2020
95-80-7	U221	UN1709
95-94-3	U207	
95-95-4	U230	NA2020
96-12-8	U066	UN2872
96-22-0		UN1156
96-33-3		UN1919
96-45-7	U116	
97-63-2	U118	UN2277
98-00-0		UN2874
98-01-1	U125	UN1199
98-07-7	U023	UN2226
98-09-9	U020	
98-13-5		UN1804
98-16-8		UN2948
98-51-1		UN2667
98-82-8	U055	
98-83-9		UN2303
98-86-2	U004	
98-87-3	U017	
98-88-4		UN1737
98-95-3	U169	UN1662
99-08-1		UN1664
99-35-4	U234	UN1354
99-55-8	U181	UN2660
99-65-0		UN1597
99-99-0		UN1664
100-00-5		UN1578
100-01-6	P077	UN1660
100-02-7	U170	UN1663
100-25-4		UN1597
100-41-4		UN1175
100-42-5		UN2055
100-44-7	P028	UN1738
100-47-0		UN2224
100-61-8		UN2294
100-63-0		UN2572
100-75-4	U179	
101-14-4	U158	
101-55-3	U030	
101-68-8		UN2489
102-81-8		UN2873
103-85-5	P093	
105-30-6		UN2053
105-46-4		UN1123
105-67-9	U101	UN2261
106-42-3		UN1307
106-44-5	U052	
106-46-7	U072	UN1592
106-47-8	P024	UN2018, UN2019
106-48-9		UN2020
106-49-0	U353	UN1708
106-50-3		UN1673
106-51-4	U197	UN2587
106-89-8	U041	UN2023

CAS No.	RCRA	Shipping Identification Number
106-92-3		UN2219
106-93-4	U067	UN1605
106-97-8		UN1011
106-99-0		UN1010
107-02-8	P003	UN1092
107-05-1		UN1100
107-06-2	U077	UN1184
107-07-3		UN1135
107-10-8	U194	UN1277
107-11-9		UN2334
107-12-0	P101	UN2404
107-13-1	U009	UN1093
107-18-6	P005	UN1098
107-19-7	P102	NA1986
107-20-0	P023	UN2232
107-30-2	U046	UN1239
107-31-3		UN1243
107-49-3	P111	
107-87-9		UN1249
108-03-2		UN2608
108-05-4		UN1301
108-10-1	U161	UN1245
108-18-9		UN1158
108-21-4		UN1220
108-23-6		UN2407
108-24-7		UN1715
108-31-6	U147	NA2215
108-38-3		UN1307
108-39-4	U052	
108-43-0		UN2020
108-44-1		UN1708
108-46-3	U201	UN2867
108-60-1	U027	
108-83-8		UN1157
108-87-2		UN2296
108-88-3	U220	UN1294
108-90-7	U037	UN1134
108-91-8		UN2357
108-94-1	U057	UN1915
108-95-2	U188	UN1671,UN2312,UN2821
108-98-5	P014	UN2337
109-06-8	U191	UN2313
109-60-4		UN1276
109-61-5		UN2740
109-66-0		UN1265
109-77-3	U149	UN2647
109-79-5		UN2347
109-86-4		UN1188
109-87-5		UN1234
109-89-7		UN1154
109-94-4		UN1190
109-99-9	U213	UN2056
110-00-9	U124	UN2389
110-12-3		UN2302
110-16-7		NA2215
110-17-8		NA9126
110-19-0		UN1213

CAS No.	RCRA	Shipping Identification Number
110-49-6		UN1189
110-54-3		UN1208
110-57-6		NA2924
110-62-3		UN2058
110-75-8	U042	
110-80-5	U359	UN1171
110-82-7	U056	UN1145
110-83-8		UN2256
110-86-1	U196	UN1282
110-89-4		UN2401
110-91-8		NA1760,UN2054
111-15-9		UN1172
111-40-0		UN2079
111-44-4	U025	UN1916
111-54-6	U114	
111-69-3		UN2205
111-84-2		UN1920
111-91-1	U024	
115-02-6	U015	
115-07-1		UN1077
115-21-9		UN1196
115-29-7	P050	NA2761
115-32-2		NA2761
116-06-3	P070	
117-80-6		NA2761
117-81-7	U028	
117-84-0	U107	
118-74-1	U127	UN2729
118-96-7		UN0209,UN1356
119-90-4	U091	
119-93-7	U095	
120-58-1	U141	
120-82-1		UN2321
120-83-2	U081	
121-14-2	U105	UN1600,UN2038
121-21-1		NA9184
121-29-9		NA9184
121-44-8		UN1296
121-45-9		UN2329
121-69-7		UN2253
121-75-5		NA2783
122-09-8	P046	
122-66-7	U109	
123-19-3		UN2710
123-31-9		UN2662
123-33-1	U148	
123-38-6		UN1275
123-42-2		UN1148
123-62-6		UN2496
123-63-7	U182	UN1264
123-72-8		UN1129
123-73-9	U053	UN1143
123-86-4		UN1123
123-91-1	U108	UN1165
124-04-9		NA9077
124-38-9		UN1013,UN1845,UN2187
124-40-3	U092	UN1032,UN1160

CAS No.	RCRA	Shipping Identification Number
124-41-4		UN1289,UN1431
124-65-2		UN1688
126-72-7	U235	
126-98-7	U152	
126-99-8		UN1991
127-18-4	U210	UN1897
127-82-2		NA9160
129-00-0		NA9184
130-15-4	U166	
131-11-3	U102	
131-52-2		UN2567
131-74-8	P009	UN0004
131-89-5	P034	
133-06-2		NA9099
134-32-7	U167	UN1650,UN2077
137-26-8	U244	NA2771
140-88-5	U113	UN1917
141-43-5		UN2491
141-78-6	U112	UN1173
141-79-7		UN1229
142-71-2		NA9106
142-82-5		UN1206
142-84-7	U110	UN2383
143-33-9	P106	UN1689
143-50-0	U142	NA2761
145-73-3	P088	
148-82-3	U150	
151-50-8	P098	UN1680
151-56-4	P054	UN1185
152-16-9	P085	
156-60-5	U079	
156-62-7		UN1403
189-55-9	U064	
193-39-5	U137	
206-44-0	U120	
218-01-9	U050	
225-51-4	U016	
287-92-3		UN1146
297-97-2	P040	
298-00-0	P071	NA2783
298-02-2	P094	
298-04-4	P039	NA2783
300-76-5		NA2783
301-04-2	U144	UN1616
302-01-2	U133	UN2029
303-34-4	U143	
305-03-3	U035	
309-00-2	P004	NA2761
311-45-5	P041	
315-18-4		NA2757
319-84-6		NA2761
329-71-5		UN0076,UN1320,UN1321
330-54-1		NA2767
333-41-5		NA2783
353-42-4		UN2965
353-50-4	U033	UN2417
357-57-3	P018	UN1570

CAS No.	RCRA	Shipping Identification Number
460-19-5	P031	UN1026
465-73-6	P060	
479-45-8		UN0208
492-80-8	U014	
494-03-1	U026	
496-72-0		NA1709
504-24-5	P008	UN2671
504-29-0		UN2671
504-60-9	U186	
506-61-6	P099	
506-64-9	P104	UN1684
506-68-3	U246	UN1889
506-77-4	P033	UN1589
506-87-6		NA9084
506-97-7		UN1716
509-14-8	P112	UN1510
510-15-6	U038	
513-49-5		UN1125
528-29-0		UN1597
532-27-4		UN1697
534-07-6		UN2649
534-52-1	P047	UN1598
540-59-0		UN1150
540-73-8	U099	UN2382
540-88-5		UN1123
541-09-3		NA9180
541-41-3		UN1182
541-53-7	P049	
541-73-1	U071	
541-85-5		UN2271
542-62-1	P013	UN1565
542-75-6	U084	
542-76-7	P027	
542-88-1	P016	UN2249
543-90-8		NA2570
544-18-3		NA9104
544-92-3	P029	UN1587
554-84-7		UN1663
556-61-1		UN2477
557-19-7	P074	UN1653
557-21-7	P121	UN1713
557-34-6		NA9153
557-41-5		NA9159
563-12-2		NA2783
563-68-8	U214	
563-80-4		UN2397
573-56-8		UN1320,UN1321
583-60-8		UN2297
584-84-9		UN2078
591-08-2	P002	
592-01-8	P021	UN1575
592-04-1		UN1636
592-85-8		UN1646
592-87-0		NA2291
593-60-2		UN1885
594-42-3	P118	UN1670
594-72-9		UN2650

CAS No.	RCRA	Shipping Identification Number
598-31-2	P017	UN1569
606-20-2	U106	UN2038
608-93-5	U183	
610-39-9		UN2038
615-53-2	U178	
621-64-7	U111	
624-83-9	P064	UN2480
625-16-1		UN1105
626-38-0		UN1105
627-13-4		UN1865
628-63-7		UN1104,UN1105
628-86-4	P065	UN0135
630-08-0		UN1016,NA9202
630-10-4	P103	
630-20-6	U208	
631-61-8		NA9079
636-21-5	U222	
640-19-7	P057	
684-16-2		UN2420
684-93-5	U177	
692-42-2	P038	
696-28-6	P036	NA1556
757-58-4	P062	NA2783
759-73-9	U176	
764-41-0	U074	NA2924
765-34-4	U126	UN2622
815-82-7		NA9111
823-40-5	U221	NA1709
924-16-3	U172	
930-55-2	U180	
933-75-5		NA2020
933-78-8		NA2020
959-98-8		NA2761
1066-30-4		NA9101
1066-33-7		NA9081
1072-35-1		NA2811
1111-78-0		NA9083
1116-54-7	U173	
1120-71-4	U193	
1185-57-5		NA9118
1194-65-6		NA2769
1300-71-6		UN2261
1300-73-8		UN1711
1303-28-2	P011	UN1559
1305-78-8		UN1910
1309-64-4		NA9201
1310-58-3		UN1813,UN1814
1310-73-2		UN1823,UN1824
1314-32-5	P113	
1314-56-3		UN1807
1314-62-1	P120	UN2862
1314-80-3	U189	UN1340
1314-84-7	P122	UN1714
1314-87-0		NA2291
1314-96-1	P107	
1319-72-8		NA2765
1319-77-3	U052	UN2076

CAS No.	RCRA	Shipping Identification Number
1320-18-9		NA2765
1321-12-6		UN1663
1327-52-2	P010	UN1553
1327-53-3	P012	UN1561
1330-20-7	U239	UN1307
1332-07-6		NA9155
1332-21-4		UN2212,UN2590
1333-74-0		UN1049,UN1966
1333-86-4		UN1362
1335-32-6	U146	
1336-36-3		UN2315
1338-23-4	U160	UN2127,UN2550
1338-24-5		NA9137
1464-53-5	U085	
1563-66-2		NA2757
1600-27-7		UN1629
1615-80-1	U086	
1634-04-4		UN2350,UN2398
1762-95-4		NA9092
1863-63-4		NA9080
1888-71-7	U243	
1918-00-9		NA2769
1928-38-7		NA2765
1928-47-8		NA2765
1928-61-6		NA2765
1929-73-3		NA2765
2008-46-0		NA2765
2032-65-7		NA2757
2303-16-4	U062	
2312-35-8		NA2765
2545-59-7		NA2765
2551-62-4		UN1080
2699-79-8		UN2191
2763-96-4	P007	
2764-72-9		NA2781
2921-88-2		NA2783
2944-67-4		NA9119
2971-38-2		NA2765
3012-65-5		NA9087
3164-29-2		NA9091
3165-93-3	U049	UN1579
3251-23-8		NA1479
3288-58-2	U087	
3386-35-9		NA9157
3689-24-5	P109	UN1703
3813-14-7		NA2765
4098-71-9		UN2290
4170-30-3	U053	UN1143
4549-40-0	P084	
5333-41-5		NA2783
5344-82-1	P026	
5893-66-3		NA2449
5972-73-6		NA2449
6009-70-7		NA2449
6369-96-6		NA2765
6369-97-7		NA2765
6484-52-2		NA1942,UN2067,UN2426

CAS No.	RCRA	Shipping Identification Number
6533-73-9	U215	
7428-48-0		NA2811
7429-90-5		UN1396
7439-97-6	U151	UN2809,NA2809
7440-01-9		UN1065
7440-16-6		UN1436
7440-21-3		UN1346
7440-23-5		UN1428
7440-36-0		UN2871
7440-37-1		UN1006,UN1951
7440-38-2		UN1558
7440-39-3		UN1399
7440-41-7	P015	UN1567
7440-58-6		UN1326,UN2545
7440-59-7		UN1046,UN1964
7446-08-4	U204	NA2811
7446-09-5		UN1079
7446-11-9		UN1829
7446-14-2		UN1794,NA2291
7446-18-6	P115	UN1707
7446-27-7	U145	
7447-39-4		UN2802
7487-94-7		UN1624
7488-56-4	U205	UN2657
7550-45-0		UN1838
7558-79-4		NA9147
7601-54-9		NA9148
7631-89-2		UN1685
7632-00-0		UN1500
7637-07-2		UN1008
7645-25-2		UN1617
7646-85-7		UN1840,UN2331
7647-01-0		UN1050,UN1789,NA1789,UN2186
7647-18-9		UN1549,UN1730,UN1731
7664-38-2		UN1805
7664-39-3	U134	UN1052,UN1790
7664-41-7		UN1005,UN2073,UN2672
7664-93-9		UN1830,UN2796
7681-49-4		UN1690
7681-57-4		NA2693
7697-37-2		NA1760,UN2031
7699-45-8		NA9156
7705-08-0		UN1773,UN2582
7718-54-9		NA9139
7719-09-7		UN1836
7719-12-2		UN1809
7722-84-1		UN2014,UN2984
7723-14-0		UN2447,UN1381
7726-95-6		UN1744
7733-02-0		NA9161
7738-94-5		NA1463,UN1755
7758-29-4		NA9148
7758-95-4		NA2291
7758-98-7		NA9109
7761-88-8		UN1493
7773-06-0		NA9089
7775-11-3		NA9145

CAS No.	RCRA	Shipping Identification Number
7778-39-4	P010	UN1553
7778-44-1		UN1573
7778-50-9		NA1479
7778-54-3		UN1748,UN2208,UN2280
7779-86-4		UN1514
7782-41-4	P056	UN1045
7782-42-5		UN2658
7782-50-5		UN1017
7782-63-0		NA9125
7782-65-2		UN2192
7782-82-3		UN2630
7782-86-7		UN1627
7783-00-8	U204	
7783-06-4	U135	UN1053
7783-07-5		UN2202
7783-18-8		NA9093
7783-35-9		UN1645
7783-41-7		UN2190
7783-46-2		NA2811
7783-49-5		NA9158
7783-50-8		NA9120
7783-54-2		UN2451
7783-56-4		UN1549
7783-60-0		UN2418
7783-70-2		UN1732
7783-79-1		UN2194
7783-80-4		UN2195
7784-34-1		UN1560
7784-40-9		UN1617
7784-41-0		UN1677
7784-42-1		UN2188
7784-46-5		UN1686
7785-84-4		NA9148
7786-34-7		NA2783
7786-81-4		NA9141
7787-47-5		UN1566
7787-49-7		UN1566
7787-55-5		UN2464
7788-98-9		NA9086
7789-00-6		NA9142
7789-06-2		NA9149
7789-09-5		UN1439
7789-30-2		UN1745
7789-42-6		NA2570
7789-43-7		NA9103
7789-61-9		UN1549
7790-91-2		UN1749
7790-94-5		UN1754
7791-12-0	U216	
7791-23-3		UN2879
7803-51-2	P096	UN2199
7803-52-3		UN2676
7803-55-6	P119	UN2859
7803-62-5		UN2203
8001-35-2	P123	NA2761
8001-58-9	U051	NA1993
8003-34-7		NA9184

CAS No.	RCRA	Shipping Identification Number
8006-61-9		UN1203
8006-64-2		UN1299
8007-45-2		UN1136,NA1136,UN1137,NA1137,NA1993
8014-95-7		UN1830,UN1831,UN2796
8030-30-6		UN1255,UN1256,UN2553
8052-42-4		NA1999
9004-66-4	U139	
10025-67-9		UN1828
10025-87-3		UN1810
10025-91-9		UN1733
10026-11-6		UN2503
10026-13-8		UN1806
10028-22-5		NA9121
10031-59-1	P115	UN1707
10035-10-6		UN1048
10039-32-4		NA9147
10043-01-3		NA1760,NA9078
10045-89-3		NA9122
10045-94-0		UN1625
10049-04-4		NA9191
10049-05-5		NA9102
10099-74-8		UN1469
10101-53-8		NA9100
10101-63-0		NA2811
10101-89-0		NA9148
10102-06-4		UN2980,UN2981
10102-18-8		UN2630
10102-43-9	P076	UN1660
10102-44-0	P078	UN1067
10102-45-1	U217	UN2727
10102-48-4		UN1617
10108-64-2		NA2570
10124-50-2		UN1678
10124-56-8		NA9148
10140-65-5		NA9147
10192-30-0		NA2693
10196-04-0		NA9090
10294-33-4		UN2692
10294-34-5		UN1741
10361-89-4		NA9148
10380-29-7		NA9110
10415-75-5		UN1627
10544-72-6	P078	UN1067
10588-01-9		NA1479
11097-69-1		UN2315
11115-74-5		NA1463,UN1755
12002-03-8		UN1585
12039-52-0	P114	
12054-48-7		NA9140
12125-01-8		UN2505
12125-02-9		NA9085
12135-76-1		UN2683
12427-38-2		UN2210,UN2968
13463-39-3	P073	UN1259
13597-99-4		UN2464
13746-89-9		UN2728
13765-19-0	U032	NA9096

CAS No.	RCRA	Shipping Identification Number
13814-96-5		NA2291
13826-83-0		NA9088
13952-84-6		UN1125
14017-41-5		NA9105
14258-49-2		NA2449
14307-35-8		NA9134
14639-97-5		NA9154
14639-98-6		NA9154
14644-61-2		NA9163
14977-61-8		UN1758
15699-18-0		NA9138
15739-80-7		UN1794,NA2291
15950-66-0		NA2020
16721-80-5		NA2923,NA2949
16752-77-5	P066	
16923-95-8		NA9162
17702-41-9		UN1868
18883-66-4	U206	
19287-45-7		UN1911
19624-22-7		UN1380
20816-12-0	P087	UN2471
20830-81-3	U059	
20859-73-8	P006	UN1397
21908-53-2		UN1641
23950-58-5	U192	
25013-15-4		UN2618
25154-54-5		UN1597
25155-30-0		NA9146
25168-15-4		NA2765
25168-26-7		NA2765
25376-45-8	U221	
25550-58-7		UN0076,UN1599
25551-13-7		UN2325
25639-42-3		UN2617
26264-06-2		NA9097
26471-62-5	U223	
26628-22-8	P105	UN1687
26952-23-8		UN2047
27137-85-5		UN1766
27176-87-0		NA2584
27323-41-7		NA9151
27774-13-6		NA9152
29191-52-4		UN2431
30525-89-4		UN2213
32534-95-5		NA2765
33213-65-9		NA2761
36478-76-9		UN2980,UN2981
37211-05-5		NA9139
39196-18-4	P045	
42504-46-1		NA9127
52628-25-8		NA9154
52652-59-2		NA2811
52740-16-6		UN1574
53467-11-1		NA2765
53469-21-9		UN2315
55488-87-4		NA9119
56189-09-4		NA2811

CAS No.	RCRA	Shipping Identification Number
61792-07-2		NA2765
68476-85-7		UN1075
X00001-02-5		UN1102,UN1103,UN1930,UN2718
X00001-09-2		UN1566
X00001-23-0		UN1365
X00001-44-5		UN1060
X00001-75-2		UN1564

APPENDIX D.4

Index of CAS Numbers by Substance Names

CAS No.	Substance Name
117-79-3	AAQ
3383-96-8	Abate
83-32-9	Acenaphthene
208-96-8	Acenaphthylene
75-07-0	Acetaldehyde
60-35-5	Acetamide
62-38-4	(Acetato-O)phenylmercury
64-19-7	Acetic acid
60-35-5	Acetic acid amide
631-61-8	Acetic acid, ammonium salt
628-63-7	Acetic acid, *n*-amyl ester
626-38-0	Acetic acid, *sec*-amyl ester
625-16-1	Acetic acid, *tert*-amyl ester
123-86-4	Acetic acid, butyl ester
105-46-4	Acetic acid, *sec*-butyl ester
540-88-5	Acetic acid, *tert*-butyl ester
543-90-8	Acetic acid, cadmium salt
1066-30-4	Acetic acid, chromium(III) salt
142-71-2	Acetic acid, copper(II) salt
141-78-6	Acetic acid, ethyl ester
142-92-7	Acetic acid, hexyl ester
123-92-2	Acetic acid, isoamyl ester
110-19-0	Acetic acid, isobutyl ester
108-21-4	Acetic acid, isopropyl ester
301-04-2	Acetic acid, lead salt
1600-27-7	Acetic acid, mercury(II) salt
79-20-9	Acetic acid, methyl ester
109-60-4	Acetic acid, propyl ester
563-68-8	Acetic acid, thallium(I) salt
108-05-4	Acetic acid, vinyl ester
557-34-6	Acetic acid, zinc salt
75-07-0	Acetic aldehyde

CAS No.	Substance Name
108-24-7	Acetic anhydride
67-64-1	Acetone
75-86-5	Acetone cyanohydrin
1752-30-3	Acetone thiosemicarbazide
75-05-8	Acetonitrile
81-81-2	3-(*alpha*-Acetonylbenzyl)-4-hydroxycoumarin and salts
98-86-2	Acetophenone
62-55-5	Acetothioamide
108-05-4	Acetoxyethylene
900-95-8	Acetoxytriphenyl stannane
53-96-3	2-Acetylaminofluorene
506-97-7	Acetyl bromide
75-36-5	Acetyl chloride
74-86-2	Acetylene
540-59-0	Acetylene dichloride
79-27-6	Acetylene tetrabromide
79-34-5	Acetylene tetrachloride
79-21-0	Acetyl hydroperoxide
50-78-2	Acetylsalicylic acid
591-08-2	1-Acetylthiourea
591-08-2	1-Acetyl-2-thiourea
107-02-8	Acrolein
79-06-1	Acrylamide
79-10-7	Acrylic acid
141-32-2	Acrylic acid, butyl ester
140-88-5	Acrylic acid, ethyl ester
96-33-3	Acrylic acid, methyl ester
107-02-8	Acrylic aldehyde
107-13-1	Acrylonitrile
814-68-6	Acryloyl chloride
814-68-6	Acrylyl chloride
50-76-0	Actinomycin D
1333-86-4	Activated carbon
124-04-9	Adipic acid
103-23-1	Adipic acid, di-2-ethylhexyl ester
111-69-3	Adiponitrile
23214-92-8	Adriamycin
X00001-22-9	Aflatoxins
106-92-3	AGE
116-06-3	Aldicarb
309-00-2	Aldrin
107-18-6	Allyl alcohol
107-11-9	Allylamine
94-59-7	5-Allyl-1,3-benzodioxole
107-05-1	Allyl chloride
106-92-3	Allyl glycidyl ether
94-59-7	4-Allyl-1,2-(methylenedioxy)benzene
2179-59-1	Allyl propyl disulfide
7429-90-5	Aluminum
1344-28-1	*alpha*-Alumina
X00001-02-5	Aluminum alkyls (NOC)
1344-28-1	Aluminum oxide
20859-73-8	Aluminum phosphide
7429-90-5	Aluminum pyro powders
X00001-01-4	Aluminum salts, soluble
10043-01-3	Aluminum sulfate

CAS No.	Substance Name
7429-90-5	Aluminum welding fumes
133-90-4	Amiben
106-50-3	*p*-Aminoaniline
90-04-0	*o*-Aminoanisole
104-94-9	*p*-Aminoanisole
117-79-3	2-Amino-9,10-anthracenedione
117-79-3	2-Aminoanthraquinone
60-09-3	4-Aminoazobenzene
62-53-3	Aminobenzene
92-67-1	4-Aminobiphenyl
3037-72-7	(4-Aminobutyl) diethoxymethylsilane
133-90-4	3-Amino-2,5-dichlorobenzoic acid
99-98-9	*p*-Aminodimethylaniline
97-56-3	4-Amino-2',3-dimethylazobenzene
87-62-7	2-Amino-1,3-dimethylbenzene
21087-64-9	4-Amino-6-(1,1-dimethylethyl)-3-methylthio-*as*-triazin-5(4H)-one
92-67-1	4-Aminodiphenyl
60-09-3	p-Aminodiphenylimide
141-43-5	2-Aminoethanol
615-05-4	3-Amino-4-methoxyaniline
90-04-0	1-Amino-2-methoxybenzene
104-94-9	1-Amino-4-methoxybenzene
120-71-8	1-Amino-2-methoxy-5-methylbenzene
99-59-2	1-Amino-2-methoxy-5-nitrobenzene
120-71-8	3-Amino-4-methoxytoluene
82-28-0	1-Amino-2-methylanthraquinone
95-53-4	2-Amino-1-methylbenzene
106-49-0	4-Amino-1-methylbenzene
2763-96-4	5-(Aminomethyl)-3-isoxazolol
2763-96-4	5-(Aminomethyl)-3(2H)-isoxazolone
134-32-7	1-Aminonaphthalene
62-53-3	Aminophen
70-69-9	4'-Aminopropiophenone
54-62-6	Aminopterin
504-29-0	2-Aminopyridine
504-24-5	4-Aminopyridine
591-08-2	N-(Aminothioxomethyl)acetamide
95-53-4	2-Aminotoluene
61-82-5	3-Amino-1,2,4-triazole
78-53-5	Amiton
3734-97-2	Amiton oxalate
61-82-5	Amitrol
61-82-5	Amitrole
7773-06-0	Ammate
7664-41-7	Ammonia
631-61-8	Ammonium acetate
1863-63-4	Ammonium benzoate
1066-33-7	Ammonium bicarbonate
7789-09-5	Ammonium bichromate
1341-49-7	Ammonium bifluoride
10192-30-0	Ammonium bisulfite
1111-78-0	Ammonium carbamate
506-87-6	Ammonium carbonate
12125-02-9	Ammonium chloride
7788-98-9	Ammonium chromate
3012-65-5	Ammonium citrate, dibasic

CAS No.	Substance Name
7789-09-5	Ammonium dichromate
13826-83-0	Ammonium fluoborate
12125-01-8	Ammonium fluoride
5972-73-6	Ammonium hydrogenoxalate
1336-21-6	Ammonium hydroxide
7803-55-6	Ammonium metavanadate
15699-18-0	Ammonium nickel sulfate
6484-52-2	Ammonium nitrate
5972-73-6	Ammonium oxalate
6009-70-7	Ammonium oxalate
14258-49-2	Ammonium oxalate
131-74-8	Ammonium picrate
16919-19-0	Ammonium silicofluoride
7773-06-0	Ammonium sulfamate
7783-20-2	Ammonium sulfate
12135-76-1	Ammonium sulfide
10196-04-0	Ammonium sulfite
3164-29-2	Ammonium tartrate
14307-43-8	Ammonium tartrate
1762-95-4	Ammonium thiocyanate
7783-18-8	Ammonium thiosulfate
7803-55-6	Ammonium vanadate
12172-73-5	Amosite asbestos
300-62-9	Amphetamine
628-63-7	*n*-Amyl acetate
626-38-0	*sec*-Amyl acetate
625-16-1	*tert*-Amyl acetate
110-62-3	Amyl aldehyde
541-85-5	*sec*-Amyl ethyl ketone
13463-67-7	Anatase
62-53-3	Aniline
X00001-03-6	Aniline homologues
29191-52-4	Anisidine
90-04-0	*o*-Anisidine
104-94-9	*p*-Anisidine
29191-52-4	Anisidine (*o*- and *p*-isomers)
134-29-2	*o*-Anisidine hydrochloride
120-12-7	Anthracene
7440-36-0	Antimony
X00001-05-8	Antimony compounds
7647-18-9	Antimony pentachloride
7783-70-2	Antimony pentafluoride
28300-74-5	Antimony potassium tartrate
7789-61-9	Antimony tribromide
10025-91-9	Antimony trichloride
7783-56-4	Antimony trifluoride
1309-64-4	Antimony trioxide
1397-94-0	Antimycin A
86-88-4	ANTU
140-57-8	Aramite
7440-37-1	Argon
12674-11-2	Aroclor 1016
11104-28-2	Aroclor 1221
11141-16-5	Aroclor 1232
53469-21-9	Aroclor 1242
12672-29-6	Aroclor 1248
11097-69-1	Aroclor 1254

CAS No.	Substance Name
11096-82-5	Aroclor 1260
1336-36-3	Aroclors
8007-45-2	Aromatic hydrocarbons, polycyclic, particulate
7440-38-2	Arsenic
1327-52-2	Arsenic acid
7778-39-4	Arsenic acid
X00001-80-9	Arsenic and compounds
X00001-12-7	Arsenic compounds, organic
X00001-07-0	Arsenic compounds, soluble
1303-32-8	Arsenic disulfide
1327-53-3	Arsenic(III) oxide
1303-28-2	Arsenic(V) oxide
1303-28-2	Arsenic pentoxide
7784-34-1	Arsenic trichloride
1327-53-3	Arsenic trioxide
1303-33-9	Arsenic trisulfide
1327-53-3	Arsenous oxide
7784-34-1	Arsenous trichloride
7784-42-1	Arsine
1332-21-4	Asbestos (friable)
12172-73-5	Asbestos, Amosite
12001-29-5	Asbestos, Chrysotile
12001-28-4	Asbestos, Crocidolite
1332-21-4	Asbestos synthetic fibers
8052-42-4	Asphalt
50-78-2	Aspirin
95-63-6	Asymmetrical trimethylbenzene
1912-24-9	Atrazine
492-80-8	Auramine base
110-86-1	Azabenzene
115-02-6	Azaserine
91-22-5	1-Azanaphthalene
446-86-6	Azathioprine
2642-71-9	Azinphos-ethyl
86-50-0	Azinphos-methyl
151-56-4	Aziridine
6923-22-4	Azodrin
7727-43-7	Barite
7440-39-3	Barium
X00001-75-2	Barium compounds
X00001-08-1	Barium compounds, soluble
542-62-1	Barium cyanide
7727-43-7	Barium sulfate
X00001-06-9	Barley dust
114-26-1	Baygon
55-38-9	Baytex
154-93-8	BCNU
8021-39-4	Beechwood creosote
17804-35-2	Benomyl
225-51-4	Benz[c]acridine
225-51-4	3,4-Benzacridine
98-87-3	Benzal chloride
55-21-0	Benzamide
56-55-3	Benz[a]anthracene
56-55-3	1,2-Benzanthracene
62-53-3	Benzenamine

CAS No.	Substance Name
71-43-2	Benzene
98-05-5	Benzenearsonic acid
842-07-9	Benzeneazo-*beta*-naphthol
98-88-4	Benzenecarbonyl chloride
106-50-3	1,4-Benzenediamine
100-21-0	1,4-Benzenedicarboxylic acid
85-44-9	1,2-Benzenedicarboxylic acid anhydride
117-81-7	1,2-Benzenedicarboxylic acid, bis(2-ethylhexyl) ester
84-74-2	1,2-Benzenedicarboxylic acid, dibutyl ester
84-66-2	1,2-Benzenedicarboxylic acid, diethyl ester
131-11-3	1,2-Benzenedicarboxylic acid, dimethyl ester
117-84-0	1,2-Benzenedicarboxylic acid, di-n-occtyl ester
1477-55-0	1,3-Benzenedimethanamine
120-80-9	1,2-Benzenediol
108-46-3	1,3-Benzenediol
123-31-9	1,4-Benzenediol
98-09-9	Benzenesulfonic acid chloride
98-09-9	Benzenesulfonyl chloride
108-98-5	Benzenethiol
92-87-5	Benzidine
81-07-2	1,2-Benzisothiazolin-3-one-1,1-dioxide, and salts
56-55-3	Benzo[a]anthracene
205-99-2	Benzo[b]fluoranthene
207-08-9	Benzo[k]fluoranthene
206-44-0	Benzo[j,k]fluorene
65-85-0	Benzoic acid
55-21-0	Benzoic acid amide
1863-63-4	Benzoic acid, ammonium salt
81-07-2	*o*-Benzoic acid sulfimide
71-43-2	Benzol
100-47-0	Benzonitrile
191-24-2	Benzo[ghi]perylene
50-32-8	Benzo[a]pyrene
50-32-8	3,4-Benzopyrene
91-22-5	Benzopyridine
106-51-4	Benzoquinone
106-51-4	*p*-Benzoquinone
98-07-7	Benzotrichloride
98-88-4	Benzoyl chloride
94-36-0	Benzoyl peroxide
218-01-9	1,2-Benzphenanthrene
100-44-7	Benzyl chloride
140-29-4	Benzyl cyanide
98-87-3	Benzyl dichloride
98-87-3	Benzylidine chloride
7440-41-7	Beryllium
7787-47-5	Beryllium chloride
X00001-09-2	Beryllium compounds
7787-49-7	Beryllium fluoride
13597-99-4	Beryllium nitrate
7787-55-5	Beryllium nitrate trihydrate
2426-08-6	BGE
319-84-6	*alpha*-BHC
319-85-7	*beta*-BHC
319-86-8	*delta*-BHC
58-89-9	*gamma*-BHC
128-37-0	BHT

CAS No.	Substance Name
92-87-5	4,4'-Bianiline
141-66-2	Bidrin
1464-53-5	2,2'-Bioxirane
92-52-4	Biphenyl
92-67-1	4-Biphenylamine
92-87-5	(1,1'-Biphenyl)-4,4'-diamine
132-64-9	2,2'-Biphenylene oxide
101-14-4	Bis(4-amino-3-chlorophenyl)methane
101-80-4	Bis(4-aminophenyl) ether
101-77-9	Bis(4-aminophenyl)methane
139-65-1	Bis(4-aminophenyl) sulfide
139-13-9	N,N-Bis(carboxymethyl)glycine
111-91-1	Bis(chloroethoxy)methane
111-91-1	Bis(2-chloroethoxy)methane
305-03-3	4-(Bis(2-chloroethyl)amino)benzenebutanoic acid
148-82-3	3-(p-Bis(2-chloroethyl)amino)phenyl-L-alanine
66-75-1	5-(Bis(2-chloroethyl)amino)uracil
111-44-4	Bis(2-chloroethyl) ether
51-75-2	Bis(2-chloroethyl)methylamine
51-75-2	N,N-Bis(2-chloroethyl)methylamine
494-03-1	N,N-Bis(2-chloroethyl)-2-naphthylamine
154-93-8	Bischloroethylnitrosourea
505-60-2	Bis(chloroethyl) sulfide
108-60-1	Bis(2-chloroisopropyl) ether
542-88-1	Bis(chloromethyl) ether
108-60-1	Bis(2-chloro-1-methylethyl) ether
534-07-6	Bis(chloromethyl) ketone
78-71-7	3,3-Bis(chloromethyl)oxetane
115-32-2	1,1-Bis(4-chlorophenyl)-2,2,2-trichloroethanol
101-61-1	4,4'-Bis(dimethylamino)diphenylmethane
90-94-8	Bis(4-dimethylaminophenyl) ketone
137-26-8	Bis(dimethylthiocarbamoyl) disulfide
103-23-1	Bis(2-ethylhexyl) adipate
117-81-7	Bis(2-ethylhexyl) phthalate
111-42-2	Bis(2-hydroxyethyl)amine
80-05-7	2,2-Bis(4-hydroxyphenyl)propane
72-43-5	2,2-Bis(p-methoxyphenyl)-1,1,1-trichloroethane
50782-69-9	S-(2-(Bis(-1-methylethyl)amino)ethyl) methylphosphonothioic acid
1163-19-5	Bis(pentabromophenyl) ether
80-05-7	Bisphenol A
1304-82-1	Bismuth telluride
X00001-10-5	Bismuth telluride, selenium doped
10102-06-4	Bis(nitrato-O)dioxouranium
36478-76-9	Bis(nitrato-O,O')dioxouranium
80-05-7	Bisphenol-A
97-18-7	Bithionol
4044-65-9	Bitoscanate
7727-43-7	Blanc fixe
35400-43-2	Bolstar
1303-96-4	Borates, tetra, sodium salts
1330-43-4	Borates, tetra, sodium salts
1303-96-4	Borax
1330-43-4	Borax
1344-90-7	Borax
12179-04-3	Borax
61028-24-8	Borax

CAS No.	Substance Name
1303-86-2	Boron oxide
10294-33-4	Boron tribromide
10294-34-5	Boron trichloride
7637-07-2	Boron trifluoride
353-42-4	Boron trifluoride compound with methyl ether (1:1)
353-42-4	Boron trifluoride, methyl ether complex
314-40-9	Bromacil
28772-56-7	Bromadiolone
7726-95-6	Bromine
506-68-3	Bromine cyanide
7789-30-2	Bromine pentafluoride
598-31-2	Bromoacetone
74-97-5	Bromochloromethane
75-27-4	Bromodichloromethane
593-60-2	Bromoethylene
75-25-2	Bromoform
74-83-9	Bromomethane
101-55-3	1-Bromo-4-phenoxybenzene
101-55-3	4-Bromophenyl phenyl ether
598-31-2	1-Bromo-2-propanone
75-63-8	Bromotrifluoromethane
357-57-3	Brucine
106-99-0	Butadiene
106-99-0	1,3-Butadiene
123-72-8	Butanal
106-97-8	Butane
109-79-5	Butanethiol
107-92-6	Butanoic acid
75-65-0	*tert*-Butanol
71-36-3	1-Butanol
78-92-2	2-Butanol
78-93-3	2-Butanone
1338-23-4	2-Butanone peroxide
123-73-9	2-Butenal
4170-30-3	2-Butenal
108-31-6	*cis*-Butenedioic anhydride
111-76-2	2-Butoxyethanol
123-86-4	*n*-Butyl acetate
105-46-4	*sec*-Butyl acetate
540-88-5	*tert*-Butyl acetate
141-32-2	Butyl acrylate
71-36-3	Butyl alcohol
71-36-3	*n*-Butyl alcohol
78-92-2	*sec*-Butyl alcohol
75-65-0	*tert*-Butyl alcohol
109-73-9	Butylamine
513-49-5	*sec*-Butylamine
13952-84-6	*sec*-Butylamine
75-64-9	*tert*-Butylamine
513-49-5	2-Butylamine
13952-84-6	2-Butylamine
128-37-0	Butylated hydroxytoluene
85-68-7	Butyl benzyl phthalate
111-76-2	Butyl cellosolve
1189-85-1	*tert*-Butyl chromate
94-80-4	Butyl 2,4-dichlorophenoxyacetate

CAS No.	Substance Name
94-79-1	*sec*-Butyl 2,4-dichlorophenoxyacetate
106-88-7	1,2-Butylene oxide
2426-08-6	*n*-Butyl glycidyl ether
138-22-7	*n*-Butyl lactate
109-79-5	Butyl mercaptan
1634-04-4	*tert*-Butyl methyl ether
591-78-6	*n*-Butyl methyl ketone
924-16-3	N-Butyl-N-nitroso-1-butanamine
89-72-5	*o-sec*-Butylphenol
140-57-8	Butylphenoxyisopropyl chloroethyl sulfite
84-74-2	*n*-Butyl phthalate
141-32-2	Butyl 2-Propenoate
98-51-1	*p-tert*-Butyltoluene
98-51-1	4-*tert*-Butyltoluene
123-72-8	Butyraldehyde
107-92-6	Butyric acid
6533-73-9	Carbonic acid, dithallium(I) salt
463-58-1	Carbonyl sulfide
75-60-5	Cacodylic acid
7440-43-9	Cadmium
543-90-8	Cadmium acetate
X00001-81-0	Cadmium and compounds
7789-42-6	Cadmium bromide
10108-64-2	Cadmium chloride
1306-19-0	Cadmium oxide
X00001-13-8	Cadmium salts
2223-93-0	Cadmium stearate
7778-44-1	Calcium arsenate
52740-16-6	Calcium arsenite
1317-65-3	Calcium carbonate (marble)
13765-19-0	Calcium chromate
156-62-7	Calcium cyanamide
592-01-8	Calcium cyanide
26264-06-2	Calcium dodecylbenzenesulfonate
1305-62-0	Calcium hydroxide
7778-54-3	Calcium hypochlorite
1305-78-8	Calcium oxide
1344-95-2	Calcium silicate
7778-18-9	Calcium sulfate
10101-41-4	Calcium sulfate dihydrate
75-20-7	Calcium carbide
76-22-2	Camphor
56-25-7	Cantharidin
105-60-2	Caprolactam
2425-06-1	Captafol
133-06-2	Captan
51-83-2	Carbachol chloride
51-79-6	Carbamic acid, ethyl ester
630-10-4	Carbamimidoselenoic acid
63-25-2	Carbaryl
1563-66-2	Carbofuran
108-95-2	Carbolic acid
1333-86-4	Carbon (activated)
75-15-0	Carbon bisulfide
1333-86-4	Carbon black
124-38-9	Carbon dioxide

CAS No.	Substance Name
75-15-0	Carbon disulfide
492-80-8	4,4'-Carbonimidoylbis(N,N-dimethylaniline)
492-80-8	4,4'-Carbonimidoylbis(N,N-dimethylbenzenamine)
630-08-0	Carbon monoxide
79-22-1	Carbonochloridic acid, methyl ester
75-44-5	Carbon oxychloride
353-50-4	Carbon oxyfluoride
558-13-4	Carbon tetrabromide
56-23-5	Carbon tetrachloride
75-44-5	Carbonyl chloride
353-50-4	Carbonyl fluoride
786-19-6	Carbophenothion
81-88-9	(9-(o-Carboxyphenyl)-6-(diethylamino)-3H-xanthen-3-ylidene) diethylammonium chloride
120-80-9	Catechol
1310-73-2	Caustic soda
13010-47-4	CCNU
21923-23-9	Celathion
110-80-5	Cellosolve
111-15-9	Cellosolve acetate
9004-34-6	Cellulose (paper fiber)
21351-79-1	Cesium hydroxide
106-51-4	Chinone
75-87-6	Chloral
133-90-4	Chloramben
305-03-3	Chlorambucil
56-75-7	Chloramphenicol
57-74-9	Chlordane
143-50-0	Chlordecone
470-90-6	Chlorfenvinfos
8001-35-2	Chlorinated camphene
55720-99-5	Chlorinated diphenyl oxide
77-13-1	Chlorinated fluorocarbon
X00001-29-6	Chlorinated phenols
55720-99-5	Chlorinated phenyl ether
7782-50-5	Chlorine
506-77-4	Chlorine cyanide
10049-04-4	Chlorine dioxide
10049-04-4	Chlorine(IV) oxide
7790-91-2	Chlorine trifluoride
24934-91-6	Chlormephos
999-81-5	Chlormequat chloride
494-03-1	Chlornaphazine
107-20-0	Chloroacetaldehyde
79-11-8	Chloroacetic acid
532-27-4	alpha-Chloroacetophenone
532-27-4	2-Chloroacetophenone
79-04-9	Chloroacetyl chloride
80-63-7	2-Chloroacrylic acid, methyl ester
542-75-6	3-Chloroallyl chloride
106-47-8	p-Chloroaniline
106-47-8	4-Chloroaniline
98-88-4	alpha-Chlorobenzaldehyde
106-47-8	4-Chlorobenzenamine
108-90-7	Chlorobenzene
510-15-6	Chlorobenzilate
2698-41-1	o-Chlorobenzylidene malononitrile

CAS No.	Substance Name
74-97-5	Chlorobromomethane
126-99-8	2-Chlorobutadiene
126-99-8	2-Chloro-1,3-butadiene
510-15-6	4-Chloro-*alpha*-(4-chlorophenyl)-*alpha*-hydroxybenzeneacetic acid, ethyl ester
59-50-7	*p*-Chloro-*m*-cresol
59-50-7	4-Chloro-*m*-cresol
124-48-1	Chlorodibromomethane
96-12-8	3-Chloro-1,2-dibromopropane
75-45-6	Chlorodifluoromethane
53469-21-9	Chlorodiphenyl (42% Chlorine)
11097-69-1	Chlorodiphenyl (54 % Chlorine)
106-89-8	1-Chloro-2,3-epoxypropane
75-00-3	Chloroethane
1622-32-8	2-Chloroethanesulfonyl chloride
107-07-3	2-Chloroethanol
75-01-4	Chloroethene
110-75-8	2-Chloroethoxyethene
13010-47-4	1-(2-Chloroethyl)-3-cyclohexyl-1-nitrosourea
627-11-2	Chloroethyl chloroformate
75-01-4	Chloroethylene
999-81-5	N-(2-Chloroethyl)-N,N,N-trimethyl ammonium chloride
110-75-8	2-Chloroethyl vinyl ether
67-66-3	Chloroform
627-11-2	Chloroformic acid, chloroethyl ester
541-41-3	Chloroformic acid, ethyl ester
108-23-6	Chloroformic acid, isopropyl ester
79-22-1	Chloroformic acid, methyl ester
109-61-5	Chloroformic acid, propyl ester
75-44-5	Chloroformyl chloride
52-68-6	Chlorofos
74-87-3	Chloromethane
107-30-2	Chloromethoxymethane
3165-93-3	4-Chloro-2-methylbenzenamine hydrochloride
100-44-7	Chloromethylbenzene
542-88-1	Chloromethyl ether
106-89-8	(Chloromethyl)ethylene oxide
107-30-2	Chloromethyl methyl ether
100-14-1	1-(Chloromethyl)-4-nitrobenzene
106-89-8	2-(Chloromethyl)oxirane
59-50-7	4-Chloro-3-methylphenol
532-27-4	Chloromethyl phenyl ketone
1558-25-4	(Chloromethyl) trichlorosilane
91-58-7	*beta*-Chloronaphthalene
91-58-7	2-Chloronaphthalene
100-00-5	1-Chloro-4-nitrobenzene
600-25-9	1-Chloro-1-nitropropane
76-15-3	Chloropentafluoroethane
3691-35-8	Chlorophacinone
95-57-8	*o*-Chlorophenol
95-57-8	2-Chlorophenol
108-43-0	3-Chlorophenol
106-48-9	4-Chlorophenol
532-27-4	2-Chloro-1-phenylethanone
100-44-7	Chlorophenylmethane
7005-72-3	4-Chlorophenyl phenyl ether
5344-82-1	1-(*o*-Chlorophenyl)thiourea

CAS No.	Substance Name
5344-82-1	(2-Chlorophenyl)thiourea
52-68-6	Chlorophos
76-06-2	Chloropicrin
126-99-8	Chloroprene
126-99-8	beta-Chloroprene
542-76-7	3-Chloropropanenitrile
107-05-1	1-Chloro-2-propene
542-76-7	3-Chloropropionitrile
106-89-8	3-Chloropropylene oxide
1331-28-8	o-Chlorostyrene
2039-87-4	o-Chlorostyrene
1331-28-8	2-Chlorostyrene
7790-94-5	Chlorosulfonic acid
1897-45-6	Chlorothalonil
71-55-6	Chlorothene
100-44-7	alpha-Chlorotoluene
95-49-8	o-Chlorotoluene
95-49-8	2-Chlorotoluene
3165-93-3	4-Chloro-o-toluidine hydrochloride
1929-82-4	2-Chloro-6-(trichloromethyl)pyridine
961-11-5	2-Chloro-1-(2,4,5-trichlorophenyl)vinyl dimethyl phosphate
75-77-4	Chloro trimethylsilane
1982-47-4	Chloroxuron
3569-57-1	3-Chlorpropyl octyl sulfoxide
2921-88-2	Chlorpyrifos
21923-23-9	Chlorthiophos
X00001-15-0	Chromates
1066-30-4	Chromic acetate
7738-94-5	Chromic acid
11115-74-5	Chromic acid
13765-19-0	Chromic acid, calcium salt
10025-73-7	Chromic chloride
10101-53-8	Chromic sulfate
X00001-16-1	Chromite ore processing
7440-47-3	Chromium
X00001-30-9	Chromium and compounds
X00001-17-2	Chromium(II) compounds
X00001-18-3	Chromium(III) compounds
X00001-20-7	Chromium(VI) compounds, certain water insoluble
X00001-19-4	Chromium(VI) compounds, water soluble
14977-61-8	Chromium oxychloride
10049-05-5	Chromous chloride
14977-61-8	Chromyl chloride
218-01-9	Chrysene
12001-29-5	Chrysotile asbestos
2650-18-2	C.I. Acid Blue 9, diammonium salt
3844-45-9	C.I. Acid Blue 9, disodium salt
4680-78-8	C.I. Acid Green 3
569-64-2	C.I. Basic Green 4
989-38-8	C.I. Basic Red 1
1937-37-7	C.I. Direct Black 38
2602-46-2	C.I. Direct Blue 6
72-57-1	C.I. Direct Blue 14
16071-86-6	C.I. Direct Brown 95
2832-40-8	C.I. Disperse Yellow 3
3761-53-3	C.I. Food Red 5

CAS No.	Substance Name
81-88-9	C.I. Food Red 15
3118-97-6	C.I. Solvent Orange 7
82-28-0	C.I. Solvent Orange 35
60-11-7	C.I. Solvent Yellow 2
97-56-3	C.I. Solvent Yellow 3
842-07-9	C.I. Solvent Yellow 14
492-80-8	C.I. Solvent Yellow 34 (Auramine)
128-66-5	C.I. Vat Yellow 4
15663-27-1	Cisplatin
3012-65-5	Citric acid, diammonium salt
2971-90-6	Clopidol
X00001-21-8	Coal dust
8007-45-2	Coal tar
65996-93-2	Coal tar pitch volatiles
8007-45-2	Coal tar pitch volatiles, benzene solubles
7440-48-4	Cobalt
10210-68-1	Cobalt carbonyl
X00001-14-9	Cobalt compounds
16842-03-8	Cobalt hydrocarbonyl
7789-43-7	Cobaltous bromide
544-18-3	Cobaltous formate
14017-41-5	Cobaltous sulfamate
10210-68-1	Cobalt tetracarbonyl
X00001-37-6	Coke oven emissions
64-86-8	Colchicine
7440-50-8	Copper
142-71-2	Copper(II) acetate
X00001-31-0	Copper and compounds
544-92-3	Copper cyanide
X00001-23-0	Cotton dust
56-72-4	Coumaphos
5836-29-3	Coumatetralyl
2971-90-6	Coyden
136-78-7	Crag herbicide
8001-58-9	Creosote
120-71-8	*p*-Cresidine
1319-77-3	Cresol(s)
108-39-4	*m*-Cresol
95-48-7	*o*-Cresol
106-44-5	*p*-Cresol
1319-77-3	Cresol (mixed isomers)
1319-77-3	Cresylic acid
108-39-4	*m*-Cresylic acid
95-48-7	*o*-Cresylic acid
106-44-5	*p*-Cresylic acid
535-89-7	Crimidine
14464-46-1	Cristobalite (silica)
12001-28-4	Crocidolite asbestos
123-73-9	Crotonaldehyde
4170-30-3	Crotonaldehyde
299-86-5	Crufomate
98-82-8	Cumene
80-15-9	Cumene hydroperoxide
135-20-6	Cupferron
142-71-2	Cupric acetate
12002-03-8	Cupric acetoarsenite
7447-39-4	Cupric chloride

CAS No.	Substance Name
3251-23-8	Cupric nitrate
5893-66-3	Cupric oxalate
7758-98-7	Cupric sulfate
10380-29-7	Cupric sulfate, ammoniated
815-82-7	Cupric tartrate
544-92-3	Cuprous cyanide
420-04-2	Cyanamide
57-12-5	Cyanide anion
57-12-5	Cyanide ion
57-12-5	Cyanides (soluble cyanide salts)
137-05-3	Cyanoacrylic acid, methyl ester
107-13-1	Cyanoethene
460-19-5	Cyanogen
506-68-3	Cyanogen bromide
506-77-4	Cyanogen chloride
506-78-5	Cyanogen iodide
75-05-8	Cyanomethane
2636-26-2	Cyanophos
675-14-9	Cyanuric fluoride
14901-08-7	Cycasin
106-51-4	1,4-Cyclohexadienedione
110-82-7	Cyclohexane
108-93-0	Cyclohexanol
108-94-1	Cyclohexanone
110-83-8	Cyclohexene
66-81-9	Cycloheximide
108-91-8	Cyclohexylamine
131-89-5	2-Cyclohexyl-4,6-dinitrophenol
121-82-4	Cyclonite
542-92-7	1,3-Cyclopentadiene
542-92-7	Cyclopentadiene
287-92-3	Cyclopentane
50-18-0	Cyclophosphamide
13121-70-5	Cyhexatin
94-75-7	2,4-D acid
94-75-7	2,4-D, salts and esters
50-76-0	Dactinomycin
75-99-0	Dalapon
115-90-2	Dasanit
20830-81-3	Daunomycin
96-12-8	DBCP
924-16-3	DBN
2650-18-2	D&C Blue no. 4
3761-53-3	D&C Red no. 5
81-88-9	D&C Red no. 19
72-54-8	DDD
72-54-8	4,4'-DDD
72-55-9	DDE
72-55-9	4,4'-DDE
50-29-3	DDT
50-29-3	4,4'-DDT
62-73-7	DDVP
17702-41-9	Decaborane
1163-19-5	Decabromodiphenyl oxide
1163-19-5	Decabromophenyl ether

CAS No.	Substance Name
143-50-0	Decachlorooctahydro-1,3,4-metheno-2H-cyclobuta[c,d]-pentalen-2-one
117-81-7	DEHP
78-34-2	Delnav
8065-48-3	Demeton
919-86-8	Demeton-S-methyl
18883-66-4	2-Deoxy-2-(3-methyl-3-nitrosoureido)-D-Glucopyranose
56-53-1	DES
94-11-1	2,4-D esters
94-79-1	2,4-D esters
94-80-4	2,4-D esters
1320-18-9	2,4-D esters
1928-38-7	2,4-D esters
1928-61-6	2,4-D esters
1929-73-3	2,4-D esters
2971-38-2	2,4-D esters
25168-26-7	2,4-D esters
53467-11-1	2,4-D esters
2238-07-5	DGE
123-42-2	Diacetone alcohol
10311-84-9	Dialifos
2303-16-4	Diallate
302-01-2	Diamine
615-05-4	2,4-Diaminoanisole
39156-41-7	2,4-Diaminoanisole sulfate
106-50-3	1,4-Diaminobenzene
92-87-5	4,4'-Diaminobiphenyl
91-94-1	4,4'-Diamino-3,3'-dichlorobiphenyl
119-90-4	4,4'-Diamino-3,3'-dimethoxy-1,1'-biphenyl
119-93-7	4,4'-Diamino-3,3'-dimethylbiphenyl
101-80-4	4,4'-Diaminodiphenyl ether
107-15-3	1,2-Diaminoethane
615-05-4	1,3-Diamino-4-methoxybenzene
615-05-4	1,4-Diaminophenyl methyl ether
25376-45-8	Diaminotoluene, mixed isomers
496-72-0	Diaminotoluene
95-80-7	2,4-Diaminotoluene
823-40-5	2,6-Diaminotoluene
7783-20-2	Diammonium sulfate
119-90-4	o-Dianisidine
7631-86-9	Diatomaceous earth
68855-54-9	Diatomaceous earth, uncalcined
333-41-5	Diazinon
5333-41-5	Diazinon
334-88-3	Diazomethane
7558-79-4	Dibasic sodium phosphate
224-42-0	Dibenz[a,j]acridine
226-36-8	Dibenz[a,h]acridine
53-70-3	Dibenz[a,h]anthracene
53-70-3	1,2:5,6-Dibenzanthracene
53-70-3	Dibenzo[a,h]anthracene
194-59-2	7H-Dibenzo[c,g]carbazole
132-64-9	Dibenzofuran
189-64-0	Dibenzo[a,h]pyrene
189-55-9	Dibenzo[a,i]pyrene
189-55-9	1,2:7,8-Dibenzopyrene
128-66-5	3,4:8,9-Dibenzopyrene-5,10-dione

CAS No.	Substance Name
94-36-0	Dibenzoyl peroxide
19287-45-7	Diborane
300-76-5	Dibrom
96-12-8	1,2-Dibromo-3-chloropropane
75-61-6	Dibromodifluoromethane
74-95-3	Dibromomethane
106-93-4	1,2-Dibromoethane
126-72-7	2,3-Dibromo-1-propanol phosphate (3:1)
102-81-8	Dibutylaminoethanol
102-81-8	2-N-Dibutylaminoethanol
102-81-8	2-(Dibutylamino)ethanol
128-37-0	2,6-Di-*tert*-butyl-*p*-cresol
4835-11-4	N,N'-Dibutylhexamethylenediamine
107-66-4	Dibutyl phosphate
84-74-2	Dibutyl phthalate
84-74-2	Di-*n*-butyl phthalate
1918-00-9	Dicamba
4342-03-4	Dicarbazine
1194-65-6	Dichlobenil
117-80-6	Dichlone
534-07-6	1,3-Dichloroacetone
7572-29-4	Dichloroacetylene
2303-16-4	S-(2,3-Dichloroallyl) diisopropylthiocarbamate
133-90-4	2,5-Dichloro-3-aminobenzoic acid
541-73-1	*m*-Dichlorobenzene
95-50-1	*o*-Dichlorobenzene
106-46-7	*p*-Dichlorobenzene
95-50-1	1,2-Dichlorobenzene
541-73-1	1,3-Dichlorobenzene
106-46-7	1,4-Dichlorobenzene
25321-22-6	Dichlorobenzene, mixed isomers
91-94-1	3,3'-Dichlorobenzidine
91-94-1	3,3'-Dichloro-(1,1'-biphenyl)- 4,4'diamine
75-27-4	Dichlorobromomethane
110-57-6	1,4-Dichloro-2-butene
764-41-0	1,4-Dichloro-2-butene
110-57-6	*trans*-1,4-Dichlorobutene
101-14-4	3,3'-Dichloro-4,4'-diaminodiphenylmethane
75-71-8	Dichlorodifluoromethane
542-88-1	*sym*-Dichlorodimethyl ether
118-52-5	1,3-Dichloro-5,5-dimethylhydantoin
58270-08-9	Dichloro(4,4-dimethyl-5-((((methylamino) carbonyl)oxy)imino)pentanenitrile)-zinc
23950-58-5	3,5-Dichloro-N-(1,1-dimethyl-2-propynyl)benzamide
75-78-5	Dichloro dimethylsilane
72-54-8	Dichlorodiphenyl dichloroethane
50-29-3	Dichlorodiphenyl trichloroethane
75-34-3	1,1-Dichloroethane
107-06-2	1,2-Dichloroethane
10140-87-1	1,2-Dichloroethanol, acetate
75-35-4	1,1-Dichloroethene
156-60-5	*trans*-1,2-Dichloroethene
10140-87-1	1,2-Dichloroethyl acetate
540-59-0	*sym*-Dichloroethylene
75-35-4	1,1-Dichloroethylene
540-59-0	1,2-Dichloroethylene

CAS No.	Substance Name
156-60-5	1,2-*trans*-Dichloroethylene
156-60-5	*trans*-1,2-Dichloroethylene
111-44-4	Dichloroethyl ether
75-43-4	Dichlorofluoromethane
102-36-3	1,2-Dichloro-4-isocyanatobenzene
108-60-1	Dichloroisopropyl ether
75-09-2	Dichloromethane
98-87-3	(Dichloromethyl)benzene
51-75-2	2,2'-Dichloro-N-methyldiethylamine
149-74-6	Dichloromethylphenylsilane
75-43-4	Dichloromonofluoromethane
594-72-9	1,1-Dichloro-1-nitroethane
120-83-2	2,4-Dichlorophenol
87-65-0	2,6-Dichlorophenol
1929-73-3	2,4-Dichlorophenoxyacetic acid, butoxyethyl ester
94-80-4	2,4-Dichlorophenoxyacetic acid, butyl ester
94-79-1	2,4-Dichlorophenoxyacetic acid, sec-butyl ester
2971-38-2	2,4-Dichlorophenoxyacetic acid, 4-chloro-2-butenyl ester
25168-26-7	2,4-Dichlorophenoxyacetic acid, isooctyl ester
94-11-1	2,4-Dichlorophenoxyacetic acid, isopropyl ester
1928-38-7	2,4-Dichlorophenoxyacetic acid, methyl ester
1320-18-9	2,4-Dichlorophenoxyacetic acid, monoester with propylene glycol butyl ether
1928-61-6	2,4-Dichlorophenoxyacetic acid, propyl ester
94-75-7	2,4-Dichlorophenoxyacetic acid, salts and esters
53467-11-1	*alpha*-[(2,4-Dichlorophenoxy)acetyl]-*omega*-butoxypoly(oxypropylene)
696-28-6	Dichloro phenylarsine
1836-75-5	2,4-Dichlorophenyl-*p*-nitrophenyl ether
27137-85-5	Dichlorophenyl trichlorosilane
26638-19-7	Dichloropropane
78-99-9	1,1-Dichloropropane
78-87-5	1,2-Dichloropropane
142-28-9	1,3-Dichloropropane
8003-19-8	Dichloropropane-dichloropropene mixture
26952-23-8	Dichloropropene
542-75-6	1,3-Dichloropropene
78-88-6	2,3-Dichloropropene
26952-23-8	2,3-Dichloropropene
75-99-0	2,2-Dichloropropionic acid
26952-23-8	Dichloropropylene
542-75-6	1,3-Dichloropropylene
78-88-6	2,3-Dichloropropylene
76-14-2	1,2-Dichloro-1,1,2,2-tetrafluoroethane
76-14-2	Dichlorotetrafluoroethane
3615-21-2	4,5-Dichloro-2-(trifluoromethyl)benzimidazole
62-73-7	2,2-Dichlorovinyl dimethyl phosphate
62-73-7	Dichlorovinyl phosphoric acid dimethyl ester
62-73-7	Dichlorvos
1194-65-6	Dicholobenil
115-32-2	Dicofol
141-66-2	Dicrotophos
1897-45-6	1,3-Dicyanotetrachlorobenzene
77-73-6	Dicyclopentadiene
102-54-5	Dicyclopentadienyl iron
60-57-1	Dieldrin
1464-53-5	Diepoxybutane

CAS No.	Substance Name
1464-53-5	1,2:3,4,-Diepoxybutane
111-42-2	Diethanolamine
109-89-7	Diethylamine
100-37-8	Diethylaminoethanol
1642-54-2	Diethylcarbamazine citrate
810-49-3	Diethyl chlorophosphate
95-06-7	Diethyldithiocarbamic acid, 2-chloroallyl ester
123-91-1	1,4-Diethylene dioxide
111-40-0	Diethylenetriamine
100-37-8	Diethylethanolamine
60-29-7	Diethyl ether
298-04-4	O,O-Diethyl S-(2-(ethylthio)ethyl) phosphorodithioate
103-23-1	Diethylhexyl adipate
117-81-7	Di(2-ethylhexyl) phthalate
1615-80-1	N,N'-Diethylhydrazine
1615-80-1	1,2-Diethylhydrazine
96-22-0	Diethyl ketone
3288-58-2	O,O-Diethyl S-methyl dithiophosphate
3288-58-2	O,O-Diethyl S-methyl phosphorodithioate
311-45-5	Diethyl (4-nitrophenyl) phosphate
311-45-5	Diethyl-*p*-nitrophenyl phosphate
56-38-2	Diethyl 4-nitrophenyl phosphorothionate
55-18-5	N,N-Diethylnitrosamine
56-38-2	Diethyl parathion
84-66-2	Diethyl phthalate
297-97-2	O,O-Diethyl O-pyrazinyl phosphorothioate
56-53-1	Diethylstilbestrol
56-53-1	*alpha,alpha'*-Diethyl-4,4'-stilbenediol
64-67-5	Diethyl sulfate
75-61-6	Difluorodibromomethane
76-11-9	1,1-Difluoro-1,2,2,2-tetrachloroethane
76-12-0	1,2-Difluoro-1,1,2,2-tetrachloroethane
71-63-6	Digitoxin
2238-07-5	Diglycidyl ether
20830-75-5	Digoxin
96-45-7	4,5-Dihydro-2-mercaptoimidazole
56-49-5	1,2-Dihydro-3-methylbenz[j]aceanthrylene
56-04-2	2,3-Dihydro-6-methyl-2-thioxo-4-(1H)-pyrimidinone
60-51-5	Dimethoate
119-90-4	3,3'-Dimethoxybenzidine
119-90-4	3,3'-Dimethoxy-(1,1'-biphenyl)-4,4'-diamine
119-90-4	3,3'-Dimethoxy-4,4'-diaminodiphenyl
109-87-5	Dimethoxymethane
357-57-3	2,3-Dimethoxystrychnidin-10-one
127-19-5	Dimethyl acetamide
127-19-5	N,N-Dimethyl acetamide
91-80-5	2-((2-(Dimethylamino)ethyl)-2-thenylamino)pyridine
60-11-7	4-Dimethylaminoazobenzene
1300-73-8	Dimethylaminobenzene
124-40-3	Dimethylamine
97-56-3	2',3-Dimethyl-4-aminoazobenzene
121-69-7	N,N-Dimethylaminobenzene
79-44-7	(Dimethylamino)carbonyl chloride
121-69-7	Dimethylaniline
121-69-7	N,N-Dimethylaniline

CAS No.	Substance Name
87-62-7	2,6-Dimethylaniline
692-42-2	Diethylarsine
75-60-5	Dimethyl arsenate
57-97-6	7,12-Dimethylbenz[a]anthracene
57-97-6	7,12-Dimethyl-1,2-benzanthracene
1330-20-7	Dimethylbenzene
108-38-3	*m*-Dimethylbenzene
95-47-6	*o*-Dimethylbenzene
106-42-3	*p*-Dimethylbenzene
95-47-6	1,2-Dimethylbenzene
106-42-3	1,4-Dimethylbenzene
119-93-7	3,3'-Dimethylbenzidine
80-15-9	*alpha,alpha*-Dimethylbenzylhydroperoxide
119-93-7	3,3'-Dimethyl-(1,1'-biphenyl)- 4,4'-diamine
1910-42-5	1,1'-Dimethyl-4,4'-bipyridinium dichloride
79-44-7	Dimethylcarbamoyl chloride
79-44-7	Dimethylcarbamyl chloride
2524-03-0	Dimethyl chlorothiophosphate
300-76-5	Dimethyl-1,2-dibromo-2,2-dichloroethyl phosphate
75-78-5	Dimethyl dichlorosilane
62-73-7	Dimethyl 2,2-dichlorovinyl phosphate
26419-73-8	2,4-Dimethyl-1,3-Dithiolane-2-carboxaldehyde, O-(methylcarbamoy) oxime
26419-73-8	O-(((2,4-Dimethyl-1,3-dithiolan-2-y)methylcarbamic acid
68-12-2	Dimethylformamide
68-12-2	N,N-Dimethylformamide
108-83-8	2,6-Dimethyl-4-heptanone
57-14-7	Dimethylhydrazine
57-14-7	*as*-Dimethylhydrazine
57-14-7	N,N-Dimethylhydrazine
57-14-7	1,1-Dimethylhydrazine
540-73-8	1,2-Dimethylhydrazine
57-14-7	Dimethyl hydrazine, unsymmetrical
52-68-6	O,O-Dimethyl (1-hydroxy-2,2,2-trichloroethyl) phosphonate
67-64-1	Dimethyl ketone
39196-18-4	3,3'-Dimethyl-1-(methylthio)-2-butanone, O-((methylamino)carbonyl) oxime
2587-90-8	O,O-Dimethyl-S-(2-methylthio)ethyl phosphorothioate
3254-63-5	Dimethyl 4-(methylthio)phenyl phosphate
79-46-9	Dimethylnitromethane
298-00-0	O,O-Dimethyl O-*p*-nitrophenyl phosphorothioate
62-75-9	Dimethylnitrosoamine
122-09-8	*alpha,alpha*-Dimethylphenethylamine
105-67-9	2,4-Dimethylphenol
60-11-7	N,N-Dimethyl-4-phenylazobenzenamine
122-09-8	*alpha,alpha*-Dimethylphenethylamine
99-98-9	Dimethyl-p-phenylenediamine
122-09-8	1,1-Dimethyl-2-phenylethanamine
105-67-9	2,4-Dimethylphenol
121-69-7	Dimethylphenylamine
3118-97-6	1-((2,4-Dimethylphenyl)azo)-2-naphthalenol
2524-03-0	Dimethyl phosphorochloridothioate
131-11-3	Dimethyl phthalate
77-78-1	Dimethyl sulfate
75-18-3	Dimethyl sulfide
52-68-6	O,O-Dimethyl (2,2,2-trichloro-1-hydroxyethyl) phosphonate

CAS No.	Substance Name
2164-17-2	N,N-Dimethyl-N'-(3-(trifluoromethyl)phenyl)urea
644-64-4	Dimetilan
52-68-6	Dimetox
148-01-6	Dinitolmide
99-65-0	m-Dinitrobenzene
528-29-0	o-Dinitrobenzene
100-25-4	p-Dinitrobenzene
528-29-0	1,2-Dinitrobenzene
99-65-0	1,3-Dinitrobenzene
100-25-4	1,4-Dinitrobenzene
25154-54-5	Dinitrobenzene, mixed
534-52-1	Dinitro-o-cresol and salts
534-52-1	4,6-Dinitro-o-cresol and salts
131-89-5	4,6-Dinitro-o-cyclohexylphenol
1582-09-8	2,6-Dinitro-N,N-dipropyl-4-(trifluoromethyl)benzenamine
534-52-1	2,4-Dinitro-6-methylphenol and salts
88-85-7	2,4-Dinitro-6-(1-methylpropyl)phenol
25550-58-7	Dinitrophenol
51-28-5	2,4-Dinitrophenol
329-71-5	2,5-Dinitrophenol
573-56-8	2,6-Dinitrophenol
148-01-6	3,5-Dinitro-o-toluamide
121-14-2	Dinitrotoluene
25321-14-6	Dinitrotoluene
121-14-2	2,4-Dinitrotoluene
606-20-2	2,6-Dinitrotoluene
610-39-9	3,4-Dinitrotoluene
51-28-5	Dinofan
88-85-7	Dinoseb
1420-07-1	Dinoterb
103-23-1	Dioctyl adipate
117-81-7	Di-sec-octyl phthalate
117-84-0	Di-n-octyl phthalate
123-91-1	Dioxane
123-91-1	p-Dioxane
123-91-1	1,4-Dioxane
78-34-2	Dioxathion
1746-01-6	Dioxin
82-66-6	Diphacinone
92-52-4	Diphenyl
122-39-4	Diphenylamine
57-41-0	Diphenylhydantoin
122-66-7	1,2-Diphenylhydrazine
101-68-8	Diphenylmethane diisocyanate
101-68-8	Diphenylmethane-4,4'-diisocyanate
101-68-8	4,4'-Diphenylmethane diisocyanate
86-30-6	N,N-Diphenyl-N-nitrosoamine
101-84-8	Diphenyl oxide
142-84-7	Dipropylamine
34590-94-8	Dipropylene glycol methyl ether
34590-94-8	Dipropylene glycol monomethyl ether
123-19-3	Dipropyl ketone
621-64-7	Di-n-propylnitrosamine
85-00-7	Diquat
2764-72-9	Diquat
85-00-7	Diquat dibromide
85-00-7	o-Diquat dibromide

CAS No.	Substance Name
1937-37-7	Direct black 38
2602-46-2	Direct blue 6
16071-86-6	Direct brown 95
7558-79-4	Disodium phosphate
10039-32-4	Disodium phosphate dodecahydrate
10140-65-5	Disodium phosphate hydrate
97-77-8	Disulfiram
298-04-4	Disulfoton
298-04-4	Disyston
6533-73-9	Dithallium(I) carbonate
514-73-8	Dithiazanine iodide
541-53-7	Dithiobiuret
541-53-7	2,4-Dithiobiuret
3689-24-5	Dithiopyrophosphoric acid, tetraethyl ester
330-54-1	Diuron
106-99-0	Divinyl
1321-74-0	Divinylbenzene
108-57-6	*m*-Divinylbenzene
79-44-7	DMCC
68-12-2	DMF
62-75-9	DMNA
51-28-5	2,4-DNP
103-23-1	DOA
27176-87-0	Dodecylbenzenesulfonic acid
117-81-7	DOP
23214-92-8	Doxorubicin
2921-88-2	Dursban
944-22-9	Dyfonate
60-00-4	EDTA
112-62-9	Emery
316-42-7	Emetine dihydrochloride
115-29-7	Endosulfan
1031-07-8	Endosulfan sulfate
959-98-8	*alpha*-Endosulfan
33213-65-9	*beta*-Endosulfan
145-73-3	Endothall
2778-04-3	Endothion
72-20-8	Endrin
7421-93-4	Endrin aldehyde
759-73-9	ENU
1395-21-7	Enzymes
106-89-8	Epichlorohydrin
51-43-4	Epinephrine
2104-64-5	EPN
106-88-7	1,2-Epoxybutane
106-89-8	1,2-Epoxy-3-chloropropane
75-21-8	1,2-Epoxyethane
96-09-3	(Epoxyethyl)benzene
75-56-9	1,2-Epoxypropane
765-34-4	2,3-Epoxy-1-propanal
556-52-5	2,3-Epoxy-1-propanol
50-14-6	Ergocalciferol
379-79-3	Ergotamine tartrate
X00001-24-1	Estrogens, conjugated
75-07-0	Ethanal
60-35-5	Ethanamide

CAS No.	Substance Name
74-84-0	Ethane
107-21-1	1,2-Ethanediol
111-54-6	1,2-Ethanediyl-bis-carbamodithioic acid
12427-38-2	1,2-Ethanediylbiscarbamodithioic acid, manganese(2 +) salt
12122-67-7	1,2-Ethanediylbiscarbamodithioic acid, zinc salt
62207-76-5	((2,2'-(1,2-Ethanediylbis(nitrilomethylidyne))bis(6 cobalt
67-72-1	Ethane hexachloride
75-05-8	Ethanenitrile
62-55-5	Ethanethioamide
75-08-1	Ethanethiol
64-17-5	Ethanol
141-43-5	Ethanolamine
75-36-5	Ethanoyl chloride
60-29-7	Ether
563-12-2	Ethion
13194-48-4	Ethoprop
13194-48-4	Ethoprophos
110-80-5	2-Ethoxyethanol
111-15-9	2-Ethoxyethyl acetate
62-44-2	N-(4-Ethoxyphenyl)acetamide
141-78-6	Ethyl acetate
140-88-5	Ethyl acrylate
64-17-5	Ethyl alcohol
75-04-7	Ethylamine
541-85-5	Ethyl *sec*-amyl ketone
100-41-4	Ethyl benzene
538-07-8	Ethylbis(2-chloroethyl)amine
74-96-4	Ethyl bromide
106-35-4	Ethyl butyl ketone
51-79-6	Ethyl carbamate
110-80-5	Ethyl Cellosolve
75-00-3	Ethyl chloride
541-41-3	Ethyl chloroformate
107-12-0	Ethyl cyanide
510-15-6	Ethyl 4,4'-dichlorobenzilate
74-85-1	Ethylene
111-54-6	Ethylenebisdithiocarbamic acid
62207-76-5	N,N'-Ethylenebis(3-fluorosalicylidineiminato) cobalt(II)
79-10-7	Ethylene carboxylic acid
107-07-3	Ethylene chlorohydrin
107-15-3	Ethylenediamine
60-00-4	Ethylenediamine tetraacetic acid
106-93-4	Ethylene dibromide
107-06-2	Ethylene dichloride
628-96-6	Ethylene dinitrate
371-62-0	Ethylene fluorohydrin
107-21-1	Ethylene glycol
628-96-6	Ethylene glycol dinitrate
110-49-6	Ethylene glycol methyl ether acetate
110-80-5	Ethylene glycol monoethyl ether
111-15-9	Ethylene glycol monoethyl ether acetate
109-59-1	Ethylene glycol monoisopropyl ether
109-86-4	Ethylene glycol monomethyl ether
110-49-6	Ethylene glycol monomethyl ether acetate
151-56-4	Ethyleneimine

CAS No.	Substance Name
68-76-8	2,3,5-Ethylenimine-1,4-benzoquinone
75-21-8	Ethylene oxide
96-45-7	Ethylene thiourea
60-29-7	Ethyl ether
106-88-7	Ethylethylene oxide
109-94-4	Ethyl formate
75-34-3	Ethylidene chloride
75-34-3	Ethylidene dichloride
16219-75-3	Ethylidenenorbornene
16219-75-3	5-Ethylidene-2-norbornene
75-08-1	Ethyl mercaptan
97-63-2	Ethyl methacrylate
62-50-0	Ethyl methanesulfonate
78-93-3	Ethyl methyl ketone
615-53-2	Ethyl methylnitrosocarbamate
2703-13-1	O-Ethyl-O-(4-(methylthio)phenyl) methylphosphonothioate
35400-43-2	O-Ethyl O-(4-(methylthio)phenyl) S-propyl phosphorodithioate
100-74-3	N-Ethylmorpholine
759-73-9	N-Ethyl-N-nitrosocarbamide
55-18-5	N-Ethyl-N-nitrosoethanamine
106-88-7	2-Ethyloxirane
56-38-2	Ethyl parathion
84-66-2	Ethyl phthalate
140-88-5	Ethyl 2-propenoate
78-10-4	Ethyl silicate
64-67-5	Ethyl sulfate
542-90-5	Ethyl thiocyanate
115-21-9	Ethyl trichlorosilane
51-79-6	Ethyl urethane
3118-97-6	Ext. D&C Red no. 14
52-85-7	Famphur
3844-45-9	FD&C Blue no. 1
4680-78-8	FD&C Green no. 1
22224-92-6	Fenamiphos
122-14-5	Fenitrothion
115-90-2	Fensulfothion
55-38-9	Fenthion
14484-64-1	Ferbam
7720-78-7	Ferrous sulfate
1185-57-5	Ferric ammonium citrate
2944-67-4	Ferric ammonium oxalate
55488-87-4	Ferric ammonium oxalate
7705-08-0	Ferric chloride
9004-66-4	Ferric dextran
7783-50-8	Ferric fluoride
10421-48-4	Ferric nitrate
1309-37-1	Ferric oxide
10028-22-5	Ferric sulfate
102-54-5	Ferrocene
10045-89-3	Ferrous ammonium sulfate
7758-94-3	Ferrous chloride
7782-63-0	Ferrous sulfate
7782-63-0	Ferrous sulfate heptahydrate
12604-58-9	Ferrovanadium

CAS No.	Substance Name
60676-86-0	Fibrous glass
4301-50-2	Fluenethil
4301-50-2	Fluenetil
2164-17-2	Fluometuron
640-19-7	Fluoracetamide
62-74-8	Fluoroacetic acid, sodium salt
206-44-0	Fluoranthene
86-73-7	Fluorene
53-96-3	N-Fluoren-2-ylacetamide
53-96-3	N-2-Fluorenylacetamide
53-96-3	N-9H-Fluoren-2-ylacetamide
X00001-32-1	Fluorides
7782-41-4	Fluorine
640-19-7	2-Fluoroacetamide
144-49-0	Fluoroacetic acid
359-06-8	Fluoroacetyl chloride
75-69-4	Fluorotrichloromethane
51-21-8	Fluorouracil
944-22-9	Fonofos
109-87-5	Formal
50-00-0	Formaldehyde
107-16-4	Formaldehyde cyanohydrin
50-00-0	Formalin
75-12-7	Formamide
23422-53-9	Formetanate hydrochloride
23422-53-9	Formetane hydrochloride
64-18-6	Formic acid
544-18-3	Formic acid, cobalt(II) salt
109-94-4	Formic acid, ethyl ester
625-55-8	Formic acid, isopropyl ester
107-31-3	Formic acid, methyl ester
557-41-5	Formic acid, zinc salt
50-00-0	Formol
74-90-8	Formonitrile
2540-82-1	Formothion
17702-57-7	Formparanate
21548-32-3	Fosthietan
77-13-1	Freon 113
3878-19-1	Fuberidazole
628-86-4	Fulminic acid, mercury salt
628-86-4	Fulminic acid, mercury(II) salt
110-17-8	Fumaric acid
1563-66-2	Furadan
110-00-9	Furan
98-01-1	2-Furancarboxaldehyde
108-31-6	2,5-Furandione
98-01-1	Furfural
110-00-9	Furfuran
98-00-0	Furfurol
98-00-0	Furfuryl alcohol
13450-90-3	Gallium trichloride
8006-61-9	Gasoline
7782-65-2	Germane
7782-65-2	Germanium tetrahydride
64-19-7	Glacial acetic acid
60676-86-0	Glass, fibrous

CAS No.	Substance Name
111-30-8	Glutaraldehyde
56-81-5	Glycerin
56-81-5	Glycerol
55-63-0	Glyceryl trinitrate
765-34-4	Glycidylaldehyde
556-52-5	Glycidol
106-89-8	Glycidyl chloride
X00001-28-5	Glycol ethers
110-80-5	Glycol monoethyl ether
X00001-06-9	Grain dust
7782-42-5	Graphite
X00001-33-2	Graphite, synthetic
86-50-0	Guthion
7778-18-9	Gypsum
10101-41-4	Gypsum
7440-58-6	Hafnium
319-84-6	HCH, hexachlorcyclohexane isomers
7440-59-7	Helium
76-44-8	Heptachlor
1024-57-3	Heptachlor epoxide
76-44-8	1,4,5,6,7,8,8,-Heptachloro-3a,4,7,7a-tetrahydro-4,7-methano-1H-indene
142-82-5	n-Heptane
110-43-0	2-Heptanone
106-35-4	3-Heptanone
123-19-3	4-Heptanone
36355-01-8	Hexabromobiphenyl
118-74-1	Hexachlorobenzene
87-68-3	Hexachlorobutadiene
87-68-3	Hexachloro-1,3-butadiene
87-68-3	1,1,2,3,4,4-Hexachloro-1,3-butadiene
58-89-9	Hexachlorocyclohexane (gamma isomer)
77-47-4	Hexachlorocyclopentadiene
77-47-4	Hexachloro-1,3-cyclopentadiene
77-47-4	1,2,3,4,5,5,-Hexachloro-1,3-cyclopentadiene
67-72-1	Hexachloroethane
67-72-1	1,1,1,2,2,2-Hexachloroethane
465-73-6	Hexachlorohexahydro-endo,endo-dimethanonaphthalene
1335-87-1	Hexachloronaphthalene
70-30-4	Hexachlorophene
1888-71-7	Hexachloropropene
1888-71-7	1,1,2,3,3,3-Hexachloro-1-propene
757-58-4	Hexaethyl tetraphosphate
684-16-2	Hexafluoroacetone
110-82-7	Hexahydrobenzene
100-75-4	Hexahydro-N-nitrosopyridine
121-82-4	Hexahydro-1,3,5-trinitro-s-triazine
680-31-9	Hexamethyl phosphoramide
680-31-9	Hexamethyl phosphoric triamide
110-54-3	Hexane
110-54-3	n-Hexane
X00001-34-3	Hexane, other than normal isomer
591-78-6	2-Hexanone
108-10-1	Hexone
108-84-9	sec-Hexyl acetate

CAS No.	Substance Name
142-92-7	sec-Hexyl acetate
107-41-5	Hexylene glycol
302-01-2	Hydrazine
79-19-6	Hydrazinecarbothioamide
563-41-7	Hydrazinecarboxamide monohydrochloride
10034-93-2	Hydrazine sulfate
10034-93-2	Hydrazinium sulfate
122-66-7	Hydrazobenzene
8007-45-2	Hydrocarbons, polycyclic aromatic, particulate
7647-01-0	Hydrochloric acid
74-90-8	Hydrocyanic acid
57-12-5	Hydrocyanic acid, ion(1-)
7664-39-3	Hydrofluoric acid
1333-74-0	Hydrogen
61788-32-7	Hydrogenated terphenyls
10035-10-6	Hydrogen bromide
7647-01-0	Hydrogen chloride
74-90-8	Hydrogen cyanide
7664-39-3	Hydrogen fluoride
7722-84-1	Hydrogen peroxide
7803-51-2	Hydrogen phosphide
7783-07-5	Hydrogen selenide
7783-06-4	Hydrogen sulfide
123-31-9	Hydroquinone
150-76-5	Hydroquinone monomethyl ether
7783-06-4	Hydrosulfuric acid
107-16-4	Hydroxyacetonitrile
108-95-2	Hydroxybenzene
90-43-7	2-Hydroxybiphenyl
78-92-2	2-Hydroxybutane
75-60-5	Hydroxydimethylarsine oxide
105-67-9	1-Hydroxy-2,4-dimethylbenzene
51-28-5	1-Hydroxy-2,4-dinitrobenzene
107-21-1	2-Hydroxyethanol
51-43-4	4-(1-Hydroxy-2-(methylamino)ethyl)-1,2-Benzenediol
95-48-7	1-Hydroxy-2-methylbenzene
106-44-5	1-Hydroxy-4-methylbenzene
123-42-2	4-Hydroxy-4-methyl-2-pentanone
2832-40-8	4'-((2-Hydroxy-5-methylphenyl)azo)acetanilide
2832-40-8	N-(4-((2-Hydroxy-5-methylphenyl)azo)phenyl)acetamide
75-86-5	2-Hydroxy-2-methylpropanenitrile
842-07-9	2-Hydroxynaphthyl-1-azobenzene
100-02-7	p-Hydroxynitrobenzene
120-80-9	o-Hydroxyphenol
123-31-9	p-Hydroxyphenol
67-63-0	2-Hydroxypropanol
57-57-8	3-Hydroxypropionic acid lactone
999-61-1	2-Hydroxypropyl acrylate
1319-77-3	Hydroxytoluene
108-39-4	m-Hydroxytoluene
95-48-7	o-Hydroxytoluene
106-44-5	p-Hydroxytoluene
4016-14-2	IGE
96-45-7	2-Imidazolidinethione
111-42-2	2,2'-Iminobisethanol
95-13-6	Indene

CAS No.	Substance Name
193-39-5	Indeno[1,2,3-cd]pyrene
7440-74-6	Indium
X00001-35-4	Indium compounds
7553-56-2	Iodine
506-78-5	Iodine cyanide
75-47-8	Iodoform
74-88-4	Iodomethane
55488-87-4	Iron ammonium oxalate
9004-66-4	Iron dextran
1309-37-1	Iron oxide
13463-40-6	Iron pentacarbonyl
X00001-36-5	Iron salts, soluble
123-92-2	Isoamyl acetate
123-51-3	Isoamyl alcohol
110-12-3	Isoamyl methyl ketone
297-78-9	Isobenzan
85-44-9	1,3-Isobenzofurandione
79-31-2	Isobutanoic acid
78-83-0	Isobutanol
110-19-0	Isobutyl acetate
78-83-1	Isobutyl alcohol
78-81-9	Isobutylamine
108-11-2	Isobutyl methyl carbinol
108-10-1	Isobutyl methyl ketone
78-84-2	Isobutyraldehyde
79-31-2	Isobutyric acid
78-82-0	Isobutyronitrile
102-36-3	Isocyanic acid, 3,4-dichlorophenyl ester
624-83-9	Isocyanic acid, methyl ester
465-73-6	Isodrin
55-91-4	Isofluorphate
26952-21-6	Isooctanol
26952-21-6	Isooctyl alcohol
78-59-1	Isophorone
4098-71-9	Isophorone diisocyanate
626-17-5	Isophthalonitrile
78-79-5	Isoprene
67-63-0	Isopropanol
42504-46-1	Isopropanolamine dodecylbenzenesulfonate
98-83-9	Isopropenylbenzene
109-59-1	Isopropoxyethanol
114-26-1	O-(2-Isopropoxyphenyl) N-methylcarbamate
108-21-4	Isopropyl acetate
67-63-0	Isopropyl alcohol
75-31-0	Isopropylamine
768-52-5	N-Isopropylaniline
98-82-8	Isopropylbenzene
80-15-9	Isopropylbenzene hydroperoxide
108-23-6	Isopropyl chloroformate
94-11-1	Isopropyl 2,4-dichlorophenoxyacetate
108-20-3	Isopropyl ether
625-55-8	Isopropyl formate
4016-14-2	Isopropyl glycidyl ether
80-05-7	4,4'-Isopropylidenediphenol
119-38-0	Isopropylmethylpyrazolyl dimethylcarbamate
93-79-8	Isopropyl 2,4,5-trichlorophenoxyacetate
120-58-1	Isosafrole

CAS No.	Substance Name
X00001-38-7	Kaolin
115-32-2	Kelthane
143-50-0	Kepone
463-51-4	Ketene
138-22-7	Lactic acid, *n*-butyl ester
78-97-7	Lactonitrile
16752-77-5	Lannate
303-34-4	Lasiocarpine
7439-92-1	Lead
301-04-2	Lead acetate
X00001-49-0	Lead and compounds
7645-25-2	Lead arsenate
7784-40-9	Lead arsenate
10102-48-4	Lead arsenate
7758-95-4	Lead chloride
1344-38-3	Lead chromate
7758-97-6	Lead chromate
13814-96-5	Lead fluoborate
7783-46-2	Lead fluoride
X00001-39-8	Lead, inorganic
10101-63-0	Lead iodide
10099-74-8	Lead nitrate
7446-27-7	Lead phosphate
1072-35-1	Lead stearate
7428-48-0	Lead stearate
52652-59-2	Lead stearate
56189-09-4	Lead stearate
1335-32-6	Lead subacetate
7446-14-2	Lead sulfate
15739-80-7	Lead sulfate
1314-87-0	Lead sulfide
592-87-0	Lead thiocyanate
21609-90-5	Leptophos
541-25-3	Lewisite
1305-78-8	Lime
1317-65-3	Limestone
58-89-9	Lindane
319-84-6	Lindane
68476-85-7	Liquified petroleum gas
14307-35-8	Lithium chromate
7580-67-8	Lithium hydride
68476-85-7	LPG
532-27-4	Mace
632-99-5	Magenta manufacture
1309-48-4	Magnesia
546-93-0	Magnesite dust
546-93-0	Magnesium carbonate dust
1309-48-4	Magnesium oxide
569-64-2	Malachite green
121-75-5	Malathion
110-16-7	Maleic acid
108-31-6	Maleic acid anhydride
108-31-6	Maleic anhydride
123-33-1	Maleic hydrazide
2757-18-8	Malonic acid, thallous salt

CAS No.	Substance Name
109-77-3	Malononitrile
12427-38-2	Maneb
7439-96-5	Manganese
X00001-40-1	Manganese compounds
12079-65-1	Manganese cyclopentadienyl tricarbonyl
12427-38-2	Manganese ethylenebis(dithiocarbamate)
1317-35-7	Manganese tetroxide
X00001-44-5	MAPP
1317-65-3	Marble
101-68-8	MBI
101-68-8	MDI
51-75-2	Mechlorethamine
137-05-3	Mecrylate
78-93-3	MEK
108-78-1	Melamine
148-82-3	Melphalan
950-10-7	Mephosfolan
2032-65-7	Mercaptodimethur
96-45-7	Mercaptoimidazoline
1600-27-7	Mercuric acetate
7487-94-7	Mercuric chloride
592-04-1	Mercuric cyanide
10045-94-0	Mercuric nitrate
21908-53-2	Mercuric oxide
7783-35-9	Mercuric sulfate
592-85-8	Mercuric thiocyanate
7782-86-7	Mercurous nitrate
10415-75-5	Mercurous nitrate
7782-86-7	Mercurous nitrate monohydrate
7439-97-6	Mercury
X00001-41-2	Mercury, alkyl compounds
X00001-51-4	Mercury and compounds
X00001-42-3	Mercury, aryl compounds
628-86-4	Mercury(II) fulminate
X00001-43-4	Mercury, inorganic compounds
141-79-7	Mesityl oxide
8022-00-2	Metasystox-R
10476-95-6	Methacrolein diacetate
79-41-4	Methacrylic acid
80-62-6	Methacrylic acid, methyl ester
97-63-2	Methacrylic acid, ethyl ester
760-93-0	Methacrylic anhydride
126-98-7	Methacrylonitrile
920-46-7	Methacryloyl chloride
30674-80-7	Methacryloyloxyethyl isocyanate
10265-92-6	Methamidophos
50-00-0	Methanal
74-82-8	Methane
62-50-0	Methanesulfonic acid, ethyl ester
558-25-8	Methanesulfonyl fluoride
74-93-1	Methanethiol
64-18-6	Methanoic acid
67-56-1	Methanol
91-80-5	Methapyrilene
950-37-8	Methidathion
2032-65-7	Methiocarb
16752-77-5	Methomyl

CAS No.	Substance Name
X00001-25-2	Methoxsalen with ultra-violet A therapy
104-94-9	p-Methoxyaniline
134-29-2	2-Methoxyaniline hydrochloride
134-29-2	2-Methoxybenzenamine hydrochloride
39156-41-7	4-Methoxy-1,3-benzenediamine, sulfate (1:1)
72-43-5	Methoxychlor
109-86-4	2-Methoxyethanol
110-49-6	2-Methoxyethyl acetate
151-38-2	Methoxyethylmercuric acetate
120-71-8	2-Methoxy-5-methylaniline
1634-04-4	2-Methoxy-2-methylpropane
99-59-2	2-Methoxy-5-nitroaniline
150-76-5	4-Methoxyphenol
39156-41-7	4-Methoxy-m-phenylenediamine sulfate
107-98-2	1-Methoxy-2-propanol
123-38-6	Methylacetaldehyde
79-20-9	Methyl acetate
74-99-7	Methylacetylene
X00001-44-5	Methylacetylene-propadiene mixture
96-33-3	Methyl acrylate
126-98-7	Methylacrylonitrile
109-87-5	Methylal
67-56-1	Methyl alcohol
74-89-5	Methylamine
105-30-6	Methylamyl alcohol
108-11-2	Methylamyl alcohol
105-30-6	2-Methylamyl alcohol
110-43-0	Methyl n-amyl ketone
100-61-8	N-Methyl aniline
95-53-4	o-Methylaniline
636-21-5	o-Methylaniline hydrochloride
75-55-8	2-Methylaziridine
95-53-4	2-Methylbenzenamine
106-49-0	4-Methylbenzenamine
108-88-3	Methylbenzene
636-21-5	2-Methylbenzenamine hydrochloride
95-80-7	4-Methyl-1,3-Benzenediamine
51-75-2	N-Methylbis(beta-chloroethyl)amine
74-83-9	Methyl bromide
504-60-9	1-Methylbutadiene
563-80-4	3-Methyl-2-butanone
563-80-4	3-Methyl butan-2-one
1634-04-4	Methyl tert-butyl ether
591-78-6	Methyl n-butyl ketone
114-26-1	Methylcarbamic acid, o-isopropoxyphenyl ester
16752-77-5	N((Methylcarbamoyl)oxy)thioacetimidic acid, methyl ester
109-86-4	Methyl cellosolve
110-49-6	Methyl cellosolve acetate
56-49-5	3-Methylchloanthrene
74-87-3	Methyl chloride
80-63-7	Methyl 2-chloroacrylate
79-22-1	Methyl chlorocarbonate
71-55-6	Methyl chloroform
79-22-1	Methyl chloroformate
107-30-2	Methyl chloromethyl ether
52-68-6	Methyl chlorophos

CAS No.	Substance Name
75-30-2	Methyl cyanide
137-05-3	Methyl 2-cyanoacrylate
108-87-2	Methylcyclohexane
25639-42-3	Methylcyclohexanol
583-60-8	o-Methylcyclohexanone
583-60-8	2-Methylcyclohexanone
12108-13-3	Methylcyclopentadienyl manganese tricarbonyl
8022-00-2	Methyl demeton
121-14-2	1-Methyl-2,4-dinitrobenzene
606-20-2	1-Methyl-2,6-dinitrobenzene
624-92-0	Methyl disulfide
101-77-9	4,4'-Methylenebis(aniline)
101-14-4	4,4'-Methylenebis(2-chloroaniline)
101-14-4	4,4'-Methylenebis(2-chlorobenzenamine)
5124-30-1	4,4'-Methylenebis(cyclohexyl isocyanate)
5124-30-1	Methylenebis(4-cyclohexyl isocyanate)
101-61-1	4,4'-Methylenebis(N,N-dimethylaniline)
101-61-1	4,4'-Methylenebis(N,N-dimethyl)benzenamine
838-88-0	4,4'-Methylenebis(2-methylaniline)
111-91-1	1,1'-(Methylenebis(oxy))bis(2-chloroethane)
101-68-8	Methylenebisphenyl isocyanate
70-30-4	2,2'-Methylenebis(3,4,6-trichlorophenol)
74-95-3	Methylene bromide
75-09-2	Methylene chloride
101-77-9	4,4'-Methylenedianiline
75-09-2	Methylene dichloride
94-59-7	1,2-Methylenedioxy-4-allylbenzene
120-58-1	1,2-Methylenedioxy-4-propenylbenzene
94-58-6	1,2-Methylenedioxy-4-propylbenzene
50-00-0	Methylene oxide
98-82-8	(1-Methylethyl)benzene
78-92-2	Methylethylcarbinol
115-07-1	Methylethylene
75-55-8	2-Methylethyleneimine
78-93-3	Methyl ethyl ketone
1338-23-4	Methyl ethyl ketone peroxide
64-00-6	3-(1-Methylethyl)phenol, methylcarbamate
107-31-3	Methyl formate
541-85-5	5-Methyl-3-heptanone
110-12-3	5-Methyl-2-hexanone
60-34-4	Methylhydrazine
74-88-4	Methyl iodide
110-12-3	Methyl isoamyl ketone
105-30-6	Methyl isobutyl carbinol
108-11-2	Methyl isobutyl carbinol
108-10-1	Methyl isobutyl ketone
624-83-9	Methyl isocyanate
107-44-8	Methylisopropoxyfluorophosphine oxide
563-80-4	Methyl isopropyl ketone
556-61-1	Methyl isothiocyanate
75-86-5	2-Methyllactonitrile
74-93-1	Methyl mercaptan
502-39-6	Methylmercuric dicyanamide
80-62-6	Methyl methacrylate
116-06-3	2-Methyl-2-(methylthio)propanal, O-((methylamino)carbonyl)oxime
563-80-4	3-Methyl methyl ethyl ketone

CAS No.	Substance Name
116-06-3	2-Methyl-2-(methylthio)propanal, O-((methylamino)carbonyl)oxime
99-55-8	2-Methyl-5-nitrobenzenamine
70-25-7	N-Methyl-N'-nitro-N-nitrosoguanidine
615-53-2	Methylnitrosocarbamic acid, ethyl ester
684-93-5	N-Methyl-N-nitrosocarbamide
4549-40-0	N-Methyl-N-nitrosoethenamine
62-75-9	N-Methyl-N-nitrosomethanamine
681-84-5	Methyl orthosilicate
75-56-9	Methyloxirane
298-00-0	Methyl parathion
108-10-1	4-Methyl-2-pentanone
3735-23-7	Methyl phenkapton
1319-77-3	Methylphenol
95-48-7	2-Methylphenol
108-39-4	3-Methylphenol
106-44-5	4-Methylphenol
25376-45-8	Methyl phenylenediamine
95-80-7	4-Methyl-1,3-phenylenediamine
80-15-9	1-Methyl-1-phenylethylhydroperoxide
98-86-2	Methyl phenyl ketone
614-78-8	2-Methylphenylthiourea
614-78-8	1-(o-Methylphenyl)thiourea
676-97-1	Methylphosphonic dichloride
2703-13-1	Methylphosphonothioic acid, O-ethyl-O-(4-methylthio)phenyl ester
2665-30-7	Methylphosphonothioic acid, O-(4-nitrophenyl)-O-phenyl ester
78-84-2	2-Methylpropanal
78-83-1	2-Methyl-1-propanol
75-65-0	2-Methyl-2-propanol
126-98-7	2-Methyl-2-propenenitrile
96-33-3	Methyl prop-2-enoate
97-63-2	2-Methyl-2-propenoic acid, ethyl ester
80-62-6	2-Methyl-2-propenoic acid, methyl ester
78-84-2	2-Methylpropionaldehyde
107-87-9	Methyl propyl ketone
109-06-8	2-Methylpyridine
54-11-5	(S)-3-(1-Methyl-2-pyrrolidinyl)pyridine and salts
681-84-5	Methyl silicate
25013-15-4	Methylstyrene
25013-15-4	Methylstyrene, aryl
98-83-9	alpha-Methyl styrene
77-78-1	Methyl sulfate
479-45-8	N-Methyl-N,2,4,6-tetranitro aniline
556-64-9	Methyl thiocyanate
56-04-2	Methylthiouracil
108-38-3	m-Methyltoluene
95-47-6	o-Methyltoluene
106-42-3	p-Methyltoluene
75-79-6	Methyl trichlorosilane
140-76-1	2-Methyl-5-vinylpyridine
78-94-4	Methyl vinyl ketone
1129-41-5	Metolcarb
21087-64-9	Metribuzin
443-48-1	Metronidazole
7786-34-7	Mevinphos

CAS No.	Substance Name
315-18-4	Mexacarbate
108-10-1	MIBK
12001-25-2	Mica
101-61-1	Michler's base
90-94-8	Michler's ketone
108-10-1	MIK
8012-95-1	Mineral oil
X00001-00-3	Mineral wool fiber
2385-85-5	Mirex
50-07-7	Mitomycin C
101-14-4	MOCA
7439-98-7	Molybdenum
X00001-46-7	Molybdenum, insoluble compounds
1313-27-5	Molybdenum(VI) oxide
X00001-45-6	Molybdenum, soluble compounds
1313-27-5	Molybdenum trioxide
1313-27-5	Molybdic acid anhydride
108-90-7	Monochlorobenzene
6923-22-4	Monocrotophos
75-04-7	Monoethylamine
100-61-8	Monomethyl aniline
60-34-4	Monomethyl hydrazine
74-89-5	Monomethylamine
110-91-8	Morpholine
7647-01-0	Muriatic acid
2763-96-4	Muscimol
505-60-2	Mustard gas
300-76-5	Naled
8030-30-6	Naphtha
X00001-59-2	Naphtha, rubber solvent
8030-31-7	Naphtha (coal tar)
91-59-8	2-Naphthalenamine
91-20-3	Naphthalene
130-15-4	1,4-Naphthalenedione
86-88-4	1-Naphthalenylthiourea
91-20-3	Naphthalin
1338-24-5	Naphthenic acid
130-15-4	1,4-Naphthoquinone
134-32-7	alpha-Naphthylamine
91-59-8	beta-Naphthylamine
134-32-7	1-Naphthylamine
91-59-8	2-Naphthylamine
63-25-2	alpha-Naphthyl methylcarbamate
63-25-2	1-Naphthyl N-methylcarbamate
86-88-4	alpha-Naphthyl thiourea
86-88-4	1-(1-Naphthyl)-2-thiourea
86-88-4	1-Naphthyl-2-thiourea
55-18-5	NDEA
1116-54-7	NDELA
621-64-7	NDPA
22224-92-6	Nemacur
7440-01-9	Neon
55-63-0	NG
563-12-2	Nialate
7440-02-0	Nickel
X00001-53-6	Nickel and compounds

CAS No.	Substance Name
15699-18-0	Nickel ammonium sulfate
13463-39-3	Nickel carbonyl
7718-54-9	Nickel chloride
37211-05-5	Nickel chloride
7718-54-9	Nickel(II) chloride
557-19-7	Nickel cyanide
557-19-7	Nickel(II) cyanide
12054-48-7	Nickel hydroxide
14216-75-2	Nickel nitrate
X00001-48-9	Nickel refining, roasting
X00001-47-8	Nickel, soluble compounds
7786-81-4	Nickel sulfate
12035-72-2	Nickel sulfide roasting
13463-39-3	Nickel tetracarbonyl
54-11-5	Nicotine and salts
65-30-5	Nicotine sulfate
1929-82-4	Nitrapyrin
7697-37-2	Nitric acid
10102-43-9	Nitric oxide
139-13-9	Nitrilotriacetic acid
100-01-6	*p*-Nitroaniline
100-01-6	4-Nitroaniline
99-59-2	5-Nitro-*o*-anisidine
100-01-6	4-Nitrobenzenamine
98-95-3	Nitrobenzene
100-14-1	4-Nitrobenzyl chloride
92-93-3	4-Nitrobiphenyl
100-00-5	*p*-Nitrochlorobenzene
1122-60-7	Nitrocyclohexane
92-93-3	4-Nitrodiphenyl
79-24-3	Nitroethane
1836-75-5	Nitrofen
10102-44-0	Nitrogen dioxide
10544-72-6	Nitrogen dioxide
51-75-2	Nitrogen mustard
538-07-8	Nitrogen mustard gas
10102-43-9	Nitrogen(II) oxide
10102-44-0	Nitrogen(IV) oxide
10544-72-6	Nitrogen(IV) oxide
10544-72-6	Nitrogen tetroxide
7783-54-2	Nitrogen trifluoride
55-63-0	Nitroglycerin
75-52-5	Nitromethane
99-59-2	5-Nitro-2-methoxyaniline
82-68-8	Nitropentachlorobenzene
554-84-7	*m*-Nitrophenol
88-75-5	*o*-Nitrophenol
100-02-7	*p*-Nitrophenol
25154-55-6	Nitrophenol, mixed
88-75-5	2-Nitrophenol
554-84-7	3-Nitrophenol
100-02-7	4-Nitrophenol
2665-30-7	O-(4-Nitrophenyl) O-phenyl methylphosphonothiate
108-03-2	1-Nitropropane
79-46-9	2-Nitropropane
1124-33-0	4-Nitropyridine-1-oxide
924-16-3	N-Nitrosodibutylamine

CAS No.	Substance Name
924-16-3	N-Nitrosodi-*n*-butylamine
1116-54-7	N-Nitrosodiethanolamine
55-18-5	N-Nitrosodiethylamine
62-75-9	N-Nitrosodimethylamine
621-64-7	N-Nitrosodi-*n*-propylamine
86-30-6	N-Nitrosodiphenylamine
156-10-5	*p*-Nitrosodiphenylamine
759-73-9	N-Nitroso-N-ethylurea
1116-54-7	2,2'-(Nitrosoimino)bisethanol
70-25-7	N-Nitroso-N-methyl-N'-nitroguanidine
684-93-5	N-Nitroso-N-methylurea
615-53-2	N-Nitroso-N-methylurethane
4549-40-0	N-Nitrosomethylvinylamine
13256-13-8	N-Nitrosomethylvinylamine
59-89-2	N-Nitrosomorpholine
16543-55-8	N-Nitrosonornicotine
156-10-5	*p*-Nitroso-N-phenylaniline
92-93-3	1-Nitro-4-phenylbenzene
86-30-6	N-Nitroso-N-phenylbenzenamine
156-10-5	4-Nitroso-N-phenylbenzenamine
135-20-6	N-Nitroso-N-phenylhydroxylamine ammonium salt
100-75-4	1-Nitrosopiperidine
100-75-4	N-Nitrosopiperidine
930-55-2	N-Nitrosopyrrolidine
16543-55-8	3-(1-Nitroso-2-pyrrolidinyl)pyridine
13256-22-9	N-Nitrososarcosine
1321-12-6	Nitrotoluene
99-08-1	*m*-Nitrotoluene
88-72-2	*o*-Nitrotoluene
99-99-0	*p*-Nitrotoluene
88-72-2	2-Nitrotoluene
99-08-1	3-Nitrotoluene
99-99-0	4-Nitrotoluene
99-55-8	5-Nitro-*o*-toluidine
76-06-2	Nitrotrichloromethane
59-89-2	NMOR
684-93-5	NMU
13256-13-8	NMVA
16543-55-8	NNN
111-84-2	Nonane
991-42-4	Norbormide
115-29-7	5-Norbornene-2,3-dimethanol, 1,4,5,6,7,7-hexachlorocyclic sulfite
100-75-4	NPIP
930-55-2	NPYR
13256-22-9	NSAR
1929-82-4	N-Serve
139-13-9	NTA
55-63-0	NTG
X00001-06-9	Oat dust
2698-41-1	OCBM
8001-35-2	Octachlorocamphene
2234-13-1	Octachloronaphthalene
57-74-9	1,2,4,5,6,7,8,8-Octachloro-3a,4,7,7a-tetrahydro-4,7-methanoindan
152-16-9	Octamethyldiphosphoramide

CAS No.	Substance Name
152-16-9	Octamethylpyrophosphoramide
111-65-9	Octane
8012-95-1	Oil, mineral
7664-93-9	Oil of vitriol
8014-95-7	Oleum
X00001-11-6	Organorhodium complex
20816-12-0	Osmium oxide
20816-12-0	Osmium tetroxide
630-60-4	Ouabain
145-73-3	7-Oxabicyclo[2,2,1]heptane-2,3-dicarboxylic acid
144-62-7	Oxalic acid
14258-49-2	Oxalic acid, ammonium salt
5893-66-3	Oxalic acid, copper(II) salt hemihydrate
6009-70-7	Oxalic acid, diammonium salt monohydrate
5972-73-6	Oxalic acid, monoammonium salt monohydrate
23135-22-0	Oxamyl
1120-71-4	1,2-Oxathiolane-2,2-dioxide
50-18-0	2H-1,3,2-Oxazaphosphorine,2(bis(2-chloroethyl)amino) tetrahydro-2 oxide
75-21-8	Oxirane
106-87-6	3-Oxiranyl-7-oxabicyclo[4.1.0]heptane
111-44-4	1,1′-Oxybis(2-chloroethane)
542-88-1	Oxybis(chloromethane)
108-60-1	2,2′-Oxybis(2-chloropropane)
60-29-7	1,1′-Oxybisethane
1163-19-5	1,1′-Oxybis(2,3,4,5,6-pentabromobenzene)
101-80-4	4,4′-Oxydianiline
58-36-6	10,10′-Oxydiphenoxarsine
2497-07-6	Oxydisulfoton
7783-41-7	Oxygen difluoride
434-07-1	Oxymetholone
10028-15-6	Ozone
136-40-3	PAP
9004-34-6	Paper fiber
80-05-7	Parabis A
8002-74-2	Paraffin wax
30525-89-4	Paraformaldehyde
123-63-7	Paraldehyde
1910-42-5	Paraquat
4685-14-7	Paraquat
2074-50-2	Paraquat bis(methyl sulfate)
1910-42-5	Paraquat dichloride
2074-50-2	Paraquat methosulfate
56-38-2	Parathion
298-00-0	Parathion-methyl
12002-03-8	Paris green
8007-45-2	Particulate polycyclic aromatic hydrocarbons
65996-93-2	Particulate polycyclic aromatic hydrocarbons
36355-01-8	PBBs
12674-11-2	PCB-1016
11104-28-2	PCB-1221
11141-16-5	PCB-1232
53469-21-9	PCB-1242
12672-29-6	PCB-1248
11097-69-1	PCB-1254
11096-82-5	PCB-1260

CAS No.	Substance Name
1336-36-3	PCBs
11096-82-5	PCBs
11097-69-1	PCBs
11104-28-2	PCBs
11141-16-5	PCBs
12672-29-6	PCBs
12674-11-2	PCBs
53469-21-9	PCBs
87-86-5	PCP
19624-22-7	Pentaborane
608-93-5	Pentachlorobenzene
76-01-7	Pentachloroethane
1321-64-8	Pentachloronaphthalene
82-68-8	Pentachloronitrobenzene
87-86-5	Pentachlorophenol
2570-26-5	Pentadecylamine
504-60-9	1,3-Pentadiene
115-77-5	Pentaerythritol
109-66-0	Pentane
96-22-0	2-Pentanone
107-87-9	2-Pentanone
79-21-0	Peracetic acid
118-74-1	Perchlorobenzene
87-68-3	Perchloro-1,3-butadiene
77-47-4	Perchlorocyclopentadiene
67-72-1	Perchloroethane
127-18-4	Perchloroethylene
594-42-3	Perchloromethyl mercaptan
1888-71-7	Perchloropropylene
7616-94-6	Perchloryl fluoride
X00001-50-3	Perlite
79-21-0	Peroxyacetic acid
8030-30-6	Petroleum distillates (naphtha)
122-60-1	PGE
62-44-2	Phenacetin
532-27-4	Phenacyl chloride
85-01-8	Phenanthrene
94-78-0	Phenazopyridine
136-40-3	Phenazopyridine hydrochloride
108-95-2	Phenol
92-84-2	Phenothiazine
X00001-26-3	Phenoxyacetic acid herbicides
122-09-8	Phentermine
62-53-3	Phenylamine
156-10-5	p-Phenylaminonitrosobenzene
92-67-1	4-Phenylaniline
60-09-3	4-(Phenylazo)benzenamine
60-11-7	4-(Phenylazo)-N,N-dimethylaniline
900-95-8	1-Phenylazo-beta-naphthol
92-52-4	Phenylbenzene
55-21-0	Phenylcarboxyamide
108-90-7	Phenyl chloride
98-07-7	Phenylchloroform
696-28-6	Phenyldichloroarsine
106-50-3	Phenylenediamine, para
106-50-3	p-Phenylenediamine
193-39-5	1,10-(1,2-Phenylene)pyrene

CAS No.	Substance Name
100-41-4	Phenylethane
98-86-2	1-Phenylethanone
101-84-8	Phenyl ether
55720-99-5	Phenyl ether, chlorinated
100-42-5	Phenylethylene
96-09-3	Phenylethylene oxide
122-60-1	Phenyl glycidyl ether
100-63-0	Phenylhydrazine
59-88-1	Phenylhydrazine hydrochloride
108-98-5	Phenyl mercaptan
62-38-4	Phenylmercuric acetate
62-38-4	Phenylmercury acetate
135-88-6	N-Phenyl-*beta*-naphthylamine
135-88-6	N-Phenyl-2-naphthylamine
96-09-3	Phenyloxirane
90-43-7	2-Phenylphenol
90-43-7	o-Phenylphenol
638-21-1	Phenylphosphine
98-82-8	2-Phenylpropane
80-15-9	2-Phenyl-2-propyl hydroperoxide
2097-19-0	Phenylsilatrane
103-85-5	Phenylthiourea
103-85-5	N-Phenylthiourea
103-85-5	1-Phenylthiourea
98-07-7	Phenyltrichloromethane
98-13-5	Phenyltrichlorosilane
57-41-0	Phenytoin
298-02-2	Phorate
4104-14-7	Phosacetim
7786-34-7	Phosdrin
947-02-4	Phosfolan
75-44-5	Phosgene
732-11-6	Phosmet
13171-21-6	Phosphamidon
7803-51-2	Phosphine
7664-38-2	Phosphoric acid
311-45-5	Phosphoric acid, diethyl p-nitrophenyl ester
3254-63-5	Phosphoric acid, dimethyl 4-(methylthio)phenyl ester
7446-27-7	Phosphoric acid, lead salt
1314-56-3	Phosphoric anhydride
3288-58-2	Phosphorodithioic acid, O,O-diethyl S-methylester
60-51-5	Phosphorodithioic acid, O,O-dimethyl S-(2(methylamino)-2-oxoethyl) ester
55-91-4	Phosphorofluoridic acid, bis(1-methylethyl) ester
56-38-2	Phosphorothioic acid, O,O-diethyl O-(p-nitrophenyl) ester
52-85-7	Phosphorothioic acid, O,O-dimethyl-O-(p-((dimethylamino)sulfonyl)phenyl) ester
2587-90-8	Phosphorothioic acid, O,O-dimethyl-S-(2-methylthio)ethyl ester
7723-14-0	Phosphorus
10025-87-3	Phosphorus oxychloride
10026-13-8	Phosphorus pentachloride
1314-80-3	Phosphorus pentasulfide
1314-56-3	Phosphorus pentoxide
1314-80-3	Phosphorus sulfide
7719-12-2	Phosphorus trichloride

CAS No.	Substance Name
7723-14-0	Phosphorus, white
7723-14-0	Phosphorus, yellow
100-21-0	*p*-Phthalic acid
85-44-9	Phthalic acid anhydride
85-68-7	Phthalic acid, benzyl butyl ester
117-81-7	Phthalic acid, di-2-ethylhexyl ester
84-74-2	Phthalic acid, dibutyl ester
84-66-2	Phthalic acid, diethyl ester
131-11-3	Phthalic acid, dimethyl ester
117-84-0	Phthalic acid, dioctyl ester
85-44-9	Phthalic anhydride
626-17-5	*m*-Phthalodinitrile
57-47-6	Physostigmine
57-64-7	Physostigmine salicylate
1918-02-1	Picloram
109-06-8	2-Picoline
131-74-8	Picratol
88-89-1	Picric acid
124-87-8	Picrotoxin
83-26-1	Pindone
142-64-3	Piperazine dihydrochloride
110-89-4	Piperidine
5281-13-0	Piperonal bis[2-(2-butoxyethoxy)ethyl] acetal
5281-13-0	Piprotal
504-24-5	4-Pyridinamine
23505-41-1	Pirimifos-ethyl
23505-41-1	Pirimiphos-ethyl
83-26-1	Pival
83-26-1	2-Pivalyl-1,3-indandione
7778-18-9	Plaster of Paris
7440-06-4	Platinum
X00001-52-5	Platinum, soluble salts
13121-70-5	Plictran
36355-01-8	Polybrominated biphenyls
1336-36-3	Polychlorinated biphenyls
11096-82-5	Polychlorinated biphenyls
11097-69-1	Polychlorinated biphenyls
11104-28-2	Polychlorinated biphenyls
11141-16-5	Polychlorinated biphenyls
12672-29-6	Polychlorinated biphenyls
12674-11-2	Polychlorinated biphenyls
1336-36-3	Polychlorobiphenyls
11097-69-1	Polychlorobiphenyls
53469-21-9	Polychlorobiphenyls
8007-45-2	Polycyclic aromatic hydrocarbons, particulate
X00001-54-7	Polytetrafluoroethylene decomposition products
X00001-55-8	Portland cement
1310-58-3	Potash, caustic
7784-41-0	Potassium arsenate
10124-50-2	Potassium arsenite
7778-50-9	Potassium bichromate
7789-00-6	Potassium chromate
151-50-8	Potassium cyanide
7778-50-9	Potassium dichromate
1310-58-3	Potassium hydroxide
7722-64-7	Potassium permanganate
506-61-6	Potassium silver cyanide

CAS No.	Substance Name
8007-45-2	PPAH
65996-93-2	PPAH
X00001-04-7	Precipitated silica
671-16-9	Procarbazine
366-70-1	Procarbazine hydrochloride
X00001-27-4	Progesterone
2631-37-0	Promecarb
23950-58-5	Pronamide
X00001-44-5	Propadiene-methyl acetylene mixture
123-38-6	Propanal
107-10-8	1-Propanamine
74-98-6	Propane
109-77-3	Propanedinitrile
107-12-0	Propanenitrile
1120-71-4	Propane sultone
1120-71-4	1,3-Propane sultone
55-63-0	1,2,3-Propanetriol, trinitrate
79-09-4	Propanoic acid
71-23-8	Propanol
67-63-0	2-Propanol
57-57-8	3-Propanolide
67-64-1	2-Propanone
2312-35-8	Propargite
107-19-7	Propargyl alcohol
106-96-7	Propargyl bromide
107-02-8	2-Propenal
79-06-1	2-Propenamide
115-07-1	Propene
107-13-1	2-Propenenitrile
79-10-7	2-Propenoic acid
141-32-2	2-Propenoic acid, butyl ester
140-88-5	2-Propenoic acid, ethyl ester
107-18-6	2-Propen-1-ol
57-57-8	*beta*-Propiolactone
123-38-6	Propionaldehyde
79-09-4	Propionic acid
123-62-6	Propionic anhydride
107-12-0	Propionitrile
114-26-1	Propoxur
109-60-4	*n*-Propyl acetate
71-23-8	Propyl alcohol
71-23-8	*n*-Propyl alcohol
107-10-8	*n*-Propylamine
109-61-5	Propyl chloroformate
115-07-1	Propylene
78-87-5	Propylene chloride
78-87-5	Propylene dichloride
107-98-2	Propylene glocol methyl ether
6423-43-4	Propylene glycol dinitrate
107-98-2	*alpha*-Propylene glycol monomethyl ether
75-55-8	Propyleneimine
75-55-8	1,2-Propylenimine
75-56-9	Propylene oxide
627-13-4	*n*-Propyl nitrate
142-84-7	N-Propyl-1-propanamine
51-52-5	Propylthiouracil
74-99-7	Propyne

CAS No.	Substance Name
107-19-7	2-Propyn-1-ol
1395-21-7	Proteolytic enzymes
2275-18-5	Prothoate
74-90-8	Prussic acid
95-63-6	Pseudocumene
X00001-54-7	PTFE decomposition products
129-00-0	Pyrene
121-29-9	Pyrethrin
121-21-1	Pyrethrin II
8003-34-7	Pyrethrins
8003-34-7	Pyrethrum
110-86-1	Pyridine
53558-25-1	Pyriminil
120-80-9	Pyrocatechol
107-49-3	Pyrophosphoric acid, tetraethyl ester
298-02-2	Phosphorodithioic acid, O,O-diethyl S-(ethylthio) methyl ester
297-97-2	Phosphorothioic acid, O,O-diethyl, O-pyrazinyl ester
14808-60-7	Quartz
1305-78-8	Quicklime
7439-97-6	Quicksilver
91-22-5	Quinoline
106-51-4	Quinone
82-68-8	Quintozene
X00001-63-8	Radionuclides
121-82-4	RDX
50-55-5	Reserpine
108-46-3	Resorcinol
7440-16-6	Rhodium
X00001-56-9	Rhodium, insoluble compounds
X00001-57-0	Rhodium, soluble compounds
299-84-3	Ronnel
632-99-5	Rosaniline
X00001-58-1	Rosin core solder pyrolysis products
83-79-4	Rotenone
1309-37-1	Rouge
X00001-59-2	Rubber solvent, naphtha
13463-67-7	Rutile
81-07-2	Saccharin
94-59-7	Safrole
107-44-8	Sarin
152-16-9	Schradan
7783-00-8	Selenious acid
7782-49-2	Selenium
X00001-60-5	Selenium compounds
7446-08-4	Selenium dioxide
7488-56-4	Selenium disulfide
X00001-10-5	Selenium doped bismuth telluride
7783-79-1	Selenium hexafluoride
7446-08-4	Selenium oxide
7791-23-3	Selenium oxychloride
7446-34-6	Selenium sulfide
630-10-4	Selenourea
7783-00-8	Selenous acid

CAS No.	Substance Name
563-41-7	Semicarbazide hydrochloride
21087-64-9	Sencor
115-02-6	L-Serine, diazoacetate (ester)
136-78-7	Sesone
63-25-2	Sevin
7803-62-5	Silane
7631-86-9	Silica, amorphous
14464-46-1	Silica, crystalline, Cristobalite
14808-60-7	Silica, crystalline, Quartz
15468-32-3	Silica, crystalline, Tridymite
1317-95-9	Silica, crystalline, Tripoli
60676-86-0	Silica, fused
7440-21-3	Silicon
409-21-2	Silicon carbide
7803-62-5	Silicon tetrahydride
7440-22-4	Silver
X00001-66-1	Silver and compounds
X00001-61-6	Silver, soluble compounds
506-64-9	Silver cyanide
7761-88-8	Silver nitrate
93-72-1	Silvex
32534-95-5	Silvex, isooctyl ester
14167-18-1	Salcomine
1305-62-0	Slaked lime
X00001-62-7	Soapstone
1310-73-2	Soda, caustic
7440-23-5	Sodium
7631-89-2	Sodium arsenate
7784-46-5	Sodium arsenite
26628-22-8	Sodium azide
10588-01-9	Sodium bichromate
1333-83-1	Sodium bifluoride
7631-90-5	Sodium bisulfite
124-65-2	Sodium cacodylate
7775-11-3	Sodium chromate
143-33-9	Sodium cyanide
136-78-7	Sodium 2,4-dichlorophenoxyethyl sulfate
10588-01-9	Sodium dichromate
25155-30-0	Sodium dodecylbenzenesulfonate
7681-49-4	Sodium fluoride
62-74-8	Sodium fluoroacetate
10124-56-8	Sodium hexametaphosphate
16721-80-5	Sodium hydrosulfide
1310-73-2	Sodium hydroxide
7681-52-9	Sodium hypochlorite
10022-70-5	Sodium hypochlorite
10022-70-5	Sodium hypochlorite pentahydrate
7681-57-4	Sodium metabisulfite
124-41-4	Sodium methoxide
124-41-4	Sodium methylate
7632-00-0	Sodium nitrite
131-52-2	Sodium pentachlorophenate
131-52-2	Sodium pentachlorphenolate
7558-79-4	Sodium phosphate, dibasic
10039-32-4	Sodium phosphate, dibasic
10140-65-5	Sodium phosphate, dibasic
7601-54-9	Sodium phosphate, tribasic

CAS No.	Substance Name
7758-29-4	Sodium phosphate, tribasic
7785-84-4	Sodium phosphate, tribasic
10101-89-0	Sodium phosphate, tribasic
10124-56-8	Sodium phosphate, tribasic
10361-89-4	Sodium phosphate, tribasic
13410-01-0	Sodium selenate
7782-82-3	Sodium selenite
10102-18-8	Sodium selenite
7757-82-6	Sodium sulfate
10102-20-2	Sodium tellurite
1303-96-4	Sodium tetraborate
1330-43-4	Sodium tetraborate
61028-24-8	Sodium tetraborate, anhydrous
1344-90-7	Sodium tetraborate decahydrate
12179-04-3	Sodium tetraborate pentahydrate
7758-29-4	Sodium tripolyphosphate
9005-25-8	Starch
2223-93-0	Stearic acid, cadmium salt
1072-35-1	Stearic acid, dilead(II) salt
7428-48-0	Stearic acid, lead salt
52652-59-2	Stearic acid, lead salt
557-05-1	Stearic acid, zinc salt
7803-52-3	Stibine
8052-41-3	Stoddard solvent
18883-66-4	STR
18883-66-4	Streptozotocin
7789-06-2	Strontium chromate
1314-96-1	Strontium sulfide
57-24-9	Strychnidin-10-one and salts
57-24-9	Strychnine and salts
60-41-3	Strychnine, sulfate
100-42-5	Styrene
96-09-3	Styrene epoxide
96-09-3	Styrene oxide
100-42-5	Styrol
1395-21-7	Subtilisins
57-50-1	Sucrose
95-06-7	Sulfallate
3689-24-5	Sulfotep
10025-67-9	Sulfur chloride
7446-09-5	Sulfur dioxide
2551-62-4	Sulfur hexafluoride
7783-06-4	Sulfur hydride
7664-93-9	Sulfuric acid
8014-95-7	Sulfuric acid
64-67-5	Sulfuric acid, diethyl ester
77-78-1	Sulfuric acid, dimethyl ester
7446-18-6	Sulfuric acid, thallium(I) salt
10031-59-1	Sulfuric acid, thallium(I) salt
10025-67-9	Sulfur monochloride
12771-08-3	Sulfur monochloride
5714-22-7	Sulfur pentafluoride
1314-80-3	Sulfur phosphide
7488-56-4	Sulfur selenide
7783-60-0	Sulfur tetrafluoride
7446-11-9	Sulfur trioxide
2699-79-8	Sulfuryl fluoride

CAS No.	Substance Name
35400-43-2	Sulprofos
8065-48-3	Systox
93-76-5	2,4,5-T
93-76-5	2,4,5-T acid
62-55-5	TAA
77-81-6	Tabun
14807-96-6	Talc
1319-72-8	2,4,5-T amines
2008-46-0	2,4,5-T amines
3813-14-7	2,4,5-T amines
6369-96-6	2,4,5-T amines
6369-97-7	2,4,5-T amines
7440-25-7	Tantalum
28300-74-5	Tartar emetic
3164-29-2	Tartaric acid, ammonium salt
14307-43-8	Tartaric acid, ammonium salt
815-82-7	Tartaric acid, copper(II) salt
1746-01-6	TCDD
95-95-4	TCP
72-54-8	TDE
584-84-9	TDI
91-08-7	2,6-TDI
3689-24-5	TEDP
X00001-54-7	Teflon decomposition products
13494-80-9	Tellurium
X00001-64-9	Tellurium compounds
7783-80-4	Tellurium hexafluoride
3383-96-8	Temephos
107-49-3	TEPP
13071-79-9	Terbufos
100-21-0	Terephthalic acid
26140-60-3	Terphenyl
61788-32-7	Terphenyls, hydrogenated
93-79-8	2,4,5-T esters
1928-47-8	2,4,5-T esters
2545-59-7	2,4,5-T esters
25168-15-4	2,4,5-T esters
61792-07-2	2,4,5-T esters
79-27-6	1,1,2,2-Tetrabromoethane
1897-45-6	2,4,5,6-Tetrachloro-1,3-benzenedicarbonitrile
76-11-9	1,1,1,2-Tetrachloro-2,2-difluoroethane
76-12-0	1,1,2,2-Tetrachloro-1,2-difluoroethane
630-20-6	1,1,1,2-Tetrachloroethane
79-34-5	1,1,2,2-Tetrachloroethane
127-18-4	1,1,2,2-Tetrachloroethene
95-94-3	1,2,4,5-Tetrachlorobenzene
1746-01-6	2,3,7,8-Tetrachlorodibenzo-*p*-dioxin
79-34-5	*sym*-Tetrachloroethane
127-18-4	Tetrachloroethylene
1897-45-6	2,4,5,6-Tetrachloroisophthalonitrile
56-23-5	Tetrachloromethane
1335-88-2	Tetrachloronaphthalene
58-90-2	2,3,4,6-Tetrachlorphenol
961-11-5	Tetrachlorvinphos
3689-24-5	Tetraethyl dithiopyrophosphate
78-00-2	Tetraethyl lead

CAS No.	Substance Name
78-10-4	Tetraethyl orthosilicate
78-00-2	Tetraethylplumbane
107-49-3	Tetraethyl pyrophosphate
78-10-4	Tetraethyl silicate
97-77-8	Tetraethylthiuram disulfide
597-64-8	Tetraethyl tin
109-99-9	Tetrahydrofuran
930-55-2	Tetrahydro-N-nitrosopyrrole
90-94-8	Tetramethyldiaminobenzophenone
90-94-8	N,N,N',N'-Tetramethyl-4,4'-diaminobenzophenone
101-61-1	Tetramethyldiaminodiphenylmethane
75-74-1	Tetramethyl lead
3333-52-6	Tetramethylsuccinonitrile
3383-96-8	O,O,O',O'-Tetramethyl-O,O'-thiodi-p-phenylene phosphorothioate
509-14-8	Tetranitromethane
757-58-4	Tetraphosphoric acid, hexaethyl ester
7722-88-5	Tetrasodium pyrophosphate
479-45-8	Tetryl
1314-32-5	Thallic oxide
7440-28-0	Thallium
X00001-69-4	Thallium and compounds
X00001-65-0	Thallium, soluble compounds
563-68-8	Thallium(I) acetate
6533-73-9	Thallium(I) carbonate
7791-12-0	Thallium(I) chloride
10102-45-1	Thallium(I) nitrate
1314-32-5	Thallium(III) oxide
12039-52-0	Thallium(I) selenide
10031-59-1	Thallium sulfate
7446-18-6	Thallium(I) sulfate
10031-59-1	Thallium(I) sulfate
6533-73-9	Thallous carbonate
7791-12-0	Thallous chloride
2757-18-8	Thallous malonate
7446-18-6	Thallous sulfate
109-99-9	THF
298-02-2	Thimet
62-55-5	Thioacetamide
139-65-1	4,4'-Thiobisaniline
139-65-1	4,4'-Thiobisbenzenamine
96-69-5	4,4'-Thiobis(6-*tert*-butyl-*m*-cresol)
505-60-2	1,1'-Thiobis(2-chloroethane)
4418-66-0	2,2'-Thiobis(4-chloro-6-methyl)phenol
97-18-7	2,2'-Thiobis(4,6-dichlorophenol)
62-56-6	Thiocarbamide
2231-57-4	Thiocarbazide
115-29-7	Thiodan
139-65-1	4,4'-Thiodianiline
39196-18-4	Thiofanox
68-11-1	Thioglycolic acid
541-53-7	Thioimidodicarbonic diamide
74-93-1	Thiomethanol
297-97-2	Thionazin
7719-09-7	Thionyl chloride
108-98-5	Thiophenol

CAS No.	Substance Name
79-19-6	Thiosemicarbazide
52-24-4	Thiotepa
62-56-6	Thiourea
137-26-8	Thiram
1314-20-1	Thorium dioxide
1314-20-1	Thorium(IV) oxide
7440-31-5	Tin
X00001-67-2	Tin, inorganic compounds
X00001-68-3	Tin, organic compounds
1332-29-2	Tin oxide
13463-67-7	Titanium dioxide
7550-45-0	Titanium tetrachloride
118-96-7	TNT
1836-75-5	TOK
119-93-7	o-Tolidine
108-88-3	Toluene
25376-45-8	Toluenediamine
95-80-7	2,4-Toluenediamine
823-40-5	2,6-Toluenediamine
496-72-0	3,4-Toluenediamine
584-84-9	Toluene diisocyanate
26471-62-5	1,3-Toluene diisocyanate
584-84-9	2,4-Toluene diisocyanate
91-08-7	2,6-Toluene diisocyanate
584-84-9	Toluene-2,4-diisocyanate
91-08-7	Toluene 2,6-diisocyanate
108-44-1	m-Toluidine
106-49-0	p-Toluidine
95-53-4	o-Toluidine
636-21-5	o-Toluidine hydrochloride
108-88-3	Toluol
95-53-4	o-Tolylamine
100-44-7	Tolylchloride
95-80-7	m-Tolylenediamine
1918-02-1	Tordon
8001-35-2	Toxaphene
57-74-9	Toxichlor
93-72-1	2,4,5-TP
93-72-1	2,4,5-TP acid
32534-95-5	2,4,5-TP acid esters
15271-41-7	Tranid
14567-73-8	Tremolite
299-75-2	Treosulfan
299-75-2	Treosulphan
108-78-1	2,4,6-Triamino-1,3,5-triazine
1031-47-6	Triamiphos
68-76-8	Triaziquone
24017-47-8	Triazofos
61-82-5	1H-1,2,4-Triazol-3-amine
75-25-2	Tribromomethane
126-73-8	Tributyl phosphate
12108-13-3	Tricarbonyl methylcyclopentadienyl manganese
52-68-6	Trichlorfon
75-87-6	Trichloroacetaldehyde
76-03-9	Trichloroacetic acid
76-02-8	Trichloroacetyl chloride
115-32-2	2,2,2-Trichloro-1,1-bis(4-chlorophenyl)ethanol

CAS No.	Substance Name
72-43-5	1,1,1-Trichloro-2,2-bis(*p*-methoxyphenyl)-ethane
120-82-1	1,2,4-Trichlorobenzene
1558-25-4	Trichloro(chloromethyl) silane
27137-85-5	Trichloro(dichlorophenyl)silane
71-55-6	1,1,1-Trichloroethane
79-00-5	1,1,2-Trichloroethane
79-01-6	Trichloroethene
79-01-6	Trichloroethylene
79-01-6	1,1,2-Trichloroethylene
115-21-9	Trichloroethylsilane
75-69-4	Trichlorofluoromethane
52-68-6	(2,2,2-Trichloro-1-hydroxyethyl)phosphonic acid, dimethyl ester
67-66-3	Trichloromethane
594-42-3	Trichloromethanesulfenyl chloride
98-07-7	(Trichloromethyl)benzene
133-06-2	N-((Trichloromethyl)thio)-4-cyclohenxene-1,2-dicarboximide
133-06-2	N-((Trichloromethyl)thio)tetrahydrophthalimide
75-69-4	Trichloromonofluoromethane
1321-65-9	Trichloronaphthalene
327-98-0	Trichloronate
76-06-2	Trichloronitromethane
96-18-4	1,2,3-Trichloropropane
15950-66-0	2,3,4-Trichlorophenol
933-78-8	2,3,5-Trichlorophenol
933-75-5	2,3,6-Trichlorophenol
95-95-4	2,4,5-Trichlorophenol
93-76-5	2,4,5-Trichlorophenoxyacetic acid
1319-72-8	2,4,5-Trichlorophenoxyacetic acid, 1-amino-2-propanol (1:1)
2545-59-7	2,4,5-Trichlorophenoxyacetic acid, butoxyethyl ester
61792-07-2	2,4,5-Trichlorphenoxyacetic acid, *sec*-butyl ester
2008-46-0	2,4,5-Trichlorophenoxyacetic acid, N,N-diethylethamine (1:1)
6369-96-6	2,4,5-Trichlorophenoxyacetic acid, N,N-dimethylmethamine
1928-47-8	2,4,5-Trichlorophenoxyacetic acid, 2-ethylhexyl ester
6369-97-7	2,4,5-Trichlorophenoxyacetic acid, N-methylmethanamine
25168-15-4	2,4,5-Trichlorophenoxyacetic acid, isooctyl ester
32534-95-5	2,4,5-Trichlorophenoxypropionic acid, isooctyl ester
93-79-8	2,4,5-Trichlorophenoxyacetic acid, isopropyl ester
3813-14-7	2,4,5-Trichlorophenoxyacetic acid, 2,2′,2″-nitrilotris(ethanol) (1:1)
13560-99-1	2,4,5-Trichlorphenoxyacetic acid, sodium salt
25167-82-2	Trichlorophenol
88-06-2	2,4,6-Trichlorophenol
609-19-8	3,4,5-Trichlorophenol
93-72-1	2-(2,4,5-Trichlorophenoxy)propionic acid
98-13-5	Trichlorophenylsilane
52-68-6	Trichlorophon
77-13-1	1,1,2-Trichloro-1,2,2-trifluoroethane
52-68-6	Trichlorphon

CAS No.	Substance Name
78-30-8	Tricresyl phosphate
78-30-8	o-Tricresyl phosphate
78-30-8	Tri-o-cresyl phosphate
78-30-8	Tri-ortho-cresyl phosphate
13121-70-5	Tricyclohexyltin hydroxide
15468-32-3	Tridymite (silica)
27323-41-7	Triethanolamine dodecylbenzenesulfonate
998-30-1	Triethoxysilane
121-44-8	Triethylamine
75-63-8	Trifluorobromomethane
1582-09-8	alpha,alpha,alpha-Trifluoro-2,6-dinitro-N,N-dipropyl-p-toluidine
98-16-8	3-Trifluoromethylaniline
98-16-8	3-(Trifluoromethyl)benzenamine
2164-17-2	3-(m-Trifluoromethylphenyl)-1,1-dimethylurea
75-63-8	Trifluoromonobromomethane
98-16-8	alpha,alpha,alpha-Trifluoro-m-toluidine
1582-09-8	Trifluralin
552-30-7	Trimellitic anhydride
88-05-1	2,4,6-Trimethylaniline
75-50-3	Trimethylamine
25551-13-7	Trimethylbenzene
95-63-6	1,2,4-Trimethylbenzene
75-65-0	Trimethyl carbinol
75-77-4	Trimethyl chlorosilane
824-11-3	Trimethylolpropane phosphite
121-45-9	Trimethyl phosphite
1066-45-1	Trimethyltin chloride
123-63-7	2,4,6-Trimethyl-1,3,5-trioxane
99-35-4	sym-Trinitrobenzene
99-35-4	1,3,5-Trinitrobenzene
88-89-1	2,4,6-Trinitrophenol
131-74-8	2,4,6-Trinitrophenol, ammonium salt
479-45-8	2,4,6-Trinitrophenylmethylnitramine
118-96-7	2,4,6-Trinitrotoluene
603-34-9	Triphenylamine
115-86-6	Triphenyl phosphate
639-58-7	Triphenyltin chloride
1317-95-9	Tripoli (silica)
126-72-7	TRIS
68-76-8	2,3,5-Tris(aziridinyl)-1,4-benzoquinone
68-76-8	2,3,5-Tris(1-aziridinyl)-2,5-cyclohexadiene-1,4-dione
52-24-4	Tris(1-aziridinyl)phosphine sulfide
555-77-1	Tris(2-chloroethyl)amine
126-72-7	Tris(2,3-dibromopropyl) phosphate
7601-54-9	Trisodium phosphate
10361-89-4	Trisodium phosphate decahydrate
10101-89-0	Trisodium phosphate dodecahydrate
72-57-1	Trypan blue
13560-99-1	2,4,5-T salts
X00001-71-8	Tungsten, insoluble compounds
X00001-72-9	Tungsten, soluble compounds
8006-64-2	Turpentine
58270-08-9	(T-4)-Zinc
66-75-1	Uracil mustard
X00001-74-1	Uranium, insoluble compounds

CAS No.	Substance Name
X00001-73-0	Uranium, soluble compounds
541-09-3	Uranyl acetate
10102-06-4	Uranyl nitrate
36478-76-9	Uranyl nitrate
51-79-6	Urethan
51-79-6	Urethane
110-62-3	Valeraldehyde
2001-95-8	Valinomycin
7803-55-6	Vanadic acid, ammonium salt
7440-62-2	Vanadium
1314-62-1	Vanadium(V) oxide
1314-62-1	Vanadium pentoxide
27774-13-6	Vanadyl sulfate
107-13-1	VCN
X00001-76-3	Vegetable oil
108-05-4	Vinyl acetate
79-06-1	Vinylamide
100-42-5	Vinylbenzene
593-60-2	Vinyl bromide
75-01-4	Vinyl chloride
107-13-1	Vinyl cyanide
106-87-6	Vinylcyclohexene dioxide
75-35-4	Vinylidene chloride
25013-15-4	Vinyltoluene
50-14-6	Vitamin D2
8030-30-6	VM&P Naphtha
81-81-2	Warfarin
129-06-6	Warfarin sodium
X00001-77-4	Welding fumes
X00001-06-9	Wheat dust
7723-14-0	White phosphorus
1344-95-2	Wollastonite
67-56-1	Wood alcohol
X00001-78-5	Wood dusts, hard woods
X00001-79-6	Wood dusts, soft woods
92-67-1	*p*-Xenylamine
1330-20-7	Xylene (mixed isomers)
108-38-3	*m*-Xylene
95-47-6	*o*-Xylene
106-42-3	*p*-Xylene
1477-55-0	*m*-Xylene-*alpha,alpha'*-diamine
28347-13-9	Xylene dichloride
105-67-9	2,4-Xylenol
1300-71-6	Xylenol
87-62-7	2,6-Xylidine
1300-73-8	Xylidine
108-38-3	*m*-Xylol
95-47-6	*o*-Xylol
28347-13-9	Xylylene dichloride
7723-14-0	Yellow phosphorus
7440-65-5	Yttrium

CAS No.	Substance Name
7440-66-6	Zinc
X00001-70-7	Zinc and compounds
557-34-6	Zinc acetate
14639-97-5	Zinc ammonium chloride
14639-98-6	Zinc ammonium chloride
52628-25-8	Zinc ammonium chloride
1332-07-6	Zinc borate
7699-45-8	Zinc bromide
3386-35-9	Zinc carbonate
7646-85-7	Zinc chloride
13530-65-9	Zinc chromate
557-21-7	Zinc cyanide
12122-67-7	Zinc ethylenebis(dithiocarbamate)
7783-49-5	Zinc fluoride
557-41-5	Zinc formate
7779-86-4	Zinc hydrosulfite
7779-88-6	Zinc nitrate
1314-13-2	Zinc oxide
127-82-2	Zinc phenolsulfonate
1314-84-7	Zinc phosphide
16871-71-9	Zinc silicofluoride
557-05-1	Zinc stearate
7733-02-0	Zinc sulfate
12122-67-7	Zineb
X00001-82-1	Zirconium compounds
13746-89-9	Zirconium nitrate
16923-95-8	Zirconium potassium fluoride
14644-61-2	Zirconium sulfate
10026-11-6	Zirconium tetrachloride
148-01-6	Zoalene

APPENDIX E

Pertinent Federal Forms

Material Safety Data Sheet

May be used to comply with
OSHA's Hazard Communication Standard,
29 CFR 1910.1200. Standard must be
consulted for specific requirements.

U.S. Department of Labor

Occupational Safety and Health Administration
(Non-Mandatory Form)
Form Approved
OMB No. 1218-0072

IDENTITY *(As Used on Label and List)*	Note: Blank spaces are not permitted. If any item is not applicable, or no information is available, the space must be marked to indicate that.

Section I

Manufacturer's Name	Emergency Telephone Number
Address *(Number, Street, City, State, and ZIP Code)*	Telephone Number for Information
	Date Prepared
	Signature of Preparer *(optional)*

Section II — Hazardous Ingredients/Identity Information

Hazardous Components (Specific Chemical Identity; Common Name(s))	OSHA PEL	ACGIH TLV	Other Limits Recommended	% *(optional)*

Section III — Physical/Chemical Characteristics

Boiling Point		Specific Gravity (H$_2$O = 1)	
Vapor Pressure (mm Hg.)		Melting Point	
Vapor Density (AIR = 1)		Evaporation Rate (Butyl Acetate = 1)	

Solubility in Water

Appearance and Odor

Section IV — Fire and Explosion Hazard Data

Flash Point (Method Used)	Flammable Limits	LEL	UEL

Extinguishing Media

Special Fire Fighting Procedures

Unusual Fire and Explosion Hazards

(Reproduce locally)   OSHA 174, Sept. 1985

Section V — Reactivity Data

Stability	Unstable		Conditions to Avoid
	Stable		

Incompatibility (*Materials to Avoid*)

Hazardous Decomposition or Byproducts

Hazardous Polymerization	May Occur		Conditions to Avoid
	Will Not Occur		

Section VI — Health Hazard Data

Route(s) of Entry:	Inhalation?	Skin?	Ingestion?

Health Hazards (*Acute and Chronic*)

Carcinogenicity:	NTP?	IARC Monographs?	OSHA Regulated?

Signs and Symptoms of Exposure

Medical Conditions
Generally Aggravated by Exposure

Emergency and First Aid Procedures

Section VII — Precautions for Safe Handling and Use

Steps to Be Taken in Case Material Is Released or Spilled

Waste Disposal Method

Precautions to Be Taken in Handling and Storing

Other Precautions

Section VIII — Control Measures

Respiratory Protection (*Specify Type*)

Ventilation	Local Exhaust		Special
	Mechanical (*General*)		Other

Protective Gloves	Eye Protection

Other Protective Clothing or Equipment

Work/Hygienic Practices

Page _____ of _____ pages
Form Approved OMB No. 2050-0072

Tier One

EMERGENCY AND HAZARDOUS CHEMICAL INVENTORY
Aggregate Information by Hazard Type

FOR OFFICIAL USE ONLY	ID #
	Date Received

Important: Read instructions before completing form

Reporting Period From January 1 to December 31, 19 _____

Facility Identification

Name _____
Street Address _____
City _____ State _____ Zip _____
SIC Code [ ][ ][ ][ ] Dun & Brad Number [ ][ ]-[ ][ ][ ]-[ ][ ][ ][ ]

Owner/Operator

Name _____
Mail Address _____
Phone ( ) _____

Emergency Contacts

Name _____
Title _____
Phone ( ) _____
24 Hour Phone ( ) _____

Name _____
Title _____
Phone ( ) _____
24 Hour Phone ( ) _____

[ ] Check if site plan is attached

	Hazard Type	Max Amount*	Average Daily Amount*	Number of Days On-Site	General Location
Physical Hazards	Fire	[ ][ ]	[ ][ ]	[ ][ ][ ]	_____
	Sudden Release of Pressure	[ ][ ]	[ ][ ]	[ ][ ][ ]	_____
	Reactivity	[ ][ ]	[ ][ ]	[ ][ ][ ]	_____
Health Hazards	Immediate (acute)	[ ][ ]	[ ][ ]	[ ][ ][ ]	_____
	Delayed (Chronic)	[ ][ ]	[ ][ ]	[ ][ ][ ]	_____

Certification (Read and sign after completing all sections)

I certify under penalty of law that I have personally examined and am familiar with the information submitted in this and all attached documents, and that based on my inquiry of those individuals responsible for obtaining the information, I believe that the submitted information is true, accurate and complete.

_____
Name and official title of owner/operator OR owner/operator's authorized representative

_____ _____
Signature   Date signed

* Reporting Ranges

Reporting Ranges Range Value	Weight Range in Pounds From...	To...
00	0	99
01	100	999
02	1000	9,999
03	10,000	99,999
04	100,000	999,999
05	1,000,000	9,999,999
06	10,000,000	49,999,999
07	50,000,000	99,999,999
08	100,000,000	499,999,999
09	500,000,000	999,999,999
10	1 billion	higher than 1 billion

Page _____ of _____ pages
Form Approved OMB No. 2050-0072

Tier Two

EMERGENCY AND HAZARDOUS CHEMICAL INVENTORY

Specific Information by Chemical

Facility Identification

Name _____

Street Address _____

City _____ State _____ Zip _____

SIC Code [____] Dun & Brad Number [____]-[____]

FOR OFFICIAL USE ONLY   ID # [____]   Date Received [____]

Owner/Operator Name

Name _____

Mail Address _____

Phone ( )

Emergency Contact

Name _____ Title _____

Phone ( ) 24 Hr. Phone ( )

Name _____ Title _____

Phone ( ) 24 Hr. Phone ( )

Reporting Period From January 1 to December 31, 19___

Important: Read all instructions before completing form

Chemical Description	Physical and Health Hazards (check all that apply)	Inventory			Storage Codes and Locations (Non-Confidential)	
		Max. Daily Amount (code)	Avg. Daily Amount (code)	No. of Days On-site (days)	Storage Code	Storage Locations
CAS [____] Chem. Name ___ ☐ Trade Secret Check all that apply: ☐ Pure ☐ Mix ☐ Solid ☐ Liquid ☐ Gas	☐ Fire ☐ Sudden Release of Pressure ☐ Reactivity ☐ Immediate (acute) ☐ Delayed (chronic)	☐	☐	☐		
CAS [____] Chem. Name ___ ☐ Trade Secret Check all that apply: ☐ Pure ☐ Mix ☐ Solid ☐ Liquid ☐ Gas	☐ Fire ☐ Sudden Release of Pressure ☐ Reactivity ☐ Immediate (acute) ☐ Delayed (chronic)	☐	☐	☐		
CAS [____] Chem. Name ___ ☐ Trade Secret Check all that apply: ☐ Pure ☐ Mix ☐ Solid ☐ Liquid ☐ Gas	☐ Fire ☐ Sudden Release of Pressure ☐ Reactivity ☐ Immediate (acute) ☐ Delayed (chronic)	☐	☐	☐		

Certification (Read and sign after completing all sections)

I certify under penalty of law that I have personally examined and am familiar with the information submitted in this and all attached documents, and that based on my inquiry of those individuals responsible for obtaining the information, I believe that the submitted information is true, accurate, and complete.

Name and official title of owner/operator OR owner/operator's authorized representative _____ Signature _____ Date signed _____

Optional Attachments (Check one)

☐ I have attached a site plan

☐ I have attached a list of site coordinate abbreviations

Form Approved OMB No.: 2070-0093

Approval Expires: 01/91

Page 1 of 5

(Important: Type or print; read instructions before completing form.)

U.S. Environmental Protection Agency

⊕EPA   TOXIC CHEMICAL RELEASE INVENTORY REPORTING FORM

EPA FORM R

Section 313, Title III of The Superfund Amendments and Reauthorization Act of 1986

PART I. FACILITY IDENTIFICATION INFORMATION	(This space for EPA use only)

1.

1.1 Does this report contain trade secret information?	1.2 Is this a sanitized copy?	1.3 Reporting Year
☐ Yes (Answer 1.2) ☐ No (Do not answer 1.2)	☐ Yes ☐ No	

2. CERTIFICATION (Read and sign after completing all sections.)

I hereby certify that I have reviewed the attached documents and that, to the best of my knowledge and belief, the submitted information is true and complete and that the amounts and values in this report are accurate based on reasonable estimates using data available to the preparers of this report.

Name and official title of owner/operator or senior management official

Signature	Date signed

3. FACILITY IDENTIFICATION

3.1

Facility or Establishment Name

Street Address

City	County

State	Zip Code

3.2 This report contains information for: (check one)

a. ☐ An entire covered facility.

b. ☐ Part of a covered facility.

3.3	Technical Contact	Telephone Number (Include area code) ( ) –
3.4	Public Contact	Telephone Number (Include area code) ( ) –

3.5	a. SIC Code	b.	c.

3.6

Latitude — Deg. Min. Sec. Longitude — Deg. Min. Sec.

Where to send completed forms:

U.S. Environmental Protection Agency
P.O. Box 70266
Washington, DC 20024–0266
Attn: Toxic Chemical Release Inventory

3.7	Dun & Bradstreet Number(s) a.	b.

3.8	EPA Identification Number (RCRA I.D. No.) a.	b.

3.9	NPDES Permit Number(s) a.	b.

3.10

Name of Receiving Stream(s) or Water Body(s)

a.

b.

c.

3.11

Underground Injection Well Code (UIC) Identification No.

4. PARENT COMPANY INFORMATION

4.1	Name of Parent Company

4.2	Parent Company's Dun & Bradstreet No.

EPA Form 9350-1 (1–88)

(Important: Type or print; read instructions before completing form.)   Page 2 of 5

EPA FORM **R** PART II. OFF-SITE LOCATIONS TO WHICH TOXIC CHEMICALS ARE TRANSFERRED IN WASTES	(This space for EPA use only)

1. PUBLICLY OWNED TREATMENT WORKS (POTW)

Facility Name

Street Address

City	County

State	Zip

2. OTHER OFF-SITE LOCATIONS – Number these locations sequentially on this and any additional page of this form you use.

☐ **Other off-site location**

EPA Identification Number (RCRA ID. No.)

Facility Name

Street Address

City	County

State	Zip

Is location under control of reporting facility or parent company? ☐ ☐
                                                                   Yes No

☐ **Other off-site location**

EPA Identification Number (RCRA ID. No.)

Facility Name

Street Address

City	County

State	Zip

Is location under control of reporting facility or parent company? ☐ ☐
                                                                   Yes No

☐ **Other off-site location**

EPA Identification Number (RCRA ID. No.)

Facility Name

Street Address

City	County

State	Zip

Is location under control of reporting facility or parent company? ☐ ☐
                                                                   Yes No

☐ Check if additional pages of Part II are attached.

EPA Form 9350-1 (1-88)

(Important: Type or print; read instructions before completing form.)

EPA FORM R
PART III. CHEMICAL SPECIFIC INFORMATION

(This space for EPA use only.)

I. CHEMICAL IDENTITY

1.1 ☐ Trade Secret (Provide a generic name in 1.4 below. Attach substantiation form to this submission.)

1.2 CAS # ☐☐☐☐☐☐☐ – ☐☐ – ☐ (Use leading zeros if CAS number does not fill space provided.)

1.3 Chemical or Chemical Category Name

1.4 Generic Chemical Name (Complete only if 1.1 is checked.)

MIXTURE COMPONENT IDENTITY (Do not complete this section if you have completed Section 1.)

2. Generic Chemical Name Provided by Supplier (Limit the name to a maximum of 70 characters (e.g., numbers, letters, spaces, punctuation)).

---

3. ACTIVITIES AND USES OF THE CHEMICAL AT THE FACILITY (Check all that apply.)

3.1 Manufacture:	a. ☐ Produce	b. ☐ Import	c. ☐ For on-site use/processing
	d. ☐ For sale/ distribution	e. ☐ As a byproduct	f. ☐ As an impurity
3.2 Process:	a. ☐ As a reactant	b. ☐ As a formulation component	c. ☐ As an article component
	d. ☐ Repackaging only		
3.3 Otherwise Used:	a. ☐ As a chemical processing aid	b. ☐ As a manufacturing aid	c. ☐ Ancillary or other use

---

4. MAXIMUM AMOUNT OF THE CHEMICAL ON SITE AT ANY TIME DURING THE CALENDAR YEAR

☐☐ (enter code)

---

5. RELEASES OF THE CHEMICAL TO THE ENVIRONMENT

You may report releases of less than 1,000 lbs. by checking ranges under A.1.		A. Total Release (lbs/yr)				B. Basis of Estimate (enter code)	
		A.1 Reporting Ranges			A.2 Enter Estimate		
		0	1–499	500–999			
5.1 Fugitive or non-point air emissions	5.1a					5.1b ☐	
5.2 Stack or point air emissions	5.2a					5.2b ☐	
5.3 Discharges to water 5.3.1 ☐	5.3.1a					5.3.1b ☐	C. % From Stormwater 5.3.1c
(Enter letter code from Part I Section 3.10 for streams(s).) 5.3.2 ☐	5.3.2a					5.3.2b ☐	5.3.2c
5.3.3 ☐	5.3.3a					5.3.3b ☐	5.3.3c
5.4 Underground Injection	5.4a					5.4b ☐	
5.5 Releases to land 5.5.1 ☐☐☐ (enter code)	5.5.1a					5.5.1b ☐	
5.5.2 ☐☐☐ (enter code)	5.5.2a					5.5.2b ☐	
5.5.3 ☐☐☐ (enter code)	5.5.3a					5.5.3b ☐	

☐ (Check if additional information is provided on Part IV–Supplemental Information.)

EPA Form 9350-1 (1-88)

EPA FORM **R**, Part III (Continued)

Page 4 of 5

6. TRANSFERS OF THE CHEMICAL IN WASTE TO OFF-SITE LOCATIONS

	A. Total Transfers (lbs/yr)				B. Basis of Estimate (enter code)	C. Type of Treatment/ Disposal (enter code)
You may report transfers of less than 1,000 lbs. by checking ranges under A.1	A.1 Reporting Ranges			A.2 Enter Estimate		
	0	1–499	500–999			
6.1 Discharge to POTW					6.1b ☐	
6.2 Other off-site location (Enter block number from Part II, Section 2.) ☐					6.2b ☐	6.2c ☐
6.3 Other off-site location (Enter block number from Part II, Section 2.) ☐					6.3b ☐	6.3c ☐
6.4 Other off-site location (Enter block number from Part II, Section 2.) ☐					6.4b ☐	6.4c ☐

☐ (Check if additional information is provided on Part IV–Supplemental Information)

7. WASTE TREATMENT METHODS AND EFFICIENCY

A. General Wastestream (enter code)	B. Treatment Method (enter code)	C. Range of Influent Concentration (enter code)	D. Sequential Treatment? (check if applicable)	E. Treatment Efficiency Estimate	F. Based on Operating Data? Yes No
7.1a ☐	7.1b ☐	7.1c ☐	7.1d ☐	7.1e ___ %	7.1f ☐ ☐
7.2a ☐	7.2b ☐	7.2c ☐	7.2d ☐	7.2e ___ %	7.2f ☐ ☐
7.3a ☐	7.3b ☐	7.3c ☐	7.3d ☐	7.3e ___ %	7.3f ☐ ☐
7.4a ☐	7.4b ☐	7.4c ☐	7.4d ☐	7.4e ___ %	7.4f ☐ ☐
7.5a ☐	7.5b ☐	7.5c ☐	7.5d ☐	7.5e ___ %	7.5f ☐ ☐
7.6a ☐	7.6b ☐	7.6c ☐	7.6d ☐	7.6e ___ %	7.6f ☐ ☐
7.7a ☐	7.7b ☐	7.7c ☐	7.7d ☐	7.7e ___ %	7.7f ☐ ☐
7.8a ☐	7.8b ☐	7.8c ☐	7.8d ☐	7.8e ___ %	7.8f ☐ ☐
7.9a ☐	7.9b ☐	7.9c ☐	7.9d ☐	7.9e ___ %	7.9f ☐ ☐
7.10a ☐	7.10b ☐	7.10c ☐	7.10d ☐	7.10e ___ %	7.10f ☐ ☐
7.11a ☐	7.11b ☐	7.11c ☐	7.11d ☐	7.11e ___ %	7.11f ☐ ☐
7.12a ☐	7.12b ☐	7.12c ☐	7.12d ☐	7.12e ___ %	7.12f ☐ ☐
7.13a ☐	7.13b ☐	7.13c ☐	7.13d ☐	7.13e ___ %	7.13f ☐ ☐
7.14a ☐	7.14b ☐	7.14c ☐	7.14d ☐	7.14e ___ %	7.14f ☐ ☐

☐ (Check if additional information is provided on Part IV–Supplemental Information.)

8. OPTIONAL INFORMATION ON WASTE MINIMIZATION

(Indicate actions taken to reduce the amount of the chemical being released from the facility. See the instructions for coded items and an explanation of what information to include.)

A. Type of modification (enter code)	B. Quantity of the chemical in the wastestream prior to treatment/disposal			C. Index	D. Reason for action (enter code)
	Current reporting year (lbs/yr)	Prior year (lbs/yr)	Or percent change		
☐☐	_____	_____	_____ %	☐☐	☐☐

EPA Form 9350-1(1-88)

(Important: Type or print; read instructions before completing form.)   Page 5 of 5

EPA FORM R
PART IV. SUPPLEMENTAL INFORMATION
Use this section if you need additional space for answers to questions in Parts I and III.
Number or letter this information sequentially from prior sections (e.g., D.E. F, or 5.54, 5.55).

(This space for EPA use only.)

ADDITIONAL INFORMATION ON FACILITY IDENTIFICATION (Part I – Section 3)

3.5 — SIC Code —

3.7 — Dun & Bradstreet Number(s)

3.8 — EPA Identification Number(s) RCRA I.D. No.)

3.9 — NPDES Permit Number(s)

3.10 — Name of Receiving Stream(s) or Water Body(s)

ADDITIONAL INFORMATION ON RELEASES TO LAND (Part III – Section 5.5)

Releases to Land	A. Total Release (lbs/yr)			B. Basis of Estimate (enter code)
	A.1 Reporting Ranges 0 1–499 500–999		A.2 Enter Estimate	
5.5___ [ ] [ ] (enter code)	5.5___ a			5.5___ b [ ]
5.5___ [ ] [ ] (enter code)	5.5___ a			5.5___ b [ ]
5.5___ [ ] [ ] (enter code)	5.5___ a			5.5___ b [ ]

ADDITIONAL INFORMATION ON OFF-SITE TRANSFER (Part III – Section 6)

	A. Total Transfers (lbs/yr)			B. Basis of Estimate (enter code)	C. Type of Treatment/ Disposal (enter code)
	A.1 Reporting Ranges 0 1–499 500–999		A.2 Enter Estimate		
6.___ Discharge to POTW	6.___ a			6.___ b [ ]	
6.___ Other off-site location (Enter block number from Part II, Section 2.) [ ]	6.___ a			6.___ b [ ]	6.___ c.
6.___ Other off-site location (Enter block number from Part II, Section 2.) [ ]	6.___ a			6.___ b [ ]	6.___ c.

ADDITIONAL INFORMATION ON WASTE TREATMENT (Part III – Section 7)

A. General Wastestream (enter code)	B. Treatment Method (enter code)	C. Range of Influent Concentration (enter code)	D. Sequential Treatment? (check if applicable)	E. Treatment Efficiency Estimate	F. Based on Operating Data? Yes No
7.___ a [ ]	7.___ b	7.___ c [ ]	7.___ d [ ]	7.___ e ___ %	7.___ f [ ] [ ]
7.___ a [ ]	7.___ b	7.___ c [ ]	7.___ d [ ]	7.___ e ___ %	7.___ f [ ] [ ]
7.___ a [ ]	7.___ b	7.___ c [ ]	7.___ d [ ]	7.___ e ___ %	7.___ f [ ] [ ]
7.___ a [ ]	7.___ b	7.___ c [ ]	7.___ d [ ]	7.___ e ___ %	7.___ f [ ] [ ]
7.___ a [ ]	7.___ b	7.___ c [ ]	7.___ d [ ]	7.___ e ___ %	7.___ f [ ] [ ]

EPA Form 9350-1 (1–88)

United States Environmental Protection Agency
Washington, DC 20460

Substantiation To Accompany Claims of Trade Secrecy Under the Emergency Planning and Community Right-To-Know Act of 1986

Form Approved
OMB No. 2050-0078
Approval expires 10-31-90

Paperwork Reduction Act Notice

Public reporting burden for this collection of information is estimated to vary from 27.7 hours to 33.2 hours per response, with an average of 28.8 hours per response, including time for reviewing instructions, searching existing data sources, gathering and maintaining the data needed, and completing and reviewing the collection of information. Send comments regarding the burden estimate or any other aspect of this collection of information, including suggestions for reducing this burden, to Chief, Information Policy Branch, PM-223, U.S. Environmental Protection Agency, 401 M Street, SW, Washington, DC 20460; and to the Office of Information and Regulatory Affairs, Office of Management and Budget, Washington, DC 20503.

Part 1. Substantiation Category

1.1 Title III Reporting Section (check only one)

☐ 303   ☐ 311   ☐ 312   ☐ 313

1.2 Reporting Year   19 _____

1.3 Indicate Whether This Form Is (check only one)

1.3a. ☐ **Sanitized**
(answer 1.3.1a below)

1.3.1a. Generic Class or Category

1.3b. ☐ **Unsanitized**
(answer 1.3.1b. and 1.3.2b. below)

1.3.1b. CAS Number

☐☐☐☐☐☐ – ☐☐ – ☐

1.3.2b. Specific Chemical Identity

Part 2. Facility Identification Information

2.1 Name

2.2 Street Address

2.3 City, State, and ZIP Code

2.4 Dun and Bradstreet Number

☐☐☐ – ☐☐☐ – ☐☐☐☐

EPA Form 9510-1 (7-88)

Page 1 of 5

Part 3. Responses to Substantiation Questions

3.1   Describe the specific measures you have taken to safeguard the confidentiality of the chemical identity claimed as trade secret, and indicate whether these measures will continue in the future.

3.2   Have you disclosed the information claimed as trade secret to any other person (other than a member of a local emergency planning committee, officer or employee of the United States or a State or local government, or your employee) who is not bound by a confidentiality agreement to refrain from disclosing this trade secret information to others?

☐ Yes   ☐ No

3.3   List all local, State, and Federal government entities to which you have disclosed the specific chemical identity. For each, indicate whether you asserted a confidentiality claim for the chemical identity and whether the government entity denied that claim.

Government Entity	Confidentiality Claim Asserted		Confidentiality Claim Denied	
	Yes	No	Yes	No

3.4   In order to show the validity of a trade secrecy claim, you must identify your specific use of the chemical claimed as trade secret and explain why it is a secret of interest to competitors. Therefore:

(i)   Describe the specific use of the chemical claimed as trade secret, identifying the product or process in which it is used. (If you use the chemical other than as a component of a product or in a manufacturing process, identify the activity where the chemical is used.)

(ii)   Has your company or facility identity been linked to the specific chemical identity claimed as trade secret in a patent, or in publications or other information sources available to the public or your competitors (of which you are aware)?

☐ Yes   ☐ No

If so, explain why this knowledge does not eliminate the justification for trade secrecy.

(iii)   If this use of the chemical claimed as trade secret is unknown outside your company, explain how your competitors could deduce this use from disclosure of the chemical identity together with other information on the Title III submittal form.

3.4 (iv) Explain why your use of the chemical claimed as trade secret would be valuable information to your competitors.

3.5   Indicate the nature of the harm to your competitive position that would likely result from disclosure of the specific chemical identity, and indicate why such harm would be substantial.

3.6 (i) To what extent is the chemical claimed as trade secret available to the public or your competitors in products, articles, or environmental releases?

3.6 (ii)   Describe the factors which influence the cost of determining the identity of the chemical claimed as trade secret by chemical analysis of the product, article, or waste which contains the chemical (e.g., whether the chemical is in pure form or is mixed with other substances).

Part 4. Certification (Read and sign after completing all sections)

I certify under penalty of law that I have personally examined the information submitted in this and all attached documents. Based on my inquiry of those individuals responsible for obtaining the information, I certify that the submitted information is true, accurate, and complete, and that those portions of the substantiation claimed as confidential would, if disclosed, reveal the chemical identity being claimed as a trade secret, or would reveal other confidential business or trade secret information. I acknowledge that I may be asked by the Environmental Protection Agency to provide further detailed factual substantiation relating to this claim of trade secrecy, and certify to the best of my knowledge and belief that such information is available. I understand that if it is determined by the Administrator of EPA that this trade secret claim is frivolous, EPA may assess a penalty of up to $25,000 per claim.

I acknowledge that any knowingly false or misleading statement may be punishable by fine or imprisonment or both under applicable law.

4.1 Name and official title of owner or operator or senior management official

4.2 Signature (All signatures must be original)   4.3 Date Signed

EPA Form 9510-1 (7-88)   Page 5 of 5

BILLING CODE 6560-50-C

APPENDIX F

Lethal Dose Equivalencies

In the definitions for "toxic" and "highly toxic" health hazards found in Appendix A of the Standard, reference is made to LD_{50} and LC_{50} values. These definitions are given under conditions that often are different from those for which data are available in the literature. For purposes of identifying these types of hazards, such data can be translated into approximate values for the conditions used in the OSHA Standard definitions. In some cases, it may ultimately be advisable to rerun the tests under the same conditions as in the definitions within the Standard.

The median lethal dose (LD_{50}), expressed in milligrams per kilogram of body weight, is the dose that was found to be fatal to one-half of the test animals under the stated conditions. The median lethal concentration in the air breathed by the animal (LC_{50}), expressed in parts per million by volume, is the concentration that was found to be fatal to one-half of the test animals.

SPECIES

For purposes of equivalencies between species, the various animals are divided into four groups, as in Table F.1. For oral and inhalation toxicity, the chemical is to be tested on albino rats weighing between 200 and 300 grams each. An exact equivalency between rats and some other species may not be possible, but some approximate translations are given here.

In the 1976 Edition of the NIOSH "Registry of Toxic Effects of Chemical Substances," a table of interspecies factors (page xviii) calculated from

Table F.1. Groups of Animal Species Showing Equivalent Toxic Response

Group 1	Group 2	Group 3	Group 4
Frog	Mouse	Chicken	Cat
Gerbil	Rat	Duck	Cattle
Hamster	Squirrel	Guinea Pig	Dog
	Mammal	Pigeon	Goat[a]
	(other)	Quail	Horse[a]
		Rabbit	Monkey
		Turkey	Pig
		Bird	Sheep[a]
		(other)	

[a]Domestic animals only

experimental data is given. Pertinent features of that table are summarized below and in Table F.1.

For both oral and inhalation toxicity, LD_{50} and LC_{50} values are assumed to be the same for all members of Group 2 as for the rat. Values given for members of Group 1 should be multiplied by a factor of two to get the estimated value for the rat. And any values given for members of Groups 3 and 4 should be divided by two to get the estimated value for the rat.

For skin toxicity, for which the test animal is specified to be an albino rabbit weighing between two and three kilograms, LD_{50} values are assumed to be the same for all members of Groups 2 and 3 as for the rabbit. Values given for members of Group 1 should be multiplied by a factor of two to get the estimated value for rabbit. Values given for members of Group 4 should be divided by two to get the estimated value for the rabbit.

It must be remembered that all these equivalency translations are approximate only, with considerable margin for error. Whenever such a computation yields a value within a factor of about five of the limit of a defined range, the judgment of the hazard classification should be considered tentative. In such cases it may be best to run the appropriate test with the animal species and conditions specified in the Standard.

E. A. C. Crouch and R. Wilson have stated that the interspecies factor K seems to vary randomly from chemical to chemical with a lognormal distribution corresponding to an uncertainty of a factor of about five for comparison between species (*Assessment and Management of Chemical Risks*, J. V. Rodricks and R. G. Tardiff, Eds., American Chemical Society, Washington, DC, 1984, p. 109). While this factor was evaluated for comparison of rats and mice with humans, and for carcinogens rather than for acute toxic response, it does give some indication of the uncertainty one might expect when translating other toxic data from one species to another.

INHALATION DOSE

For inhalation toxicity, the chemical is to be tested on albino rats weighing between 200 and 300 grams each with continuous inhalation for one hour. However, for a large portion of the data reported in the literature, the inhalation period is much different from the one hour specified in the Standard.

In such cases, it would seem appropriate to use the product of concentration and exposure time to give the total dose administered. Henry's Law says that the equilibrium solubility of a gas component in a liquid (the blood) is proportional to its concentration in the gas mixture, at a constant total pressure of the gas (the atmosphere). If the diffusion of the component through the lung tissue to the blood is a passive process (i.e., the cells in the lung tissue do not actively "push it through" to the blood), then the diffusion rate is also proportional to the difference in the concentrations of the component in the air and in the blood, from Fick's Law. Combining these two factors, the total amount of the test substance that dissolves in the blood is expected to be approximately proportional to the product of the exposure time and the concentration in the air.

However, this will not always be a valid translation for at least two reasons. First, if the concentration of the material in the air is high enough, the diffusion rate and the equilibrium concentration in the blood may not be proportional to the concentrations as indicated above. Second, death from a short-time high dose rate may occur by a different physiological mechanism than from a long-time low dose rate.

However, in the absence of appropriate one-hour data, the concentration that corresponds to the same total dose gives a reasonable approximation. For example, if the inhalation toxic dose for rats is reported to be $LC_{50} = 65$ ppm for a four-hour exposure, it might seem to be a "highly toxic" substance. This would translate to 260 ppm (4×65) on a one-hour exposure basis, though, which would indicate a "toxic" rather than a "highly toxic" substance.

APPENDIX G

Generic Written Hazard Communication Program

The following is a sample generic written hazard communication program. Some of the statements in this written program may not apply to a given facility, and it may be appropriate to add additional items in some cases. With this caveat, the suggested language may be adopted for use in any facility as desired.

WRITTEN HAZARD COMMUNICATION PROGRAM FOR
[*NAME OF FACILITY*]

Introduction

The purpose of this program is to ensure that [*name of facility*] is in compliance with the OSHA Hazard Communication Standard (HCS), 29 CFR 1910.1200.

The [*title of individual*] is the general coordinator of the HCS program, operating as the representative of [*title of high-level manager or executive*] who is ultimately responsible for the HCS program.

Each employee at [*name of facility*] will be informed of the HCS and its requirements, his or her rights under the Standard, the nature of material hazards in the workplace, and how to avoid harm from undue exposure to those hazards.

Further information about this written program, applicable MSDSs, or the Hazard Communication Standard in general, is available from [*title of individual and his or her location and telephone number*].

405

I. The Hazardous Materials List

The [*title of individual*] will maintain a list of all hazardous materials used at [*name of facility*]. Manufacturers and suppliers will be relied upon to determine whether each material is hazardous or not. If neither an MSDS nor assurance that one is not needed is furnished with any material when received from the supplier, that material will not be placed into use until such documentation is available.

The list will be updated upon receipt of any hazardous material not previously listed. The master list of hazardous materials is maintained at [*specific location within facility*]. Subordinate lists containing only those materials in use in a particular area will also be maintained and are located at [*list of specific locations*].

II. Material Safety Data Sheets

The [*title of individual*] will maintain a file containing an MSDS for every item on the list of hazardous materials. The MSDSs used will be those supplied by the manufacturers or other suppliers. The [*title of individual*] will maintain a subfile of MSDSs at each location where a subordinate list of materials is maintained. All employees will have access to the lists of materials and to the MSDSs at all times.

Each time a material is ordered for the first time, the purchasing department shall as a matter of course request an MSDS for the material as a condition of purchase. If one is not received with or prior to the shipment, the material shall be impounded in a secure area until one is received, and the [*title of individual*] shall dispatch a letter requesting immediate transmittal of either an MSDS or a letter disclaiming its necessity. If necessary, a second letter shall be dispatched within ten days following the first. If there is still no adequate response, the material shall be returned to the sender and the purchase order voided.

Each time a material is reordered, the purchasing department shall as a matter of course request an updated MSDS for the material if one is available. This request shall include the date shown on the most recent MSDS on file.

Whenever a complimentary sample of a material is received for evaluation from a manufacturer or distributor, the same procedure shall be followed as for purchased materials. No material shall be placed into use without either an MSDS or a letter of disclaimer.

Whenever material is purchased from a local wholesale or retail dealer, whether with petty cash or with a purchase order or requisition, the material shall not be accepted unless it is accompanied by an appropriate MSDS.

III. Labels and Other Forms of Warning

The [title of individual] will ensure that all hazardous materials in the facility are properly labeled when they arrive. Information on the labels will be checked against that on the MSDS for consistency. Any product that is not labeled in accordance with the HCS requirements will be refused and returned to the sender, except for those exempted from HCS labeling requirements.

Any containers into which materials are transferred for in-house use, other than for immediate transfer for use by the employee filling the container, shall be labeled consistently with the label on the original container. All labels for in-house use on containers holding materials whose compositions are a result of operations within the facility shall be approved for correctness of form by [title of individual].

All process vessels in the [name of facility] shall have pockets with transparent windows affixed to their outer surface at locations that can be readily seen by employees working with the equipment, and temporary labels consistent with those on the original container(s) shall be inserted in those pockets.

Process piping shall be labeled consistently with labeling requirements for other containers. The labels used for this purpose shall be inserted into transparent protective pockets which shall be attached to or hung from the pipe within three feet of each valve, union joint, or other access fitting.

The [title of individual] shall check at least monthly to assure that all containers of material are properly labeled, and that none of the original labels have been defaced.

IV. Employee Training and Information

Each employee who works with or may be exposed to hazardous materials shall be informed of the provisions of the Hazard Communication Standard, including the location and availability of the hazardous material lists, the MSDS files, and the written hazard communication program. In addition, the essential features of this information shall be posted on bulletin boards located in all areas where hazardous materials are stored or used within [name of facility]. Such postings shall indicate the specific locations of the MSDSs as well as the names of materials for which newly acquired or updated MSDSs are available.

Each employee who works with or may be exposed to hazardous materials shall receive training on the hazardous properties and safe use of those materials. Recipients of this training shall include employees who occasionally may be exposed as well as those who are regularly exposed. Additional training shall be provided for employees whenever a new hazard is introduced into their work areas. Hazardous material training is to be conducted by [*title of individual, department, or contract vendor*]. A copy of course outline and materials is appended to this written program. The training will emphasize the following elements.

- Identity of hazardous materials in the work area
- Means of identifying hazardous materials
- Physical and health hazards associated with the materials
- Symptoms of overexposure to hazardous materials
- Procedures to protect against hazards under normal use conditions
- Procedures to protect against hazards in emergency conditions
- First aid procedures where appropriate
- Procedures in case of spill or leak of hazardous materials

The [*title of individual*] will maintain records of training received by all employees and will schedule the initial training as well as any additional training needed as functions change and as new hazards are introduced into particular work areas.

V. Nonroutine Tasks

The [*title of individual*] will monitor all work orders for maintenance and other nonroutine tasks. In coordination with the supervisor responsible for the task, the [*title of individual*] will schedule and oversee appropriate training in hazard recognition and avoidance for the workers to be assigned to the task.

Those persons assigned to handling, packaging, and shipping hazardous waste from the facility shall also be responsible for cleaning up spills and coordinating efforts of outside agencies (e.g., medical, fire, and police departments) in the event of a major emergency. They shall be provided special training in the added hazardous conditions that can be experienced in such emergency situations. They shall also be given detailed training on the hazards involved in the waste management system and how to protect themselves and others from undesirable effects of those hazards.

VI. Outside Contractors' Employees

The [*title of individual*] shall be advised of all contracted work to be done by outside firms and shall coordinate hazard training and information with the contractors or their representatives. Outside contractors shall each be:

- given copies of this written program
- shown where the MSDSs are kept
- provided a summary of material hazards present in the areas in which their employees will be working in the course of fulfilling the contracts.

In addition, contractors will be requested to review with the appropriate supervisor(s) at [*name of facility*] all information regarding material hazards to be introduced by the activities of the contractor. Such supervisor(s) will then be responsible for transmitting such hazard information to employees who work in the affected area(s) and might be exposed to the hazards.

INDEX